Environmental Contaminants

Health Care and Environmental Contamination

Environmental Contaminants

Health Care and Environmental Contamination

Edited by

Alistair B.A. Boxall
University of York, Heslington, United Kingdom

Rai S. Kookana
CSIRO Land and Water, Glen Osmond, SA, Australia

ELSEVIER

Elsevier
Radarweg 29, PO Box 211, 1000 AE Amsterdam, Netherlands
The Boulevard, Langford Lane, Kidlington, Oxford OX5 1GB, United Kingdom
50 Hampshire Street, 5th Floor, Cambridge, MA 02139, United States

ISBN: 978-0-444-63857-1
ISSN: 2405-5654

Publisher: Candice Janco
Acquisition Editor: Anneka Hess
Editorial Project Manager: Emily Thomson
Production Project Manager: Vijayaraj Purushothaman
Cover Designer: Greg Harris

Typeset by SPi Global, India

Contents

Contributors

Numbers in Parentheses indicate the pages on which the author's contributions begin.

W. Baylor Steele (41), Baylor University, Waco, TX, United States

Thomas G. Bean (87), University of Maryland, College Park, MD, United States

Alistair B.A. Boxall (1, 123), University of York, Heslington, United Kingdom

Bryan W. Brooks (41), Baylor University, Waco, TX, United States

Laura J. Carter (69), University of York, Heslington, United Kingdom

Karin P.M. Helwig (199), Glasgow Caledonian University, Glasgow, Scotland

Chad A. Kinney (69), Colorado State University-Pueblo, Pueblo, CO, United States

Klaus Kümmerer (225), Leuphana University of Lüneburg, Lüneburg, Germany

Rai S. Kookana (1, 21), CSIRO, Adelaide, SA, Australia

Richard Murray-Smith (239), Regulatory Science Associates, Kip Marina, Inverkip, Renfrewshire, Scotland

Tiago S. Oliveira (7), 5820 Fair Oaks Blvd 133 95608, Carmichael, CA

Amy Pruden (137), Civil and Environmental Engineering, Virginia Tech, Blacksburg, VA, United States

Barnett A. Rattner (87), US Geological Survey, Patuxent Wildlife Research Center, Beltsville, MD, United States

Andrew C. Singer (149), NERC Centre for Ecology & Hydrology, Wallingford, United Kingdom

David Taylor (239), wca, Brunel House, Volunteer Way, Faringdon, Oxfordshire

Mike Williams (21), CSIRO, Adelaide, SA, Australia

Amy Wolff (167), Melbury Environmental Ltd

Chapter 1

Environmental Contaminants and Health Care: An Introduction

Alistair B.A. Boxall* and Rai S. Kookana†
**University of York, Heslington, United Kingdom*
†*CSIRO, Adelaide, SA, Australia*

Health care is a large sector and will grow even further due to a growing global middle class and aging population in many parts of the world. The health-care chain includes centers where pharmaceuticals and health-care products are developed and manufactured and facilities such as hospitals, dispensaries, diagnostic centers, and nursing homes where these products are used. Residential homes and training facilities (e.g., medical schools) are also a part of health-care chain since health-care products are consumed and disposed of at these locations.

Chemicals are an essential part of the health-care system, as chemicals offer the tools with which we diagnose, prevent and treat diseases (e.g., pharmaceuticals), and maintain a clean and healthy environment in hospitals and clinics (e.g., disinfectants). It is inevitable that the range of chemicals used in health-care will be released to the natural environment at different stages of their life cycle.

In most countries, with the exception of regulations regarding some basic water quality parameters, the wastes from hospitals and health-care systems are not required to be treated to remove health-care chemicals. Municipal wastewater treatment plants are not designed to remove these chemicals and at best result in incomplete removal and lead to formation of transformation products, which may still be a concern in the environment. In some instances, the transformation products may even be more toxic than the parent compounds.

Many of the chemicals used in health care, such as pharmaceuticals and disinfectants, are bioactive compounds by design and hence have the potential to cause adverse impacts on nontarget organisms in the natural environment.

Environmental Contaminants, Vol. 11. https://doi.org/10.1016/B978-0-444-63857-1.00001-2

Indeed, one of worst environmental disasters (the Asian vulture crisis) in recent history has been associated with the use of a pharmaceutical. In the early 2000s, it was noted in the Indian subcontinent that number of some species of vultures (*Gyps* spp.) had declined significantly with between a 95% and 99% reduction in the populations being reported. The observed decline in the vulture populations was attributed to the veterinary use of an inexpensive nonsteroidal antiinflammatory drug, diclofenac. The vultures were exposed to the diclofenac when they scavenged on dead animals that had been treated with high doses of diclofenac (see Chapter 6 for details). This not only led to ecological impacts but also is thought to have indirectly impacted the health of the human population in the region (see Chapter 7 for details). Now that diclofenac has been restricted from veterinary use in the region and that rehabilitation programs have been put in place, the vulture populations are showing a recovery.

In this volume, we provide a comprehensive overview of nature and sources of contaminants that are associated with health-care system—keeping in mind the entire chain from development and production of health-care products to their use and disposal in waste streams and ultimately in the environment. The volume not only offers an account of the current state of the knowledge in terms of environmental exposure and effects of health-care contaminants but also suggests solutions and guidance both in terms of their source management and removal from the environment. The volume also discusses the needs in terms of regulations and policy to minimize the adverse impacts of such contaminants in the environment. In developing the volume, we have asked authors who are world leaders in their field and are drawn from the research and business sectors to provide a high-level overview of a topic. In doing this, we hope that this volume will be a valuable resource for the nonspecialist.

In terms of hospitals as a source of pharmaceuticals and personal care products (PPCPs) that include thousands of high-volume chemicals, Olivera highlights in Chapter 2 the scale of usage and release of these chemicals in the health sector. Health-care facilities are estimated to account for 20%–25% of all human medicine usage with some hospitals using as high as a ton (1000 kg) of medicines each year. Hospital effluents have been reported to contain tens to several hundreds of microgram per liter concentrations of antibiotics, cytostatics, and other drugs. In some sewer networks with low flows, the hospital effluents can contribute >80% of total pharmaceuticals and personal care products in the network load. Therefore, there is a need to locally treat the effluents from hospitals, and some treatment options have proved to be effective.

Once released into the waste stream, the contaminants from health-care system are subject to several processes that affect how they move around the natural environment and how long they persist in the environment such as sorption, transformation, and degradation (both abiotic and biotic). In

Chapter 3, Williams and Kookana discuss the processes governing the fate and behavior of health-care contaminants in both aquatic and terrestrial ecosystems and how the physicochemical properties of the chemicals together with the prevailing environmental conditions determine their ultimate fate and thus their exposure pathways. For example, many of the antibiotics and disinfectants (such as quaternary ammonium compounds) have high affinity for solid phases, and hence, they tend to accumulate in sludge or sediments. Therefore, application of sludge (biosolids) on land may serve as an important pathway of their dispersal in the terrestrial environment.

In Chapter 4, Brooks and Baylor discuss how the field of ecotoxicology has evolved and discuss some of the factors and processes affecting the uptake and effects of pharmaceuticals in organisms in the environment. Toward the end of the chapter, they discuss how knowledge of drivers of ecotoxicity can be used to support the design of more benign pharmaceuticals.

Considering that biosolids and wastewater are increasingly being reused in crop production, Carter and Kinney (Chapter 5) discuss the potential release of health-care contaminants via these pathways to the terrestrial environment and consequent ecotoxicological impact of pharmaceuticals on organisms such as earthworms and plants. They highlight that nearly 50% of total sludge produced in the United States and 65% of that in France are applied on land. In the United States, the total discharge of some antimicrobial agents (triclosan and triclocarban) into the environment through biosolids and wastewater may be as high as 1×10^6 kg/year. In such situations, uptake of contaminants by earthworms and plants may contribute to biomagnification in terrestrial food web, and thus, their food-chain effects need attention.

The food-chain effects of such contaminants in wildlife are further discussed by Bean and Rattner in Chapter 6, where they highlight the serious lack of thorough understanding of exposure pathways and risks of health-care contaminants to wildlife, partly due to the challenges associated with such studies. They suggest that given the great diversities in physiologies among species, a combination of in vivo, in vitro, and in silico approaches may be required to fill the knowledge gaps for exposure, hazard, and risk.

In Chapter 7, Boxall discusses the potential effects of residues of pharmaceuticals in the environment on human health. This chapter shows how humans can be exposed to pharmaceuticals though the consumption of contaminated water and food. Available data indicate that levels of exposure are lower than safe levels so that direct toxicological impacts on humans are unlikely. Much more work is however needed on this topic. The chapter also discussed how residues of pharmaceuticals can also indirectly affect human health by affecting the delivery of ecosystem services and, in the case of antimicrobial compounds, selecting for antimicrobial resistance (AMR) in bacteria in the environment.

The impacts of antimicrobials on AMR are described by Pruden in Chapter 8. In this chapter, she explains how AMR is one of the greatest human health challenges of the 21st century. While significant efforts have been placed toward combatting antibiotic resistance in the clinical setting, there is increasing recognition that environmental sources of AMR are also important. The chapter examines the interplay of clinical and environmental sources of antibiotic resistance and discusses management options that may be effective for limiting the spread of resistance and advancing the life span of antibiotics for protecting human health.

Certain health-care products are used in large quantities over a short period of time to deal with pandemics and epidemics. Singer (Chapter 9) discusses the environmental and human health risks posed by society's medical response to pandemics and epidemics using the influenza pandemic 2009 as a case study. He shows how stockpiles of antiviral drugs had to be procured by different countries in preparedness. In the environmental context, antiviral drugs have been found to be highly stable and not easily degradable. Furthermore, questions have been raised about their presence in waste stream potentially compromising the efficiency of wastewater treatment plant. Given the range of viral diseases (e.g., Zika, dengue, chikungunya, SARS, and Ebola) that have been reported at different parts of world, the impact of antiviral drug use during pandemics and epidemics deserves particular attention.

Waste from health-care system presents a particular challenge, especially in developing countries, where proper systems for their safe management are not in place. This topic is covered in details by Wolff in Chapter 10. She highlights that >50% of the world's population is at particular risk from improperly treated medical waste. The chapter discusses the basic principles of safe management of heath waste and the options that may be available to do so. The chapter also demonstrates how legislation shapes and guides the health-care waste sector and how European legislation works in the United Kingdom, thus explaining the inextricable link between legislation and an evolving and safer health-care waste sector.

Product stewardship can also play a major role in the reduction of health-care-specific contaminants in the environment. Using pharmaceuticals as an example, Helwig discusses this in detail in Chapter 11 and stresses that together with green pharmacy, stewardship has the potential to deliver positive environmental, health, and financial outcomes. However, she calls for a reexamination of our attitude toward health and our relationship with pharmaceuticals.

The whole life cycle of a compound or a health-care product has to be taken into account when making risk management and risk reduction decisions. In Chapter 12, Kümmerer identifies opportunities that are provided by green and sustainable pharmacy to manage environmental contaminants from health care, considering life-cycle assessment. Suggestions have been made for action by various stakeholders in the health-care chain, including manufacturers, doctors, pharmacists, and patients.

In many regions of the world, regulations exist for managing the impacts of pharmaceuticals and other chemicals used in healthcare. These regulations are reviewed by Murray Smith in Chapter 13. In this chapter, he reviews regulatory systems in place in different regions of the world for managing impacts of manufacturing plants, for assessing the environmental safety of new active ingredients, and for managing and treating pharmaceutical wastes.

Rather than targeting a specialist group of environmental scientists, we have kept the volume free from jargon as much as possible, in order to facilitate the ready uptake of the knowledge and information by a diverse group of stakeholders associated with the health-care sector. We hope that you will find the information in this book easy to understand and useful so that in the long term, we can work together to reduce the environmental impacts of contaminants that are associated with health-care sector while ensuring the sector still has access to these chemicals whose use is vital for society.

Chapter 2

Environmental Contamination From Health-Care Facilities

Tiago S. Oliveira
5820 Fair Oaks Blvd 133 95608 Carmichael CA

INTRODUCTION

Health care is an essential component of a modern and developed society. Health-care networks typically include institutions providing care to individuals at different stages of life and health conditions and performing R&D in the medical field. These health-care services occur in facilities such as nursing homes, specialized hospitals (e.g., psychiatric hospitals and maternity hospitals), general hospitals, and university hospitals. These facilities range in size with the largest having over 1000 beds available for patient treatment.

While performing the necessary activities to meet the community health needs, wastes are generated. These wastes are typically managed following country-specific regulations (see Chapter 10). Health-care facilities use large quantities of water. Estimates indicate the consumption of up to 1200 L/bed per day occurs in hospitals in industrialized countries. This water consumption results in the generation of wastewaters that can carry microbiological (e.g., bacteria and viruses), inorganic (e.g., heavy metals), and organic (e.g., pharmaceuticals) pollutants.

In many countries, regulations require wastewaters to meet certain quality standards prior to release into the aquatic environment. In most cases, these regulations impose limits on physicochemical parameters such as temperature, pH, total suspended solids, biochemical oxygen demand, chemical oxygen demand, and conductivity. In some countries, specific sources are regulated and require monitoring of additional parameters (e.g., absorbable organic halogens, total and free chlorine, detergents, disinfectants, tensioactives, oil and grease, and toxicity) to meet established limits prior to release into the sewer systems (Carraro et al., 2016 and references therein).

Nonregulated pollutants such as pharmaceuticals and personal care products (PPCPs) have been reported in aquatic environments, and their potential risk to living organisms has been investigated (some of these impacts are

Environmental Contaminants, Vol. 11. https://doi.org/10.1016/B978-0-444-63857-1.00002-4

discussed in more detail in Chapter 4). The toxicity tests performed to date indicate that a number of PPCPs (individually and in mixtures) at environmentally relevant concentrations pose a risk to fish and lower-trophic-level organisms (Cizmas et al., 2015 and references therein). The exposure to increased PPCP concentrations might result in genetic lesions (e.g., Parolini et al., 2013), organ and reproductive abnormalities (e.g., Galus et al., 2013a, b), and behavioral changes (e.g., Brodin et al., 2013). Particular attention has been dedicated to specific therapeutic categories such as antibacterials and antiinfectives since their presence in the aquatic environment might produce antibiotic-resistant bacteria and antibiotic-resistant genes (Verlicchi et al., 2015; see Chapter 8).

After being administered, pharmaceuticals are excreted by patients in variable proportions as nonmetabolized, metabolized, or conjugated with inactivating substances (Verlicchi et al., 2010; Orias and Perrodin, 2013; Al Aukidy et al., 2014). These excreted substances are commonly discharged without pretreatment into municipal wastewater treatment plants (WWTPs) that seem unequipped to remove these nonregulated contaminants (Verlicchi et al., 2012a).

Considering the potential risks associated with the release of PPCPs into the environment, this chapter intends to (a) assess the relevance of health-care facilities (mostly general hospitals) as a source of these nonregulated compounds via effluent, (b) assess WWTP removal efficiency of these nonregulated compounds, (c) assess potential risks associated with release of PPCPs into the environment, and (d) present recommendations to improve the management of these effluents. As PPCPs include thousands of commercially available compounds, this chapter focuses on the therapeutic categories most typically found in health-care facility effluents (analgesics, antibacterials and antiinfectives, antidiabetic, antiepileptic, antiinflammatories, contrast media agents and cytostatics, diuretics, psychoanaleptics, steroid hormones, and β-blockers).

PPCPs IN HEALTH CARE FACILITIES EFFLUENTS

Health-care facilities are estimated to account for 20%–25% of all human medicine usage (Lerdau et al., 2013) with active ingredient consumption being dependent on the type of treatment provided and the size of the facility (Herrmann et al., 2015). As an example, in Germany, consumption estimates for a psychiatric hospital, a nursing home, and a general hospital ranged between 32 and 1000 kg/year. The active ingredients mostly consumed in the psychiatric hospital and the nursing home targeted the nervous system, and the active ingredients mostly consumed in the general hospital included contrast media agents used in the radiological department (Herrmann et al., 2015). The availability of active ingredient consumption data is limited in many regions making the calculation of predicted concentrations difficult.

The characterization of PPCPs in health-care facilities effluents is typically performed through measured concentrations. Measured concentrations are obtained by the collection of effluent samples and the subsequent sample preparation and analysis by analytic instrumentation (e.g., GC-MS, LC-MS/MS) in a laboratory setting. This approach has been widely used by research groups in different geographic regions (e.g., Asia, Lin et al., 2008; Europe, Kovalova et al., 2012; Verlicchi et al., 2012b; Santos et al., 2013; and North America, Oliveira et al., 2015) (Fig. 1). The total PPCPs measured in hospital effluents can reach up to low milligram-per-liter concentrations with 12 therapeutic categories being regularly measured. These therapeutic categories comprise ≥94% of the total PPCP concentrations measured. Prevalent therapeutic categories include contrast media agents and cytostatics, analgesics, and antibacterials and antiinfectives. These categories can, in some cases, reach individually >40% of the total PPCP concentrations (Fig. 1). Other relevant therapeutic categories include antiepileptics, antiinflammatories, psychoanaleptics, and β-blockers (Fig. 1).

The concentrations of PPCPs in health-care facilities are the result of a combination of three main factors: the administered quantity, the percentage

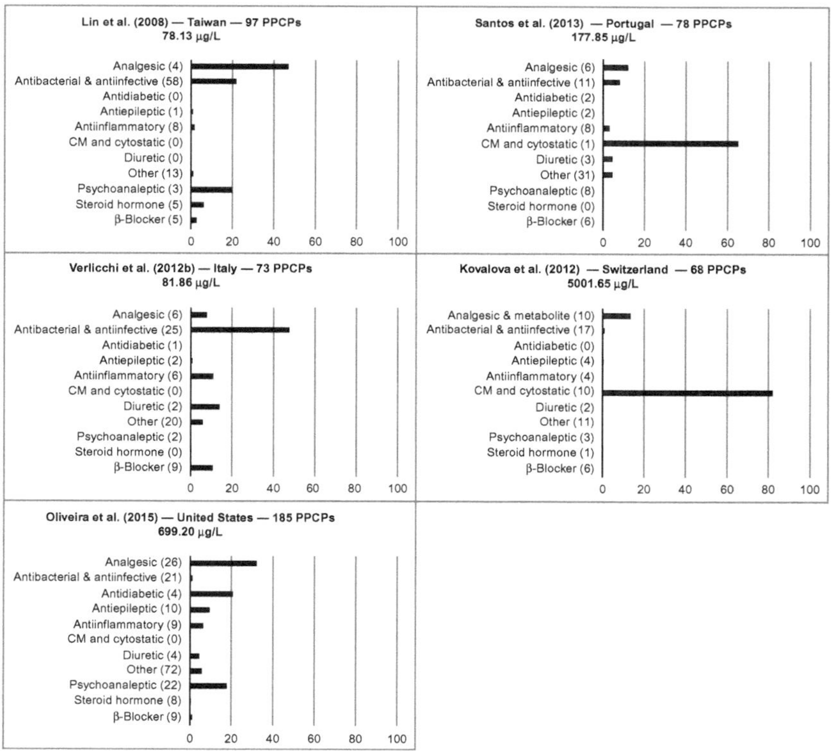

FIG. 1 Therapeutic categories most commonly measured in hospital effluent in different geographic locations. Numbers in brackets correspond to the number of PPCPs screened in the therapeutic category. Lin et al. (2008) graphic presents total median percent, and all other graphics present total average percent.

of the substance excreted by humans, and the chemical characteristics (Verlicchi et al., 2010). PPCPs in health-care facilities effluents are typically measured at concentrations <10 μg/L, and there are variations in concentrations detected across geographic regions (Fig. 2). In general hospitals, PPCPs that are regularly being measured at concentrations >10 μg/L include acetaminophen, 4,5-methylbenzotriazole, ciprofloxacin, diatrizoate, gabapentin, ibuprofen, iomeprol, iopamidol, iopromide, ioxitalamic acid, and metformin. Specialized hospitals and wards use a different range of pharmaceuticals than general hospitals. Predicted concentrations for intravenous antibiotics originating in an intensive care unit demonstrate the relevance of classes such as cephalosporins (cefazolin and ceftriaxone), carbapens (meropenem), and penicillins (ampicillin) with individual concentrations in the effluent >200 μg/L (Lopes de Souza et al., 2009). Concentrations of cytostatics originating in an oncological inpatient care ward demonstrate the relevance of 5-fluorouracil with concentrations >100 μg/L (Lenz et al., 2007).

The measured PPCP concentrations can be used to estimate loads originating from health-care facilities. Estimates of individual PPCP daily loads in

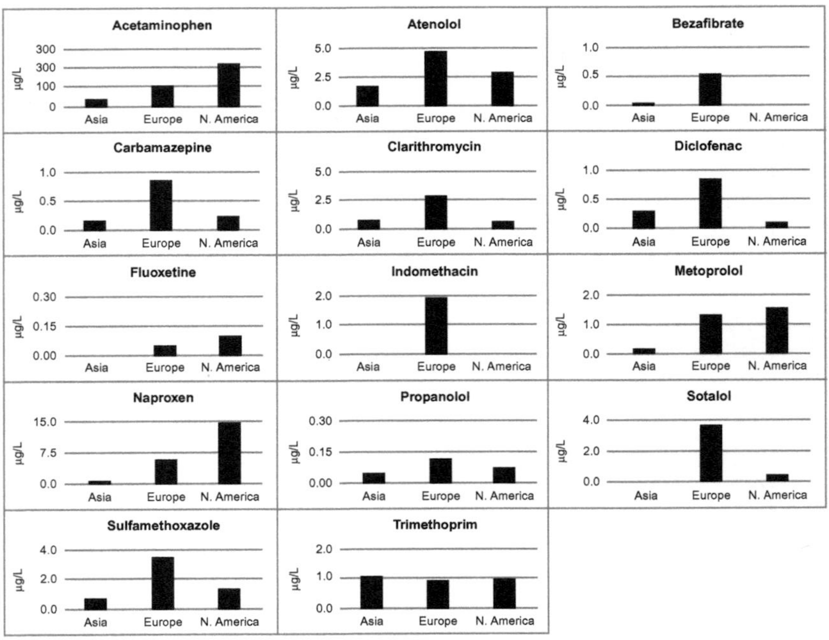

FIG. 2 Concentration of PPCPs screened in hospital effluents in different geographic regions. Asia data represent the PPCP median concentration measured by Lin et al. (2008) in hospitals in Taiwan; Europe data correspond to the maximum average measured in hospitals in Italy (Verlicchi et al., 2012b), Portugal (Santos et al., 2013), and Switzerland (Kovalova et al., 2012); the United States data correspond to the maximum average measured concentrations in hospitals in New York City (Oliveira et al., 2015).

effluents generated at a University Hospital in Switzerland ranged between 0.04 (diclofenac) and 143 g/day (acetaminophen) (Daouk et al., 2016). Other authors estimated total PPCP mass loads for hospitals with varying sizes and types ranging between 1.5 (maternity hospital with 96 beds, Portugal) and $\leq$310 g/day (university hospital with 1456 beds, Portugal, and general hospital with 250 beds, the United States) (Santos et al., 2013; Oliveira et al., 2015). These estimates might underestimate the PPCP total loads since they include few or none of the contrast media agents that have been measured at significant concentrations in hospitals providing radiological services (Kovalova et al., 2012). Before reaching the WWTP, PPCP loads are subject to dilution in sewer networks. Sewer networks with higher inflows ($\geq$10,000 m^3/day) result in a lower PPCP individual contribution from health-care facilities (<15%) (Ort et al., 2010; Oliveira et al., 2015). Sewer networks operating with lower inflow (<10,000 m^3/day) can result in a higher individual PPCP contribution from health-care facilities reaching as high as >80% for some PPCPs (Oliveira et al., 2015). Typically, concentrations of most individual PPCPs at the WWTP influent are <5 μg/L, which correspond to a $\geq$ two times dilution from hospital effluent concentration.

PPCPs IN WWTPs

WWTPs play an important role in the removal of PPCPs from wastewaters. The extent to which PPCPs are removed is compound-specific (being dependent on biodegradability and physicochemical properties—water solubility, adsorption, and volatilization) and is dependent on the WWTP characteristics (e.g., on whether the WWPT employs primary, secondary, and/or tertiary treatments), operational conditions (e.g., hydraulic and sludge retention time, pH, and temperature), reactor (conventional activated sludge (CAS), membrane biological reactor (MBR), and sequencing batch reactor (SBR)), and environmental characteristics (irradiation, precipitation, and temperature) (Verlicchi et al., 2012a, 2015; Luo et al., 2014).

Municipal WWTPs

In most cases, health-care facilities discharge effluents into municipal sewer networks without any pretreatment or dedicated WWTP. The effluents discharged into municipal sewer networks are typically routed to WWTPs that employ secondary treatment (CAS and SBR) with some using tertiary treatment (mostly UV). PPCP removal efficiencies are WWTP-specific even when similar technologies are being used (e.g., steroid hormones, SBR + UV, and psychoanaleptics, CAS) (Fig. 3). Lowest average therapeutic category removal is typically obtained for antiepileptics and psychoanaleptics. Individual PPCP removal efficiencies for municipal WWTPs-CAS are presented in Fig. 4. WWTPs-CAS typically present removal efficiencies below 80% for

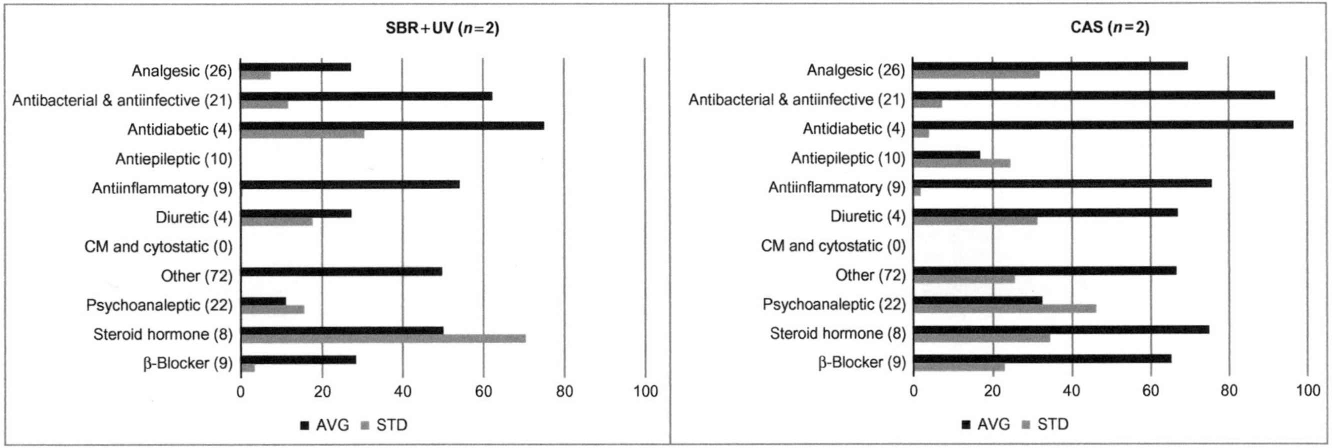

FIG. 3 Percent removal (average and standard deviation) by therapeutic category in four WWTPs in the United States—Oliveira et al. (2015). Numbers in brackets correspond to the number of PPCPs screened in the therapeutic category. No data available for contrast media agents and cytostatics.

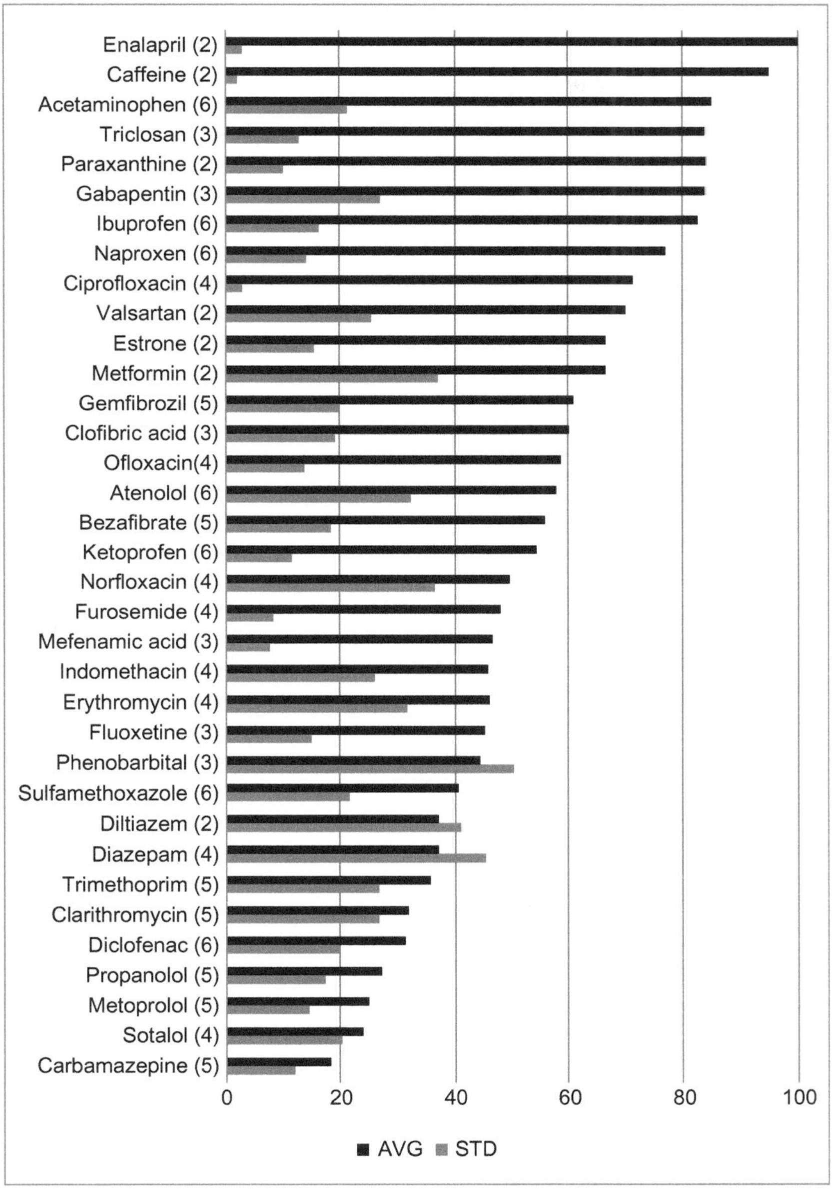

FIG. 4 PPCP percent removal (average and standard deviation) at a WWTP employing CAS and typically measured in health-care facility effluents. The number in brackets corresponds to the number of publications reporting removal (Gros et al., 2010; USEPA, 2010; Verlicchi et al., 2012a,b; Santos et al., 2013; Oliveira et al., 2015).

the majority of PPCPs assessed. PPCPs in the same therapeutic category can be removed to different degrees (e.g., antiepileptics such as carbamazepine, phenobarbital, and gabapentin and antiinflammatories such as diclofenac, indomethacin, mefenamic acid, ketoprofen, naproxen, and ibuprofen). For certain PPCPs, the reported removal efficiencies are zero or even negative (Joss et al., 2005; Gros et al., 2009; Jelic et al., 2011). The negative values can reflect poor matching of influent samples to effluent samples when the monitoring is done and/or reflect an increase in the concentration of the analyte parent compound during treatment that is due to the conversion of glucuronides and other conjugated metabolites of the pharmaceutical ingredient back to the parent compound through enzymatic processes (e.g., carbamazepine, ciprofloxacin, clarithromycin, diclofenac, erythromycin, iopamidol, phenazone, propranolol, sulfadiazine, and sulfamethoxazole). Additionally, negative PPCP removals may occur due to their release from particles (Kosma et al., 2014 and references therein; Verlicchi et al., 2015 and references therein). Typically, most PPCP concentrations in the WWTP effluent are <1 μg/L, and the PPCP mass load reduction at a WWTP is variable and can reach >95% (Verlicchi et al., 2012b; Santos et al., 2013; Oliveira et al., 2015).

Hospital Effluent—Other Treatment Options

Hospital effluents have been subject to many dedicated treatment combinations at the pilot/lab-scale plants and at full-scale plants (Verlicchi et al., 2015 and references therein). Tested combinations include the use of different infrastructures, reactors, sludge retention times, and chemicals paired or not with irradiation. To date, the best performing dedicated treatments for PPCP removal resulted from the application of MBR as a secondary treatment and powdered activated carbon (PAC) as a tertiary treatment. The combination of MBR+PAC for the treatment of hospital effluents resulted in removals >80% for a significant number of PPCPs investigated (MBR, 31 out of 67 PPCPs presented removals >80%; PAC, 23 mg/L, 37 out of 57 PPCPs presented removals >80%) (Verlicchi et al., 2015 and references therein). Other dedicated hospital effluent treatments presenting good removal of PPCPs at a reasonable cost include the use of the following choice criteria: 1.08 O_3/g DOC, 23 mg/L PAC, and 2400 J/m^3 (Kovalova et al., 2013). In addition to the combination of MBR+PAC, it has been reported that MBR+UV also presents good abatement of the most prevalent therapeutic categories in hospital effluent. When using MBR+UV, contrast media agents presented a removal of 66% of the total influent load (Kovalova et al., 2013). Like some antiepileptics, contrast media agents are among the PPCPs with lowest removal efficiencies in WWTPs.

Other hospital effluent dedicated treatment tests assessed the removal of PPCPs by adjusting parameters in existent WWTPs (e.g., sludge retention times). Longer sludge retention times allow the adaptation of slower-growing microorganisms and allow better suspended solid separation. MBR sludge retention times between >30 and 50 days present variable removal efficiencies with modest (<50%) to low (<25%) removal efficiencies being reported for several PPCPs (e.g., carbamazepine, gabapentin, iopamidol, metoprolol, ritalinic acid, indomethacin, phenazone, roxithromycin, propranolol, sotalol, iodixanol, iohexol, iomeprol, ioversol, and oxazepam (Verlicchi et al., 2015 and references therein)). Some of these PPCPs present similar removal efficiencies to the ones reported for municipal WWTPs indicating that longer sludge retention times must be used in combination with other treatment options including tertiary treatment (e.g., PAC, UV, O_3, and advanced oxidation processes) for better performance. Ultimately, the choice of the best technology to treat hospital effluent should achieve the highest reduction of its ecotoxicological effects.

PPCPs EMISSION AND ECOLOGICAL RISKS

The inefficient removal of PPCPs by WWTP results in effluent with ecotoxicological relevance that can ultimately reach the aquatic environment. The environmental risk assessment of effluents (health-care facilities and WWTPs) is generally based on the risk quotient (RQ) for a substance that is defined as the ratio between the compound concentration in the effluent and its predicted no-effect concentration (PNEC) in the environment. The PNEC is calculated at a minimum with acute ecotoxicity data (theoretic or experimental) for species from three different trophic levels (e.g., algae, invertebrates, and fish) (Orias and Perrodin, 2013). The available PNECs for PPCP ecotoxicity have been compiled by several authors (e.g., Lin et al., 2008; Escher et al., 2011; Verlicchi et al., 2012a; Orias and Perrodin, 2013; Al Aukidy et al., 2014). Currently, there is still a small number of PNECs available for PPCPs, and these present a number of limitations that prevent full reliability for assessment. These limitations are related with the fact that they are sometimes predicted based on structure-activity relationships and they are based on the results of a few experiments that only focus on subsets of end points, development stages, and species. As such, the available PNECs are only suitable for screening-level estimates (Oliveira et al., 2015).

Typically, the risk is classified high if the $RQ \geq 1$, medium if $1 < RQ < 0.1$, and low if $RQ \leq 0.1$ (Verlicchi et al., 2012b; Santos et al., 2013). Based on measured effluent concentrations, research groups reported high-ecological risk for the analgesic acetaminophen; the antibacterials and antiinfectives such as azithromycin, clarithromycin, erythromycin, metronidazole, ofloxacin, and sulfamethoxazole; the antiinflammatory ibuprofen; the contrast media agent iopromide; and the psychoanaleptic fluoxetine (Verlicchi et al., 2012b; Santos et al., 2013). Based on measured WWTP effluent concentration, Oliveira et al. (2015) reported high-ecological risk for three additionally

PPCPs measured at high concentrations in hospital effluents—the antiinflammatory naproxen, the antiepileptic gabapentin, and the psychoanaleptic caffeine. Some of these risks are described in more detail in Chapter 4.

CONCLUSION

Health-care facilities have been identified as a source of PPCPs into the built and natural environments in different geographic locations. The total PPCP concentrations measured in health-care effluents can reach the milligram-per-liter value with 12 therapeutic categories being regularly measured. The most prevalent therapeutic categories (>40% of the total PPCPs) include analgesics, antibacterials and antiinfectives, and contrast media agents and cytostatics. Individually, PPCP concentrations are typically measured at values <10 μg/L with higher concentrations being measured for pharmaceuticals highly prescribed and/or highly excreted—some examples include acetaminophen, ciprofloxacin, gabapentin, ibuprofen, iomeprol, iopamidol, iopromide, and metformin. Total PPCP mass loads have been estimated for general hospitals <310 g/day for general hospitals. Health-care PPCP contributions to the WWTP influent have been estimated to a maximum of 59% being dependent on the number/size of health-care facilities connected to the network and the sewer size. At lower inflows (<10,000 m^3/day), the individual PPCPs originating from health-care facilities can reach >80%.

Health-care facilities commonly discharge effluents into sewer networks without any pretreatment or dedicated WWTP. In municipal WWTPs, health-care effluents undergo secondary and at times tertiary treatment with individual PPCP concentration typically <1 μg/L at WWTP outfall and removal efficiencies <80% for most PPCPs. Removal efficiencies are variable between WWTPs using similar treatments and between PPCPs in the same therapeutic category. In municipal WWTPs, lowest removal efficiencies are typically obtained for antiepileptics and psychoanaleptics.

Several health-care effluent dedicated treatment options have been tested. The highest removal efficiencies (>80%) were obtained with a combination of MBR with either PAC or UV with additional choice criteria (e.g., O_3). In addition to the therapeutic categories mentioned above (antiepileptics and psychoanaleptics), contrast media are also inefficiently removed in most WWTPs. The MBR + UV combination produced the best removal efficiencies for contrast media agents (66%). Other dedicated treatment tests assessed the removal of PPCPs by adjusting parameters such as increasing sludge retention times. This adjustment was found not to improve PPCP removal when compared with municipal WWTP performance and is most likely more effective when paired with tertiary treatment options (e.g., PAC, UV, O_3, and advanced oxidation processes).

The inefficient removal of PPCPs by WWTPs results in effluent that ultimately reaches the aquatic environment. RQs for some substances in health-care facility effluent and WWTP effluent (impacted by health-care facilities) present a high risk to aquatic organisms for a number of PPCPs belonging to several therapeutic categories (analgesics such as acetaminophen; antibacterials and antiinfectives such as azithromycin, clarithromycin, erythromycin, metronidazole, ofloxacin, and sulfamethoxazole; antiepileptics such as gabapentin; antiinflammatories such as ibuprofen and naproxen; contrast media agents such as iopromide; and psychoanaleptics such as caffeine and fluoxetine). RQs are still reported for a limited number of PPCPs screened and measured in health-care facility effluents since PNECs are lacking. As such, future research on biological response (short-term and long-term effects and synergistic and antagonist effects) is essential.

To minimize the impact of PPCPs, the health-care facility effluents must be managed taking into consideration the source, the sewer network, the treatment level, and the catchment. If technically and economically feasible, it is recommended to have dedicated WWTPs operating with tertiary conditions when there are two or more health-care facilities (>250 beds) discharging into the same sewer network and the inflows are $<10,000\,m^3$/day. When no dedicated WWTP is available, it is recommended to have the health-care facilities treat the effluent prior to discharge into sewer networks. It is also recommended to adjust operational parameters at the municipal WWTP (e.g., sludge retention times and wastewater retention times) and to upgrade treatment conditions (PAC, O_3, UV, and advanced oxidation processes) when the catchment is characterized by low flow.

REFERENCES

Al Aukidy, M., Verlicchi, P., Voulvoulis, N., 2014. A framework for the assessment of the environmental risk posed by pharmaceuticals originating from hospital effluents. Sci. Total Environ. 493, 54–64.

Brodin, T., Fick, J., Jonsson, M., Klaminder, J., 2013. Dilute concentration of a psychiatric drug alter behavior of fish from natural populations. Science 339, 814–815. https://doi.org/10.1126/science.1226850.

Carraro, E., Bonetta, S., Bertino, C., Lorenzi, E., Bonetta, S., Gilli, G., 2016. Hospital effluents management: chemical, physical, microbiological risk and legislation in different countries. J. Environ. Manag. 168, 185–199.

Cizmas, L., Sharma, V.K., Gray, C.M., McDonald, T.J., 2015. Pharmaceuticals and personal care products in waters: occurrence, toxicity, and risk. Environ. Chem. Lett. https://doi.org/10.1007/s10311-015-0524-4.

Daouk, S., Chèvre, N., Vernaz, N., Widmer, C., Daali, Y., Fleury-Souverain, S., 2016. Dynamics of active pharmaceutical ingredients loads in a Swiss university hospital wastewater and prediction of the related environmental risk for the aquatic ecosystems. Sci. Total Environ. 537, 244–253.

Escher, B., Baumgartner, R., Killer, M., Treyer, K., Lienert, J., McArdell, C., 2011. Environmental toxicology and risk assessment of pharmaceuticals from hospital wastewater. Water Res. 45, 75–92.

Galus, M., Jeyaranjaan, J., Smith, E., Li, H., Metcalfe, C., Wilson, J.Y., 2013a. Chronic effects of exposure to a pharmaceutical mixture and municipal wastewater in zebrafish. Aquat. Toxicol. 132–133, 212–222. https://doi.org/10.1016/j.aquatox.2012.12.016.

Galus, M., Kirischian, N., Higgins, S., Purdy, J., Chow, J., Rangaranjan, S., Li, H., Metcalfe, C., Wilson, J.Y., 2013b. Chronic, low concentration exposure to pharmaceuticals impacts multiple organ systems in zebrafish. Aquat. Toxicol. 132–133, 200–211.

Gros, M., Petrović, M., Barceló, D., 2009. Tracing pharmaceutical residues of different therapeutic classes in environmental waters by using liquid chromatography/quadrupole-linear ion trap mass spectrometry and automated library searching. Anal. Chem. 81, 898–912.

Gros, M., Petrović, M., Barceló, D., 2010. Removal of pharmaceuticals during wastewater treatment and environment assessment using hazard indexes. Environ. Int. 36, 15–26.

Herrmann, M., Olsson, O., Fiehn, R., Herrel, M., Kümmerer, K., 2015. The significance of different health institutions and their respective contributions of active pharmaceutical ingredients to wastewater. Environ. Int. 85, 61–76.

Jelic, A., Gros, M., Ginebreda, A., Cespedas-Sánchez, R., Ventura, F., Petrović, M., Barceló, D., 2011. Occurrence, partition and removal of pharmaceuticals in sewage water and sludge during wastewater treatment. Water Res. 45, 1165–1176.

Joss, A., Keller, E., Alder, A., Göbel, A., McArdell, C.S., Ternes, T., Siegrist, H., 2005. Removal of pharmaceuticals and fragrances in biological wastewater treatment. Water Res. 39, 3139–3152.

Kosma, C.I., Lambropoulou, D.A., Albanis, T.A., 2014. Investigation of PPCPs in wastewater treatment plants in Greece: occurrence, removal and environmental risk assessment. Sci. Total Environ. 466–467, 421–438.

Kovalova, L., Siegrist, H., Singer, H., Wittmer, A., McArdell, C., 2012. Hospital wastewater treatment by membrane bioreactor: performance and efficiency for organic micropollutant elimination. Environ. Sci. Technol. 46, 1536–1545.

Kovalova, L., Siegrist, H., Von Gunten, U., Eugster, J., Hagenbuch, M., Wittmer, A., et al., 2013. Elimination of micropollutants during post-treatment of hospital wastewater with powdered activated carbon, ozone and U. Environ. Sci. Technol. 47, 7899–7908.

Lenz, K., Mahnik, S.N., Weissenbacher, N., Mader, R.M., Krenn, P., Hann, S., Koellensperger, G., Uhl, M., Knasmüller, S., Ferk, F., Bursch, W., Fuerhacker, M., 2007. Monitoring, removal and risk assessment of cytostatic drugs in hospital wastewater. Water Sci. Technol. 56 (12), 141–149.

Lerdau, M., Foster, W.G., Budnik, L.T., Zhu, B.-Z., Wang, A., 2013. Micropollutant point sources in the built environment: identification and monitoring of priority pharmaceutical substances in hospital effluents. J. Environ. Anal. Toxicol. 3, 4.

Lin, A.Y.-C., T-H, Y., Lin, C.-F., 2008. Pharmaceutical contamination in residential, industrial, and agricultural waste streams: risk to aqueous environments in Taiwan. Chemosphere 74, 131–141.

Lopes de Souza, S.M., Carvalho de Vasconcelos, E., Dziedzic, M., 2009. Environmental risk assessment of antibiotics: an intensive care unit analysis. Chemosphere, 962–967.

Luo, Y., Guo, W., Ngo, H.H., Nghiem, L.D., Hai, F.I., Zhang, J., Liang, S., Wang, X.C., 2014. A review on the occurrence of micropollutants in the aquatic environment and their fate and removal during wastewater treatment. Sci. Total Environ. 473–474, 619–641.

Oliveira, T.S., Murphy, M., Mendola, N., Wong, V., Carlson, D., Waring, L., 2015. Characterization of pharmaceuticals and personal care products in hospital effluent and waste water influent/effluent by direct-injection LC-MS-MS. Sci. Total Environ. 518–519, 459–478.

Orias, F., Perrodin, Y., 2013. Characterization of the ecotoxicity of hospital effluents: a review. Sci. Total Environ. 454–455, 250–276.

Ort, C., Lawrence, M.G., Reungoat, J., Eaglesham, G., Carter, S., Keller, J., 2010. Determining the fraction of pharmaceutical residues in wastewater originating from a hospital. Water Res. 44, 605–615.

Parolini, M., Pedriali, A., Binelli, A., 2013. Application of a biomarker response index for ranking the toxicity of five pharmaceuticals and personal care products (PPCP) to the bivalve Dreissena polymorpha. Arch. Environ. Contam. Toxicol. 64, 439–447.

Santos, L., Gros, M., Rodriguez-Mozaz, S., Delerue-Matos, C., Pena, A., Barceló, D., Montenegro, C., 2013. Contribution of hospital effluents to the load of pharmaceuticals in urban wastewaters: identification of ecologically relevant pharmaceuticals. Sci. Total Environ. 461–462, 302–316.

USEPA, 2010. Treating Contaminants of Emerging Concern—A Literature Review Database. EPA-821-R-08-002.

Verlicchi, P., Al Aukidy, M., Galletti, A., Petrović, M., Barceló, D., 2012a. Hospital effluent: investigation of the concentrations and distribution of pharmaceuticals and environment risk assessment. Sci. Total Environ. 430, 109–118.

Verlicchi, P., Al Aukidy, M., Zambello, E., 2015. What we learned from worldwide experiences on the management and treatment of hospital effluent?—An overview and a discussion on perspectives. Sci. Total Environ. 514, 467–491.

Verlicchi, P., Al Aukidy, M., Zambelo, E., 2012b. Occurrence of pharmaceutical compounds in urban wastewater: removal, mass load and environmental risk after a secondary treatment—a review. Sci. Total Environ. 429, 123–155.

Verlicchi, P., Galletti, A., Petrovic, M., Barceló, D., 2010. Hospital effluents as a source of emerging pollutants: an overview of micropollutants and sustainable treatment options. J. Hydrol. 389, 416–428.

Chapter 3

Fate and Behavior of Environmental Contaminants Arising From Health-Care Provision

Mike Williams and Rai S. Kookana
CSIRO, Adelaide, SA, Australia

INTRODUCTION

A range of chemicals are required to be used for effective health-care. They can include therapeutic agents such as active pharmaceutical ingredients (APIs), including antibiotics, disinfectants for controlling transfer of infectious diseases within health-care facilities (nosocomial infections), and chemicals that are required for procedures that specifically take place within health-care facilities, such as surgery and medical imaging. A vast majority of these chemicals are organic chemicals, which will be the focus of this chapter. These contaminants have a diverse range of physicochemical properties including K_{OW} (a measure of a compound's hydrophobicity), ionizable functional groups and their respective acid/base dissociation constant (pK_a) values, water solubility, volatility, chemical functional groups, and overall structure, which will determine their fate and behavior in the environment. This indicates that a number of fate processes will be applicable to health-care contaminants in the environments into which they are released, which is applicable not only among the many contaminant classes but also for a particular contaminant present in various receiving environments.

ENVIRONMENTAL FATE PROCESSES

As discussed in the previous chapter, most health-care contaminants are expected to initially enter the aquatic environment through discharges from wastewater treatment plants (WWTPs), which receive wastewater from both domestic and health-care facility sources. Discharges from WWTPs can also

Environmental Contaminants, Vol. 11. https://doi.org/10.1016/B978-0-444-63857-1.00003-6

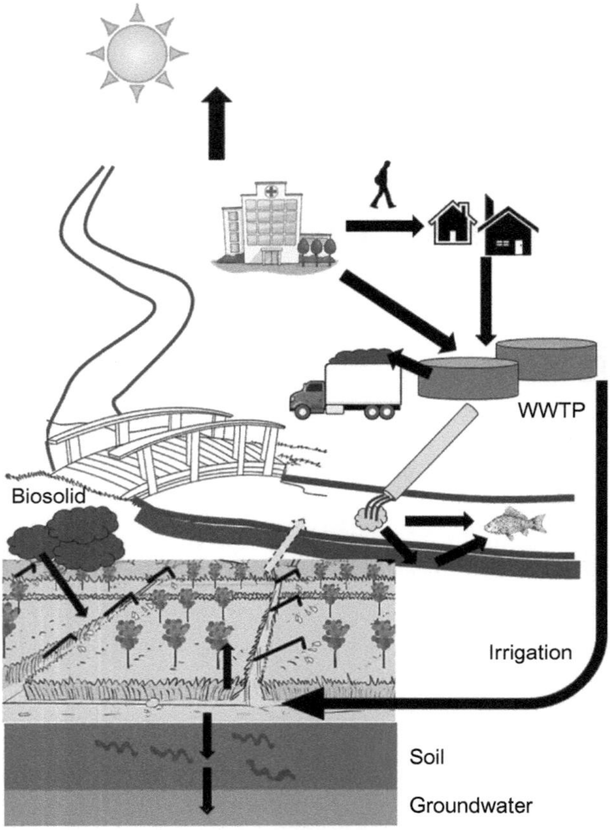

FIG. 1 Overview of potential pathways of health-care contaminants into the environment and their distribution within the environment (depicted by arrows).

lead to terrestrial contamination through transfer of biosolids or wastewater for fertilization and irrigation of agricultural land (Fig. 1). When a chemical associated with health care is released into ecosystems, a number of physical, chemical, and biological factors of the system and the chemical can influence their fate (Table 1). The fate of a contaminant, or chemical released into the environment, can relate to its transport or phase transfer (e.g., from liquid to solid or gas phases) within the receiving environment. A chemical contaminant can maintain its original chemical structure, or it can be transformed so that its chemical structure is partially modified or even mineralized, where it is broken down into basic chemical components such as CO_2 and H_2O.

Phase transfer processes include partitioning or sorption (transfer from liquid to solid phase), desorption (transfer from solid to liquid phase), volatilization (transfer from liquid/solid to air), and biological uptake (transfer from

TABLE 1 Major Fate Processes of Health-Care Contaminants in the Environment that may Affect Phase Transfer (Movement Among Environmental Compartments) or Transformation (Degradation of Chemical Structure) of the Contaminant

Contaminant Fate	Process	Key Properties Relevant to the Process	Contaminant Example	Notable Property of the Contaminant
Phase transfer	Sorption	Moderate to high log K_{OW} (or D_{OW}[a]), for example, >3 → increased sorption Ionization; unionized (neutral) or positive (cation) → increased sorption; negative (anion) → decreased sorption	Benzalkonium chloride (QAC) Ciprofloxacin (quinolone antibiotic)	Log K_{OW} = 2.9 Log K_{OW} = 0.28 Both are cations
	Uptake (organism)	Unionized (neutral) → increased uptake Moderate K_{OW} (or D_{OW}), for example, ~3 → increased uptake (plant) Moderate to high log K_{OW} (or D_{OW}), for example, >3 → increased uptake (dermal)	Carbamazepine (API) Propofol (anesthetic)	Log K_{OW} = 2.5 Log K_{OW} = 3.8 (Both unionized)
	Volatilization	High Henry's law constant value, for example, $>10^{-5}$ atm m^3/mol Low water solubility	Isoflurane (inhalation anesthetic)	H = 0.03 atm m^3/mol
	Dilution	Water solubility (>environmental concentration)	All (discharged into aquatic environment)[b]	Not applicable

Continued

TABLE 1 Major Fate Processes of Health-Care Contaminants in the Environment that may Affect Phase Transfer (Movement Among Environmental Compartments) or Transformation (Degradation of Chemical Structure) of the Contaminant—Cont'd

Contaminant Fate	Process	Key Properties Relevant to the Process	Contaminant Example	Notable Property of the Contaminant
Transformational	Hydrolysis	Contains a functional group that reacts with H_2O (or other nucleophiles in solution)	Amoxicillin (β-lactam antibiotic)	Contains multiple functional groups that react with H_2O (H^+ and OH^-)
	Photolysis	Contain functional groups that react with photons (direct) Contain functional groups that react with excited species (e.g., $OH^{\cdot}$)[c]	Sulfamethoxazole[d] (sulphonamide antibiotic)	Contains photolabile isoxazole ring for direct photolysis
	Biodegradation	Contain functional groups susceptible to microbially mediated processes Bioavailable (low sorption)	Phenylphenol (phenolic disinfectant)	Phenols susceptible to hydroxylase activity

[a] K_{OW} *adjusted for ionization state of ionizable compounds.*
[b] *Health-care contaminants are likely to be discharged at concentrations well below their water solubility values, which are generally >mg/L.*
[c] *Species such as* $OH^{\cdot}$ *are nonspecific and expected to be reactive with many contaminants.*
[d] *Photolysis of sulfonamides are pH-dependent due to ionization changing chemistry of compounds.*

liquid/solid/air to within a biological organism). A process such as dilution can mean a contaminant remains unchanged within a particular phase, such as liquid, with its concentration being reduced through addition of the same, uncontaminated phase. Phase transfer of a health-care contaminant can be an important factor in determining the mobility and bioavailability of a compound. For example, a high degree of sorption and a correspondingly low degree of desorption can significantly reduce the biological uptake, or bioavailability, of a contaminant.

Environmental processes that can transform a chemical include abiotic processes, such as photolysis and hydrolysis, where chemical reactions drive the transformation of the contaminant. During photolysis, compounds may be transformed by direct or indirect photolysis. Direct photolysis is where a compound is transformed by absorbing photons directly without the need of an intermediate reactive species. Conversely, indirect photolysis requires an intermediate species that is excited by direct photolysis (e.g., $OH^{\bullet}$ radicals) for the photolysis reaction to occur. Also, biotic processes can significantly modify the chemical structure through metabolic modification of the original contaminant or even mineralization. The importance of each environmental fate process can vary greatly among environmental compartments into which a contaminant is released. For example, chemical hydrolysis (reaction with water) and photolysis (reaction with sunlight or reactive species produced from sunlight) may be more important in aquatic environments, relative to terrestrial environments. Conversely, sorption to solids and microbial biodegradation may be more important processes within terrestrial environments. This does not mean, however, that certain processes will not occur in either aquatic or terrestrial environments, and often, a number of processes are important in influencing the overall fate of a contaminant.

PHYSICOCHEMICAL PROPERTIES OF HEALTHCARE CONTAMINANTS

The physicochemical properties of the contaminant itself will also have an important bearing on the contaminant's susceptibility to various fate processes within a receiving environment (Table 1). Some of these properties include the preference of the contaminant to remain in water (or hydrophilicity), its water solubility, the presence of ionizable functional groups, and its tendency to volatilize into the air. The hydrophilicity or, conversely, lipophilicity of a contaminant influences whether it remains within solution, binds to particulate matter, or crosses biological membranes for uptake into organisms (Schwarzenbach et al., 2003). A measure of hydrophilicity is the octanol-water partition coefficient (K_{OW}), with a higher value indicating a greater preference for partitioning into lipophilic octanol. Chemicals with higher K_{OW} values tend to have a greater tendency for sorption to environmental solids and to pass through biological membranes (Cunningham, 2004).

Further to this, many organic contaminants derived from health-care activities have ionizable functional groups, which are dependent on the pH of the environment to determine whether they exist in their ionized or charged state. The pH at which a chemical is charged is centered around its acid/base dissociation coefficient (pK_a) value (Cunningham, 2004). A charged form of a chemical will generally have a higher water solubility and lower K_{OW} compared with its uncharged state. Other factors that can affect water solubility of a contaminant include the concentration of salt in solution, where high salt concentrations (such as in marine environments) can reduce the water solubility of contaminants (Turner and Rawling, 2001). These factors often have a greater significance for contaminants with very low water solubilities.

Physicochemical properties of health-care contaminants and their use patterns mean that they are released into aquatic, terrestrial, and atmospheric environments (Fig. 1). Volatile chemicals can be released directly into the atmosphere, although volatilization can also occur from aquatic and terrestrial environments. Long-range transport can occur in the atmosphere, with chemical processes such as photolysis or reaction with radicals (especially $OH^{\bullet}$) formed from photolysis of other species being the most important fate processes (Schwarzenbach et al., 2003).

RECEIVING ENVIRONMENTS

Important processes that can influence the fate and distribution of health-care contaminants within the aquatic environment include transformation through photolysis, hydrolysis, or biological degradation (Fig. 2). Sorption to suspended particulate matter and bed sediments, uptake by aquatic organisms, dilution, and transport within the water column are fate processes that can redistribute health-care contaminants throughout the aquatic environment. Similar fate and transport processes can also be found within the terrestrial environment, although the relative importance of sorption and biodegradation may be greater in soil environments (Fig. 3). Also, water-associated processes, such as hydrolysis and transport in solution, are highly dependent on the relative amount of water present, while photolysis will only be significant on the soil surface. Within aquatic and terrestrial environments, the biogeochemical properties of the system (such as temperature, pH, solar radiation, organic carbon and clay content, microbial communities, and hydrology) are as important as the physicochemical properties of the contaminant (such as water solubility, K_{OW}, molecular size/shape, and ionizability), and the interplay between the two will determine the overall fate of the contaminant (Schwarzenbach et al., 2003). For example, ionized compounds will have a correspondingly lower pH-adjusted log K_{OW} (or D_{OW}) than their unionized K_{OW} and will also have a higher water solubility, and this degree of ionization is dependent on the pH of ionization (indicated by the acid dissociation constant, pK_a) of ionizable functional group(s) (Cunningham, 2004). The pH of

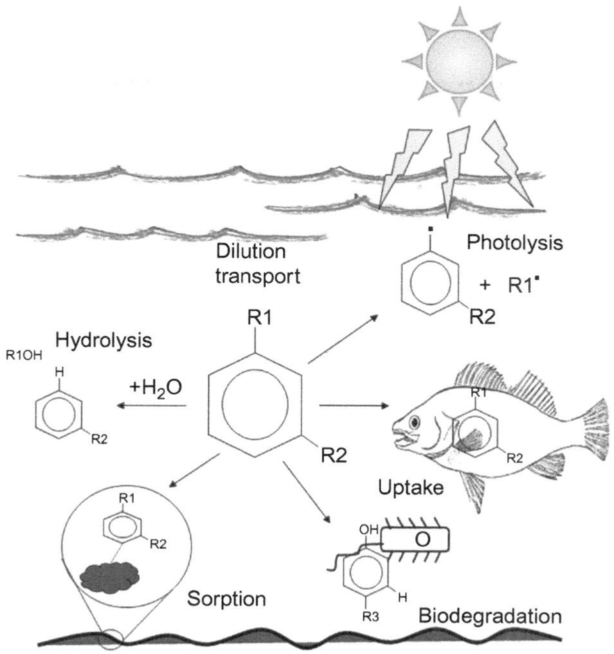

FIG. 2 Major fate processes for health-care contaminants in the aquatic environment, including the water column and sediments.

the environment into which contaminants are discharged will therefore have an important bearing on the fate of ionizable APIs. Also, contaminants with a K_{OW} (or D_{OW} for ionizable compounds) of <1 are highly unlikely to sorb onto particulate matter or accumulate in aquatic organisms (Cunningham, 2004). Furthermore, other environmental factors including salinity, temperature, and suspended and dissolved particulate matter will have an important influence on the fate of contaminants within the aquatic environment (Schwarzenbach et al., 2003). The organic carbon and clay content of soils and sediments will determine the extent of association with particulate matter, as much as the K_{OW}, water solubility, and ionization state (where applicable) of the contaminant. Aquatic and terrestrial environments into which health-care contaminants are discharged would be expected to vary considerably, with a broad range of these biogeochemical properties within different systems also expected to be highly variable.

Aquatic Environment

Health-care contaminants can enter sewerage networks directly from health-care facilities, although the majority of such contaminants are those that are

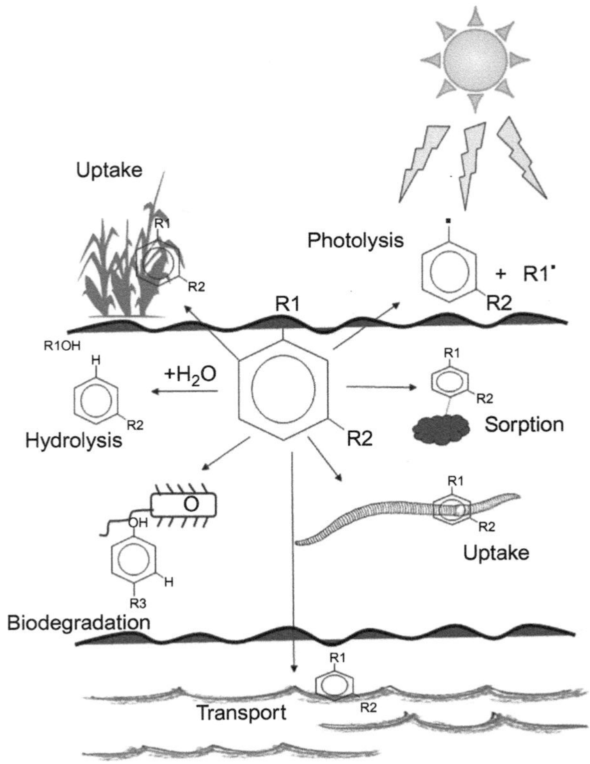

FIG. 3 Major fate processes for health-care contaminants in the terrestrial environment, from the surface to the groundwater.

highly specific to health-care facilities, including high-level disinfectants and sterilants, anesthetics, and restricted-use antibiotics (Le Corre et al., 2012). The majority of health-care contaminants are released into sewerage through domestic sources due to ongoing therapy or metabolism/excretion following discharge or overlap of consumer and health-care chemicals, such as APIs and disinfectants (Ort et al., 2010; Le Corre et al., 2012). Mitigation of contaminant concentrations can also occur prior to wastewater treatment through metabolism within patients or dilution in sewerage networks (Lienert et al., 2007; Verlicchi et al., 2015).

Fate Processes Impacting Wastewater Treatment

Wastewater may be treated onsite at a health-care facility or, more commonly, discharged to a centralized municipal WWTP, which typically only makes up a fraction of the total WWTP influent flows (Ort et al., 2010; Verlicchi et al., 2015). Municipal WWTPs are specifically designed to substantially reduce concentrations of nutrients and pathogens in sewage, although the enhanced biological activity and associated suspended flocs mean that biodegradation and sorption are also key removal processes for nontarget contaminants. The effectiveness of contaminant removal is generally a function of the type of treatment process used by the WWTP, as well as the physicochemical properties of the contaminant (Gardner et al., 2013; Rojas et al., 2013; Verlicchi et al., 2015). Removal rates of contaminants such as APIs (including cytotoxic, antibiotic, and antiviral APIs) have been previously well studied, with the range of physicochemical properties represented within this group of contaminants having a correspondingly broad range of removal rates within WWTPs (Kasprzyk-Hordern et al., 2009; Gros et al., 2010; Gardner et al., 2013; Rojas et al., 2013). Health-care contaminants that are susceptible to biodegradation include some antiviral, anesthetic, and general APIs and phenolic and aldehydic disinfectants, and they are efficiently removed during WWTP treatment (Rudel et al., 1998; Eiroa et al., 2005; Jewell et al., 2014). Quaternary ammonium compounds (QACs), used widely as disinfectants, have also been identified as being susceptible to biodegradation in laboratory-based investigations. The high affinity of QACs to particulate matter, however, can mitigate biodegradation making sorption more likely to dominate as a removal process over biodegradation (Clara et al., 2007; Kreuzinger et al., 2007; Zhang et al., 2015). While some cytostatic agents (e.g., methotrexate and cytarabine) have been identified as being susceptible to biodegradation, cytostatics as a class of contaminants generally have inherently low biodegradability (Buerge et al., 2006; Garcia-Ac et al., 2009; Besse et al., 2012; Franquet-Griell et al., 2016).

Sorption to suspended particulate matter within WWTPs is an important process for many health-care contaminants, particularly those with ionizable functional groups. For example, contaminants with a high water solubility and low to moderate K_{OW} values, such as quinolone and tetracycline antibiotics and QACs, can still have a very high affinity to particulate matter due to their ionized functional groups and, in some cases, lipophilic regions of their chemical structures (Guerra et al., 2014; Van Doorslaer et al., 2014; Zhang et al., 2015). The quinolone antibiotics can exist as multicharged species, enhancing their degree of interaction with particulate matter. Tetracycline antibiotics and the structurally related anthracycline cytostatic APIs can also form complexes with metal ions and particulate matter, leading to sorption being enhanced beyond what might be predicted from their physicochemical properties (Parolo et al., 2008). Despite containing charged functional groups under environmental conditions, however, some APIs (such as antivirals) have only a minimal affinity to particulate matter.

A number of contaminants can also be efficiently removed from WWTPs through chemical degradation, including high-use β-lactam ring-containing penicillin antibiotics, which are susceptible to hydrolysis (Hirsch et al., 1999). This is reflected in penicillins generally not being detected (or only detected at very low concentrations) in WWTP effluents despite their comparatively high usage rates (Watkinson et al., 2007; Kummerer, 2009).

Although WWTPs have enhanced conditions for removal of organic contaminants, a large number of health-care contaminants are only partially removed during WWTP treatment. A number of antiviral, antibiotic, cytotoxic, and general APIs, along with contrast media, are found in effluents discharged from WWTPs due to their resistance to removal. Some WWTPs have enhanced disinfection processes, including ozonation and chlorination, which can also increase the rate of removal of some contaminants, although complete removal is still not always achieved. Contrast media, such as organo-gadolinium complexes used in magnetic resonance imaging (MRI), and the API carbamazepine are so resistant to removal (usually 0%–10%) under most circumstances that they can be used to trace WWTP discharges in receiving aquatic environments (Liu et al., 2014; Hatje et al., 2016).

WWTP effluents are typically released directly into marine or freshwater receiving systems, carrying many of the health-care contaminants that have not been removed during the treatment process.

Fate in the Aquatic Ecosystem

When a health-care contaminant is discharged into an aquatic ecosystem, the concentration of the contaminant will immediately be greatly reduced by dilution, unless WWTP effluent makes up the majority of the environmental flow. Although dilution does not transform a contaminant, it can mitigate environmental risks to some extent by lowering the concentration of exposure to aquatic organisms. Depending on the hydrologic (flow) and geochemical characteristics of a receiving system, health-care contaminants can then be transported within this system, either dissolved in solution or bound to particulate matter. Because of the number of variables that contaminant transport relies on, this can be highly transient even within a single receiving system (Schwientek et al., 2016).

As in WWTPs, biodegradation and sorption are still likely to be important drivers of fate in aquatic ecosystems. A greater degree of variability (in terms of stability of environmental parameters) and lower levels of biological activity and particulate matter, however, can make sorption and biodegradation rates comparatively less important in aquatic ecosystems, with the exception of the benthic zone. Many health-care contaminants that strongly associate with particulate matter in WWTPs can still be discharged in effluent, albeit at much lower concentrations than in influent, and also sorb to particulate matter in the receiving environment. As a consequence, health-care contaminants such as antibiotics and, in particular, QACs are widespread in sediments of both freshwater and marine environments, which receive WWTP discharges (Simon, 2005; Kreuzinger et al., 2007; Martinez-Carballo et al., 2007; Li and Brownawell, 2010; Yang et al., 2010; Gibs et al., 2013; Xue et al., 2013; Ruan et al., 2014; Zhang et al., 2015). Strong and irreversible sorption of contaminants to sediments may make them less available for biodegradation, even where they may be susceptible to biodegradation (Li and Brownawell, 2010; Van Doorslaer et al., 2014).

Processes such as photolysis, induced directly or indirectly by exposure to solar irradiance, may become more important in aquatic ecosystems in comparison with highly turbid WWTP waters. A number of antibiotics (especially quinolones, sulfonamides, and tetracyclines) have chemical structures that can absorb photons from solar irradiance making them susceptible to chemical transformation. Direct photolysis of these antibiotics can occur within hours, although irradiation intensity, water chemistry, and pH (among other variables) play a role in the rate of degradation in the environment (Kummerer, 2009; Challis et al., 2014). Even where dissolved organic matter (DOM) present in aquatic ecosystems diminishes the impact of direct photolysis, indirect photolysis through reaction of reactive DOM species can still play an important role in aquatic degradation of antibiotics (Challis et al., 2014).

TERRESTRIAL ENVIRONMENT

Waste streams from WWTPs are not only directed into aquatic systems. Effluents and biosolids from WWTPs are increasingly becoming a globally valued resource, with respect to the need for water reuse and nutrient application in agriculture. Health-care-derived contaminants can therefore enter the terrestrial environment from WWTPs or from waters receiving WWTP effluent, with the degree of wastewater treatment being an important factor in determining the loads of contaminants present in irrigation water or biosolids.

Fate in the Terrestrial Ecosystem

The dominant fate process in the terrestrial environment is related to the extent of contaminant interaction with soil (sorption and desorption), which has an influence on other critical fate parameters, such as biodegradation and transport. Desorption is also especially pertinent to health-care contaminants bound to biosolids. Since desorption is an indication of the strength of the bond between a contaminant and particulate matter, the extent of desorption from land-applied biosolids will be important in determining the amount of a health-care contaminant being released into the terrestrial environment. The longer a contaminant interacts with particulate matter, the less likely the contaminant is to desorb, which is often referred to as "aging" or "hysteresis" (Bowman and Sans, 1985; Pignatello and Xing, 1996; Delgado-Moreno and Gan, 2013).

Also, the extent of desorption of a contaminant into soil solution is related to the amount of the contaminant available for biological access within the soil environment. This will give an indication of whether a contaminant will be susceptible to biodegradation. For example, QACs can have a strong and irreversible affinity with particulate matter making them less available for biodegradation, despite their inherent biodegradability, with aging expected to exacerbate this situation (Li and Brownawell, 2010). Conversely, sulfonamide antibiotics are anionic and have low K_{OW} (and D_{OW}) values meaning they are readily desorbed from particulate matter and, therefore, more susceptible to biodegradation in soils (Gao and Pedersen, 2005; Accinelli et al., 2007; Wu et al., 2009; Liu et al., 2010). Particulate-bound contaminants may still be biologically available due to ingestion and digestion of sorbed contaminants by soil organisms such as earthworms (Wang et al., 2014).

With respect to biological transformation of health-care contaminants in the terrestrial environment, transformation through microbial activity is likely to play a dominant role in terrestrial systems (Schwarzenbach et al., 2003). While biodegradation is a major transformation pathway in terrestrial environments, however, the extent of biodegradation is likely to be highly variable as in aquatic systems. Environmental factors that affect the physicochemical properties of a contaminant and its bioavailability in the soil environment are important but not the only ones to consider. Environmental factors that affect the activity and diversity of microbial communities present in soil, such as temperature, moisture and oxygen content of soils, and contaminant concentrations, will also play a critical role in biodegradation rates. Either the biodegradation of a health-care contaminant can be due to direct metabolic utilization by microbes or it may be indirectly degraded through production of enzymes used to degrade other chemicals present in the soil (also known as "cometabolism") (Barra Caracciolo et al., 2015). Direct utilization of a contaminant may require a degrading microbial community to first adapt the ability to degrade it before contamination takes place, which is known as a lag period (Schwarzenbach et al., 2003).

Another important consideration in the biodegradation of contaminants relates to the amount of oxygen present. Movement down a soil profile leads to a rapid decline in oxygen levels, which, along with carbon levels, has a major effect on the composition and diversity of microbial communities (Hansel et al., 2008). It is generally found that degradation of organic contaminants under high oxygen, or aerobic, conditions is substantially more rapid than under low oxygen, or anaerobic, conditions. Oxygen levels have a critical role in determining the metabolic pathways that microorganisms use, while the functional groups within the chemical structure can influence susceptibility to metabolic breakdown under either aerobic or anaerobic conditions (Schwarzenbach et al., 2003).

Wastewater and Biosolids Application on Land

Many health-care contaminants, such as antibiotics and QACs, are present in biosolids due to their high affinity to particulate matter in WWTPs, although a number of APIs with relatively low affinity for particulate matter are also detected in biosolids (Topp et al., 2008; Clarke and Smith, 2011). This is due to their relatively high loads in water during WWTP treatment still leading to some degree of partitioning to biosolids, with leaching of these contaminants from the biosolids during, for example, rainfall events (Topp et al., 2008). Even where a contaminant has a higher degree of affinity to particulate matter, they can still be transported within the terrestrial environment through colloid transport processes, which can occur both on the soil surface and in the subsurface (de Jonge et al., 2004).

Direct exposure of the terrestrial environment to health-care contaminants can occur through irrigation of crops with wastewater or with wastewater-impacted surface waters, with wastewater irrigation increasingly becoming a viable option for many dryland regions (Chefetz et al., 2008; Siemens et al., 2008; Kookana et al., 2014; Qin et al., 2015). This has the potential to introduce health-care-derived contaminants present in wastewater effluents and particularly those with resistance to degradation, lower affinity for solids, and higher water solubility. Health-care contaminants with these characteristics are more likely to be mobile in terrestrial environments and can accumulate in soils to a greater extent than biosolid-associated APIs (Wu et al., 2015). More mobile contaminants are likely to be transported through soil to an extent where a number, including APIs (carbamazepine and meprobamate) and antibiotics (trimethoprim and sulfamethoxazole), were detected in groundwater where wastewater irrigation had occurred (Snyder et al., 2004; Kinney et al., 2006). There is also increasing evidence that uptake of a number of APIs into plants can occur where APIs are present in soils. Although APIs of greater mobility are more likely to be present in the bioavailable fraction of the soil environment, maximal uptake of neutral contaminants into plants occurs at moderate log K_{OW} values (Sicbaldi et al., 1997; Carter et al., 2014), while ionization can reduce plant bioavailability in soil (Goldstein et al., 2014).

Health-care waste generated in health-care facilities is either disposed of through incineration or treated to remove pathogenic microbes (e.g., autoclaving) and disposed of in landfill, which may introduce a range of other contaminants directly to the terrestrial environment (Windfeld and Brooks, 2015). For example, polyvinyl chloride (PVC) is a plastic used extensively in its plasticized form within health-care facilities due to a number of desirable properties, including flexibility and impermeability. Up to half the weight of plasticized PVC can be composed of plasticizers (especially phthalic acid esters or PAEs) and stabilizers, which are easily leachable from the PVC (Chiellini et al., 2013). Such compounds have important implications for ecotoxicity, including modifying endocrine function (Magdouli et al., 2013). PAEs have phthalic acid as a common backbone for their structure but can vary widely in physicochemical properties depending on attached functional groups. As with other contaminants, the fate of a particular PAE is dependent on its chemical structure, with water solubility, lipophilicity, and volatility varying widely among individual PAEs. PAEs are widespread environmental contaminants, being detected in aquatic, terrestrial, and atmospheric environments due to their high-volume use and chemical properties (Net et al., 2015).

ATMOSPHERE

Although the majority of contaminants derived from health-care facilities have physicochemical properties making volatilization in the environment unlikely (e.g., low value of Henry's law constant), a number of compounds may still be expected to be present in the atmosphere following discharge into the environment. Halogenated agents containing chlorine, bromine, and/or fluorine, including halothane isoflurane, desflurane, and sevoflurane, are commonly used inhalation anesthetics (Sulbaek Andersen et al., 2010). These anesthetics are commonly delivered to a patient along with O_2 and CO_2 carrier gases and another anesthetic, N_2O. Due to the halogenated moieties contained in these compounds, they have the ability to absorb infrared radiation more effectively than CO_2 (Sulbaek Andersen et al., 2010; Vollmer et al., 2015).

The halogenated anesthetics are characterized by having a low degree of metabolism in the body and are rapidly cleared from the body through the lungs. These discharges are vented directly into the atmosphere to protect health-care workers in the same room as the patient (Vollmer et al., 2015).

It has been estimated that the number of anesthesia-based procedures worldwide (~200 million) would have the equivalent global warming potential of 1 million motor cars (Sulbaek Andersen et al., 2010). Halogenated anesthetics, especially halothane and isoflurane, also have the potential to degrade atmospheric ozone although their ozone-depleting potential is considerably less than other industrial chemicals (Kummerer, 2004; Vollmer et al., 2015). N_2O also has climate-forcing potential, but quantifying its atmospheric inputs is difficult due to high levels of natural N_2O emissions, and, because of these natural emissions, it is unlikely that health-care-related releases of N_2O will substantially contribute to overall greenhouse gas emissions (Vollmer et al., 2015). Despite the relatively low contributions of anesthetics to global climate forcing, local management of emissions, through recycling exhausts, reduced consumption, or preferential use of some anesthetics over others, is still desirable in reducing atmospheric inputs (Vollmer et al., 2015).

A substantial proportion of health-care waste is disposed of via incineration, which has the potential to release contaminants, such as mercury or incineration by-products, such as dioxins and furans (Windfeld and Brooks, 2015). This is especially important where emission controls are not in place due to the lack of regulation or availability of control technologies (Patwary et al., 2011; Windfeld and Brooks, 2015).

DISPOSAL OF POORLY TREATED HEALTHCARE WASTE

The preceding discussion has focused on the situation that exists in high-income countries (HICs), where treated discharges from WWTPs are the primary source of health-care-derived contaminants in the environment due to the majority (>90%) of the population having connectivity with sewers

(Kookana et al., 2014). Rapidly increasing wealth in many low- to medium-income countries (LMICs) means that access to health care is also rapidly increasing although regulation and management of wastes is not necessarily keeping up with this demand (Prem Ananth et al., 2010; Kookana et al., 2014; Porter and Grills, 2016). Also exacerbating this problem is the increase in medical tourism from HIC to access lower-cost health care in LMICs (Imison and Schweinsberg, 2013). In LMICs, which represent ~80% of the global population, sewer connectivity is generally poor (Kookana et al., 2014). This means that the proportion of health-care contaminants entering the environment through centralized treatment processes is likely to be considerably less in LMICs with more widespread discharge into aquatic environments (compared with "point-source" WWTP discharges). Where connectivity is available, however, sewerage waste streams may also bypass treatment systems or have WWTP treatment levels that are poor or nonexistent so that highly concentrated effluents are discharged into the aquatic environment (Kookana et al., 2014). In both cases, the fate processes in the receiving environment are where the majority of transport and transformation of health-care contaminants occurs. Similarly, terrestrial environments can also directly receive poorly treated wastewater and sludge for agricultural irrigation and fertilization (Kookana et al., 2014).

Nonliquid health-care waste in LMICs is often disposed through incineration, and with pollution control devices being rarely employed in such scenarios, the release of toxic emissions is likely to be considerably higher than in HICs (Prem Ananth et al., 2010). Furthermore, with low adherence to regulatory requirements in LMICs, solid waste materials from health-care facilities can also be disposed of directly to landfill (Patwary et al., 2011). Aside from the introduction of solid wastes and associated contaminants into landfill as is also the case in HICs, improperly treated waste streams from health care in LMICs raises important implications for direct exposure to contaminants such as pathogens (Prem Ananth et al., 2010; Windfeld and Brooks, 2015).

Poor treatment of liquid and nonliquid health-care wastes containing contaminants such as antimicrobials and pathogens, alongside low levels of environmental sanitation, is raising concerns globally of the potential for such conditions to favor the formation and environmental transfer of antimicrobial resistance (AMR) (Larsson, 2014; WHO, 2014).

CONCLUSIONS

- Many contaminants derived from health care are also likely to be sourced from the broader community, including agents used specifically within health-care facilities such as anesthetics (and associated agents) and contrast media. Antiinfectives (antibiotics, antivirals, and antifungals) and disinfectants, also used within the broader community, are more likely to be used to a greater extent within health-care settings.

- Wastewater discharges are the main source of entry into the environment, where connectivity with sewerage networks exists. Removal within WWTPs, through biodegradation and sorption to particulate matter, is highly variable and dependent on the physicochemical properties of the contaminant.
- Contaminants discharged into the aquatic environment can still undergo biodegradation and sorption, although this may be less rapid than within WWTPs. Highly mobile contaminants are more likely to be discharged into aquatic environments, but contaminants with a high affinity to particulate matter can still pass through WWTPs, due to incomplete sorption, and associate with sediments in receiving aquatic environments.
- Photolysis (direct and indirect) may become a more important fate process, under suitable conditions, for susceptible contaminants discharged from WWTPs into aquatic ecosystems.
- Entry of contaminants into the terrestrial environment can occur through land application of biosolids, although the mobility in terrestrial environments of many particulate-bound contaminants is not well defined. Contaminants can also enter terrestrial environments through wastewater irrigation, where desorption from biosolids is less critical in defining their bioavailability. Bioavailable contaminants are more vulnerable to biodegradation, although this also implies greater accessibility to other soil organisms, such as plants. There has been increasing interest in transfer to plant crops, although this has largely focused on only select number of APIs.
- With the exception of some classes of antibiotics (e.g., those used in veterinary applications), there is considerably less information available relating to the presence and fate of health-care contaminants in the terrestrial environment. This relates to not only discharges from WWTPs but also where landfill applications of health-care waste occur, which can introduce into the terrestrial environment other classes of contaminants not discussed within this chapter.
- Landfill wastes and poorly treated sewerage discharges (to aquatic and terrestrial environments) are especially critical in lower- to middle-income countries (LMICs) where waste management and regulation has not kept pace with rapid increases in access to health care. The risks to human and ecosystem health are poorly defined in such situations, and an understanding of classes of contaminants and their respective fate needs to be considered in a much broader context than has been discussed in this chapter.

REFERENCES

Accinelli, C., Koskinen, W.C., Becker, J.M., Sadowsky, M.J., 2007. Environmental fate of two sulfonamide antimicrobial agents in soil. J. Agric. Food Chem. 55 (7), 2677–2682.

Barra Caracciolo, A., Topp, E., Grenni, P., 2015. Pharmaceuticals in the environment: biodegradation and effects on natural microbial communities: a review. J. Pharm. Biomed. Anal. 106, 25–36.

Besse, J.P., Latour, J.F., Garric, J., 2012. Anticancer drugs in surface waters what can we say about the occurrence and environmental significance of cytotoxic, cytostatic and endocrine therapy drugs? Environ. Int. 39 (1), 73–86.

Bowman, B.T., Sans, W.W., 1985. Partitioning behaviour of insecticides in soil-water systems: II. Desorption hysteresis effects. J. Environ. Qual. 14 (2), 270–273.

Buerge, I.J., Buser, H.R., Poiger, T., Muller, M.D., 2006. Occurrence and fate of the cytostatic drugs cyclophosphamide and ifosfamide in wastewater and surface waters. Environ. Sci. Technol. 40 (23), 7242–7250.

Carter, L.J., Harris, E., Williams, M., Ryan, J.J., Kookana, R.S., Boxall, A.B.A., 2014. Fate and uptake of pharmaceuticals in soil-plant systems. J. Agric. Food Chem. 62 (4), 816–825.

Challis, J.K., Hanson, M.L., Friesen, K.J., Wong, C.S., 2014. A critical assessment of the photodegradation of pharmaceuticals in aquatic environments: defining our current understanding and identifying knowledge gaps. Environ. Sci.-Process. Impacts 16 (4), 672–696.

Chefetz, B., Mualem, T., Ben-Ari, J., 2008. Sorption and mobility of pharmaceutical compounds in soil irrigated with reclaimed wastewater. Chemosphere 73, 1335–1343.

Chiellini, F., Ferri, M., Morelli, A., Dipaola, L., Latini, G., 2013. Perspectives on alternatives to phthalate plasticized poly(vinyl chloride) in medical devices applications. Prog. Polym. Sci. 38 (7), 1067–1088.

Clara, M., Scharf, S., Scheffknecht, C., Gans, O., 2007. Occurrence of selected surfactants in untreated and treated sewage. Water Res. 41 (19), 4339–4348.

Clarke, B.O., Smith, S.R., 2011. Review of 'emerging' organic contaminants in biosolids and assessment of international research priorities for the agricultural use of biosolids. Environ. Int. 37 (1), 226–247.

Cunningham, V.L., 2004. Special characteristics of pharmaceuticals related to environmental fate. In: Kummerer, K. (Ed.), Pharmaceuticals in the Environment, second ed. Springer-Verlag, Berlin, pp. 13–24.

de Jonge, L.W., Kjaergaard, C., Moldrup, P., 2004. Colloids and colloid-facilitated transport of contaminants in soils: an introduction. Vadose Zone J. 3 (2), 321–325.

Delgado-Moreno, L., Gan, J., 2013. A stable isotope dilution method for measuring bioavailability of organic contaminants. Environ. Pollut. 176, 171–177.

Eiroa, M., Kennes, C., Veiga, M.C., 2005. Simultaneous nitrification and formaldehyde biodegradation in an activated sludge unit. Bioresour. Technol. 96 (17), 1914–1918.

Franquet-Griell, H., Ventura, F., Boleda, M.R., Lacorte, S., 2016. Do cytostatic drugs reach drinking water? The case of mycophenolic acid. Environ. Pollut. 208, 532–536.

Gao, J.A., Pedersen, J.A., 2005. Adsorption of sulfonamide antimicrobial agents to clay minerals. Environ. Sci. Technol. 39 (24), 9509–9516.

Garcia-Ac, A., Segura, P.A., Gagnon, C., Sauve, S., 2009. Determination of bezafibrate, methotrexate, cyclophosphamide, orlistat and enalapril in waste and surface waters using on-line solid-phase extraction liquid chromatography coupled to polarity-switching electrospray tandem mass spectrometry. J. Environ. Monit. 11 (4), 830–838.

Gardner, M., Jones, V., Comber, S., Scrimshaw, M.D., Coello-Garcia, T., Cartmell, E., Lester, J., Ellor, B., 2013. Performance of UK wastewater treatment works with respect to trace contaminants. Sci. Total Environ. 456–457, 359–369.

Gibs, J., Heckathorn, H.A., Meyer, M.T., Klapinski, F.R., Alebus, M., Lippincott, R.L., 2013. Occurrence and partitioning of antibiotic compounds found in the water column and bottom sediments from a stream receiving two wastewater treatment plant effluents in Northern New Jersey, 2008. Sci. Total Environ. 458, 107–116.

Goldstein, M., Shenker, M., Chefetz, B., 2014. Insights into the uptake processes of wastewater-borne pharmaceuticals by vegetables. Environ. Sci. Technol. 48 (10), 5593–5600.

Gros, M., Petrovic, M., Ginebreda, A., Barcelo, D., 2010. Removal of pharmaceuticals during wastewater treatment and environmental risk assessment using hazard indexes. Environ. Int. 36 (1), 15–26.

Guerra, P., Kim, M., Shah, A., Alaee, M., Smyth, S.A., 2014. Occurrence and fate of antibiotic, analgesic/anti-inflammatory, and antifungal compounds in five wastewater treatment processes. Sci. Total Environ. 473, 235–243.

Hansel, C.M., Fendorf, S., Jardine, P.M., Francis, C.A., 2008. Changes in bacterial and archaeal community structure and functional diversity along a geochemically variable soil profile. Appl. Environ. Microbiol. 74 (5), 1620–1633.

Hatje, V., Bruland, K.W., Flegal, A.R., 2016. Increases in anthropogenic gadolinium anomalies and rare earth element concentrations in San Francisco Bay over a 20 year record. Environ. Sci. Technol. 50 (8), 4159–4168.

Hirsch, R., Ternes, T., Haberer, K., Kratz, K.L., 1999. Occurrence of antibiotics in the aquatic environment. Sci. Total Environ. 225 (1-2), 109–118.

Imison, M., Schweinsberg, S., 2013. Australian news media framing of medical tourism in low- and middle-income countries: a content review. BMC Public Health 13, 12.

Jewell, K.S., Wick, A., Ternes, T.A., 2014. Comparisons between abiotic nitration and biotransformation reactions of phenolic micropollutants in activated sludge. Water Res. 48, 478–489.

Kasprzyk-Hordern, B., Dinsdale, R.M., Guwy, A.J., 2009. The removal of pharmaceuticals, personal care products, endocrine disruptors and illicit drugs during wastewater treatment and its impact on the quality of receiving waters. Water Res. 43 (2), 363–380.

Kinney, C.A., Furlong, E.T., Werner, S.L., Cahill, J.D., 2006. Presence and distribution of wastewater-derived pharmaceuticals in soil irrigated with reclaimed water. Environ. Toxicol. Chem. 25 (2), 317–326.

Kookana, R.S., Williams, M., Boxall, A.B.A., Larsson, D.G.J., Gaw, S., Choi, K., Yamamoto, H., Thatikonda, S., Zhu, Y.G., Carriquiriborde, P., 2014. Potential ecological footprints of active pharmaceutical ingredients: an examination of risk factors in low-, middle- and high-income countries. Philos. Trans. R. Soc. B-Biol. Sci. 369 (1656), 16.

Kreuzinger, N., Fuerhacker, M., Scharf, S., Uhl, M., Gans, O., Grillitsch, B., 2007. Methodological approach towards the environmental significance of uncharacterized substances-quaternary ammonium compounds as an example. Desalination 215 (1–3), 209–222.

Kummerer, K., 2004. Emissions from medical care units. In: Kummerer, K. (Ed.), Pharmaceuticals in the Environment; Sources, Fate, Effects and Risks, second ed. Springer-Verlag, Berlin, pp. 27–44.

Kummerer, K., 2009. Antibiotics in the aquatic environment – A review – Part I. Chemosphere 75 (4), 417–434.

Larsson, D.G.J., 2014. Antibiotics in the environment. Ups. J. Med. Sci. 119 (2), 108–112.

Le Corre, K.S., Ort, C., Kateley, D., Allen, B., Escher, B.I., Keller, J., 2012. Consumption-based approach for assessing the contribution of hospitals towards the load of pharmaceutical residues in municipal wastewater. Environ. Int. 45, 99–111.

Li, X., Brownawell, B.J., 2010. Quaternary ammonium compounds in urban estuarine sediment environments – a class of contaminants in need of increased attention? Environ. Sci. Technol. 44 (19).

Lienert, J., Gudel, K., Escher, B.I., 2007. Screening method for ecotoxicological hazard assessment of 42 pharmaceuticals considering human metabolism and excretory routes. Environ. Sci. Technol. 41 (12), 4471–4478.

Liu, F., Ying, G.G., Yang, J.F., Zhou, L.J., Tao, R., Wang, L., Zhang, L.J., Peng, P.A., 2010. Dissipation of sulfamethoxazole, trimethoprim and tylosin in a soil under aerobic and anoxic conditions. Environ. Chem. 7 (4), 370–376.

Liu, Y., Blowes, D.W., Groza, L., Sabourin, M.J., Ptacek, C.J., 2014. Acesulfame-K and pharmaceuticals as co-tracers of municipal wastewater in a receiving river. Environ. Sci.-Process. Impacts 16 (12), 2789–2795.

Magdouli, S., Daghrir, R., Brar, S.K., Drogui, P., Tyagi, R.D., 2013. Di 2-ethylhexylphtalate in the aquatic and terrestrial environment: a critical review. J. Environ. Manag. 127, 36–49.

Martinez-Carballo, E., Sitka, A., Gonzalez-Barreiro, C., Kreuzinger, N., Furhacker, M., Scharf, S., Gans, O., 2007. Determination of selected quaternary ammonium compounds by liquid chromatography with mass spectrometry. Part I. Application to surface, waste and indirect discharge water samples in Austria. Environ. Pollut. 145 (2), 489–496.

Net, S., Sempere, R., Delmont, A., Paluselli, A., Ouddane, B., 2015. Occurrence, fate, behavior and ecotoxicological state of phthalates in different environmental matrices. Environ. Sci. Technol. 49 (7), 4019–4035.

Ort, C., Lawrence, M.G., Reungoat, J., Eaglesham, G., Carter, S., Keller, J., 2010. Determining the fraction of pharmaceutical residues in wastewater originating from a hospital. Water Res. 44 (2), 605–615.

Parolo, M.E., Savini, M.C., Valles, J.M., Baschini, M.T., Avena, M.J., 2008. Tetracycline adsorption on montmorillonite: pH and ionic strength effects. Appl. Clay Sci. 40 (1-4), 179–186.

Patwary, M.A., O'Hare, W.T., Sarker, M.H., 2011. Assessment of occupational and environmental safety associated with medical waste disposal in developing countries: a qualitative approach. Saf. Sci. 49 (8-9), 1200–1207.

Pignatello, J.J., Xing, B.S., 1996. Mechanisms of slow sorption of organic chemicals to natural particles. Environ. Sci. Technol. 30 (1), 1–11.

Porter, G., Grills, N., 2016. Medication misuse in India: a major public health issue in India. J. Public Health (Oxford, England) 38 (2), e150–7.

Prem Ananth, A., Prashanthini, V., Visvanathan, C., 2010. Healthcare waste management in Asia. Waste Manag. 30 (1), 154–161.

Qin, Q., Chen, X., Zhuang, J., 2015. The fate and impact of pharmaceuticals and personal care products in agricultural soils irrigated with reclaimed water. Crit. Rev. Environ. Sci. Technol. 45 (13), 1379–1408.

Rojas, M.R., Leung, C., Bonk, F., Zhu, Y., Edwards, L., Arnold, R.G., Saez, A.E., Klecka, G., 2013. Assessment of the effectiveness of secondary wastewater treatment technologies to remove trace chemicals of emerging concern. Crit. Rev. Environ. Sci. Technol. 43 (12), 1281–1314.

Ruan, T., Song, S., Wang, T., Liu, R., Lin, Y., Jiang, G., 2014. Identification and composition of emerging quaternary ammonium compounds in municipal sewage sludge in China. Environ. Sci. Technol. 48 (8), 4289–4297.

Rudel, R.A., Melly, S.J., Geno, P.W., Sun, G., Brody, J.G., 1998. Identification of alkylphenols and other estrogenic phenolic compounds in wastewater, septage, and groundwater on Cape Cod, Massachusetts. Environ. Sci. Technol. 32 (7), 861–869.

Schwarzenbach, R.P., Gschwend, P.M., Imboden, D.M., 2003. Environmental Organic Chemistry, second ed. John Wiley & Sons, Inc., New Jersey, USA.

Schwientek, M., Guillet, G., Rügner, H., Kuch, B., Grathwohl, P., 2016. A high-precision sampling scheme to assess persistence and transport characteristics of micropollutants in rivers. Sci. Total Environ. 540, 444–454.

Sicbaldi, F., Sacchi, G.A., Trevisan, M., Del Re, A.A.M., 1997. Root uptake and xylem translocation of pesticides from different chemical classes. Pestic. Sci. 50 (2), 111–119.

Siemens, J., Huschek, G., Siebe, C., Kaupenjohann, M., 2008. Concentrations and mobility of human pharmaceuticals in the world's largest wastewater irrigation system, Mexico City-Mezquital Valley. Water Res. 42 (8-9), 2124–2134.

Simon, N.S., 2005. Loosely bound oxytetracycline in riverine sediments from two tributaries of the Chesapeake Bay. Environ. Sci. Technol. 39 (10), 3480–3487.

Snyder, S.A., Leising, J., Westerhoff, P., Yoon, Y., Mash, H., Vanderford, B., 2004. Biological and physical attenuation of endocrine disruptors and pharmaceuticals: implications for water reuse. Ground Water Monitor. Res. 24 (2), 108–118.

Sulbaek Andersen, M.P., Sander, S.P., Nielsen, O.J., Wagner, D.S., Sanford, T.J., Wallington, T.J., 2010. Inhalation anaesthetics and climate change. Br. J. Anaesth. 105 (6), 760–766.

Topp, E., Monteiro, S.C., Beck, A., Coelho, B.B., Boxall, A.B.A., Duenk, P.W., Kleywegt, S., Lapen, D.R., Payne, M., Sabourin, L., Li, H.X., Metcalfe, C.D., 2008. Runoff of pharmaceuticals and personal care products following application of biosolids to an agricultural field. Sci. Total Environ. 396 (1), 52–59.

Turner, A., Rawling, M.C., 2001. The influence of salting out on the sorption of neutral organic compounds in estuaries. Water Res. 35 (18), 4379–4389.

Van Doorslaer, X., Dewulf, J., Van Langenhove, H., Demeestere, K., 2014. Fluoroquinolone antibiotics: an emerging class of environmental micropollutants. Sci. Total Environ. 500–501, 250–269.

Verlicchi, P., Al Aukidy, M., Zambello, E., 2015. What have we learned from worldwide experiences on the management and treatment of hospital effluent? – An overview and a discussion on perspectives. Sci. Total Environ. 514, 467–491.

Vollmer, M.K., Rhee, T.S., Rigby, M., Hofstetter, D., Hill, M., Schoenenberger, F., Reimann, S., 2015. Modern inhalation anesthetics: potent greenhouse gases in the global atmosphere. Geophys. Res. Lett. 42 (5), 1606–1611.

Wang, F., Ji, R., Jiang, Z., Chen, W., 2014. Species-dependent effects of biochar amendment on bioaccumulation of atrazine in earthworms. Environ. Pollut. 186, 241–247.

Watkinson, A.J., Murby, E.J., Costanzo, S.D., 2007. Removal of antibiotics in conventional and advanced wastewater treatment: implications for environmental discharge and wastewater recycling. Water Res. 41 (18), 4164–4176.

WHO, 2014. Antimicrobial Resistance: Global Report on Surveillance, Geneva. World Health Organisation (WHO), Switzerland.

Windfeld, E.S., Brooks, M.S.L., 2015. Medical waste management – A review. J. Environ. Manag. 163, 98–108.

Wu, C.X., Spongberg, A.L., Witter, J.D., 2009. Sorption and biodegradation of selected antibiotics in biosolids. J. Environ. Sci. Health A – Toxic/Hazard. Subst. Environ. Eng. 44 (5), 454–461.

Wu, X., Dodgen, L.K., Conkle, J.L., Gan, J., 2015. Plant uptake of pharmaceutical and personal care products from recycled water and biosolids: a review. Sci. Total Environ. 536, 655–666.

Xue, B.M., Zhang, R.J., Wang, Y.H., Liu, X., Li, J., Zhang, G., 2013. Antibiotic contamination in a typical developing city in south China: occurrence and ecological risks in the Yongjiang River impacted by tributary discharge and anthropogenic activities. Ecotoxicol. Environ. Saf. 92, 229–236.

Yang, J.-F., Ying, G.-G., Zhao, J.-L., Tao, R., Su, H.-C., Chen, F., 2010. Simultaneous determination of four classes of antibiotics in sediments of the Pearl Rivers using RRLC–MS/MS. Sci. Total Environ. 408 (16), 3424–3432.

Zhang, C., Cui, F., Zeng, G.M., Jiang, M., Yang, Z.Z., Yu, Z.G., Zhu, M.Y., Shen, L.Q., 2015. Quaternary ammonium compounds (QACs): a review on occurrence, fate and toxicity in the environment. Sci. Total Environ. 518, 352–362.

Chapter 4

Ecotoxicological Perspectives on Health Care and the Environment

Bryan W. Brooks and W. Baylor Steele
Baylor University, Waco, TX, United States

HEALTH CARE AND THE ENVIRONMENT: PHARMACEUTICALS ARE INDICATORS OF URBANIZATION

We live on an urban planet with more people now residing in cities. By 2050, 70% of human populations will inhabit urban areas, and a majority of these populations will be located near or along coastlines (www.un.org). In 2014, Asia-Pacific was home to over half of the global population. By 2030, 22 megacities (cities with over 10 million inhabitants) will exist in Asia alone (www.un.org). Such high population densities result in elevated demand for food, energy, water, and other resources, including chemicals, which are then concentrated in these urban regions. In fact, the food-energy-water nexus represents a global megatrend that will stress diverse global systems in the coming years, particularly when changing climate and associated sea level rise and extreme disasters introduce uncertainty to how cities prepare for and respond to these challenges.

By 2025, 50% of the global population will live in drought-susceptible regions (www.state.gov), which dictates judicious management of water resources. Water reuse, or the use of wastewater for beneficial applications, including industrial, drinking, ecological habitat (e.g., stream flow), and agriculture, presents a critically important opportunity to advance more sustainable water resource and chemical management. For example, concentration of human populations in cities results in subsequent generation of solid and liquid waste streams, including sewage. Unfortunately, 80% of the global sewage goes untreated (www.un.org) and is returned to the environment where it is then reused for various purposes, including food production. Targeted lists of chemical contaminants are often examined in human tissues and to assure air, water, and food safety, yet these efforts are differentially

Environmental Contaminants, Vol. 11. https://doi.org/10.1016/B978-0-444-63857-1.00004-8

implemented among and within countries depending on gradients of capacity. Further, lists of targeted chemical indicators employed in surveillance and monitoring efforts, even if they are effectively delivered, represent a miniscule fraction of the exposome people and ecosystems realize in urbanizing regions. For example, environmental introductions of antibiotics and antimicrobially resistant AMR microorganisms, one of the leading global health threats (www.who.int) examined later in this book in Chapter 8, can increase when water security is compromised by increased antibiotic use and decreased waste treatment, but antibiotics and AMR have not been historically examined during environmental assessments. Thus, a common management philosophy of "If you can measure it, you can manage it" is increasingly challenged by urbanization, particularly in developing countries, and illustrated by chemicals from health care in the environment.

Environment and health implications of global megatrends, including environmental quality intersections with the food-energy-water nexus, present palpable challenges to and unique opportunities for achieving the United Nations Sustainable Development Goals (www.globalgoals.org). Human pharmaceuticals in the environment are simply indicators of an urbanizing planet and the increasingly urban water cycle described above. In this chapter, we examine how robust ecotoxicological studies of pharmaceuticals in surface waters have been hampered by historical ecotoxicology tools originally developed for other organic contaminants. We then present efforts for developing and applying new ecotoxicological approaches to support hazard and risk assessments of pharmaceuticals in the aquatic environment. We conclude by relaying perspectives on opportunities pharmaceuticals provide to identify other chemicals of concern and to advance broader sustainability efforts.

HISTORICAL BIOACCUMULATION AND ECOTOXICOLOGY TOOLS: INADEQUATE FOR PHARMACEUTICALS?

Ecotoxicology During Retrospective Studies of Water Quality

Environmental assessments of chemicals and media (e.g., soil, water, and sediment) are designed to provide evidence-based approaches to support decision-making regarding protection of public health and the environment. These efforts, which are either prospective (i.e., before a chemical enters commerce) or retrospective (i.e., after a chemical enters the environment), include exposure assessment, ecotoxicity and hazard assessment, and risk characterization. Though historical environmental assessment frameworks tended to bifurcate with human health or ecological protection goals, more recent efforts are trending to perform coupled assessments of chemical risks (Perkins et al., 2013). In fact, such coupled health and ecological assessment activities align with the broader One Health concept. Unfortunately, limited empirical toxicology and ecotoxicology data exist for the majority of chemicals in commerce, estimated to be approximately 68,000 substances (Rovida and Hartung, 2009). Thus, understanding the ecotoxicology of environmental contaminants, including products from health care, depends on a combination of empirical data, when available, and modeled predictions. Because environmental fate and behavior of substances used in health care were examined elsewhere in this volume, here, we limit our treatment of these topics and instead focus on bioaccumulation and ecotoxicity (i.e., the toxicological effects on organisms in the environment).

Some of the earliest ecotoxicology efforts focused on aquatic systems due to significant environmental degradation (e.g., the Cuyahoga River burning in 1969) and the prevalence of point source discharges (e.g., directly from a pipe) of sewage from industrial and residential areas. Because wastewater effluent discharges include complex mixtures of numerous contaminants, regulatory frameworks employing whole effluent toxicity (WET) monitoring were initially developed in the United States as part of the implementation of a water quality-based approach under the Clean Water Act (Grothe et al., 1996). WET testing, the most commonly employed retrospective ecotoxicology application, includes examining samples from a municipal or industrial effluent discharge through laboratory ecotoxicology studies with common aquatic organisms, including invertebrate (e.g., *Ceriodaphnia dubia*) and fish (e.g., *Pimephales promelas*) models. If WET results indicate that no unacceptable toxicity is expected in a receiving system (e.g., a river, lake, or estuary), then effluent water quality released by a discharger is considered compliant for toxicity by the regulatory authority (e.g., state or federal government) (Grothe et al. 1996).

Initial experimental design consideration for WET assays included that the model organisms could be reliably maintained in a lab and that the assays could be performed quickly, which decreased response times and resources spent on environmental compliance. Thus, acute ecotoxicity assays (e.g., 48 h) with mortality as the toxicological endpoint were initially used for WET testing. Subsequent activities in the United States and other countries have expanded these WET-type assays to (1) examine other organisms and matrices (e.g., sediments and soils), (2) include 7-day subchronic experimental designs where sublethal responses (e.g., invertebrate reproduction and juvenile fish growth) are often applied to large discharges (e.g., >1 million gallons/day in the United States), and (3) assess potential ambient toxicity in surface waters (e.g., WET-type assays of samples collected directly from a river, lake, or estuary). In the United States, WET testing largely examines 48 h survival and 7-day reproduction of the cladoceran *C. dubia* and 48 h survival and 7-day growth of the juvenile fathead minnow (*P. promelas*) model.

It is important to recognize that these WET type of experimental designs, organisms, and response variables are also commonly used to examine the ecotoxicity of new chemicals, to study historic substances with limited data, and to derive environmental quality criteria (e.g., regulatory "bright lines" for compliance) for existing substances in the environment. Thus, laboratory ecotoxicity evaluations of individual chemicals, effluents, surface waters, and terrestrial habitats were intended to serve as "canaries in a coal mine" by protecting, but not necessarily predicting, environmental quality of aquatic and terrestrial ecosystems (Dickson et al., 1992; La Point and Waller, 2000). Chemicals associated with health-care activities, which can originate from manufacturing facilities, institutional practices (e.g., in hospitals), and personal usage, are collected by industrial and municipal sewerage systems. Other common routes of environmental introductions of human pharmaceuticals include runoff from soils, to which biosolids or effluents from water reclamation plants are applied, and from on-site (e.g., septic) wastewater treatment technologies. Some of the highest reported values of human pharmaceuticals in the environment were reported downstream from manufacturing facilities in India (Larsson et al., 2007), which promoted development of AMR in microorganisms (Kristiansson et al., 2011). Unfortunately, implementation of WET-type effluent monitoring and ecotoxicological assessments of ambient surface waters are lacking in India and most other countries.

Bioaccumulation During Prospective Studies of Chemicals

In parallel to retrospective evaluations of effluent discharges employing standardized ecotoxicity assays, surface waters, and then terrestrial habitats, prospective ecotoxicological tools were developed to support hazard and risk assessments of new chemicals, ideally before these substances enter commerce. The structure of regulatory systems developed to perform these prospective activities differs among classes of chemicals (i.e., pesticides vs industrial chemicals) and countries. Approaches for regulatory environmental assessments of human pharmaceuticals are summarized elsewhere (Straub and Hutchinson, 2012). However, the computational and biological tools, including experimental designs, model organisms, ecotoxicological endpoints, and predictive models, employed to perform such assessments are often identical to those models and endpoints employed for effluents and surface waters and are strikingly similar around the world. For example, one common scheme includes performing an assessment to make a regulatory determination if a chemical is persistent, bioaccumulative, and toxic (PBT) (Abelkop et al., 2015). If a regulatory agency determines that a chemical is a PBT substance, then it is typically identified as a chemical to preemptively exclude (or retrospectively ban) from commerce. Though PBT assessments are hazard-based approaches due to uncertainties predicting global exposure scenarios, the persistent and bioaccumulative portions of PBT were selected to specifically consider exposure and thus to complement toxicity determinations.

Usage patterns and intrinsic chemical properties and resulting biological activities of pharmaceuticals present unique challenges when applying such historical ecotoxicological assessment tools to assess and reduce environmental contamination from health care (Brooks et al., 2003, 2009a, 2012b; Ankley et al., 2007; Brooks, 2014). Whereas usage of some substances (e.g., maintenance drugs) are expected to be continuous, a number of societal and health factors influence shifts in use and environmental risk profiles of specific medicines (Gaw and Brooks, 2016). Continuous usage patterns of many substances by urban populations effectively result in consistent chemical inputs to wastewater treatment plants and subsequent release of medicines and their metabolites and degradates to the environment from effluent discharges. Such practices can present long-term exposure scenarios to aquatic life from human medicines and other consumer products.

Historically, the paradigm for environmental management of effluent discharges was "the solution to pollution is dilution." Under this concept, dilution of effluent by instream flows would decrease the amount of contaminant concentrations in the environment and thus associated exposure scenarios and environmental risks. However, a number of aquatic ecosystems are considered effluent-dominated or effluent-dependent rivers or estuaries when the majority or even the entirety of instream flows downstream of effluent discharge(s) are composed of wastewater effluent of diverse quality (Brooks et al., 2006). Building initially on recommendations by Brooks et al. (2006), more recent efforts identified effluent-dominated systems as important management units in the face of climate change (Luthy et al., 2015). Of particular relevance to these urban systems, limited to no instream dilution exists, which in turn increases chemical exposure in surface waters. In fact, as instream dilution decreases, effective exposure duration increases (Ankley et al., 2007) because introduction rates from effluent discharges exceed a chemical's rate of degradation. For example, such observations are readily observed in the Trinity River of Texas, the United States, where instream flows downstream from Dallas/Ft. Worth are almost entirely (>98%) derived from effluent discharges (Brooks et al., 2006). Because the time it takes for water to travel over 300 km from Dallas to Lake Livingston, an important drinking water source for Houston, is much less than the typical cutoff time value employed for persistence determinations (e.g., >2 months; Abelkop et al., 2015), chemicals not traditionally determined to be persistent are effectively so from an exposure duration perspective in these surface waters. Thus, relevance of employing existing regulatory cutoff values for persistence (e.g., >2 months) appears decidedly inadequate to represent realistic exposure conditions of pharmaceuticals and other consumer chemicals in urbanizing surface waters.

In addition to understanding surface water exposure scenarios in the field, it becomes important for ecotoxicologists to understand the extent to which a chemical bioaccumulates (i.e. uptake from all routes of exposure, including diet) and elicits toxicity in the environment. Much like determinations of persistence during PBT assessments, identification of bioaccumulative chemicals also employs cutoff values, which again reflect regulatory policy, instead of science, decisions. Specifically, bioaccumulative chemical determinations are typically made when experimentally derived bioconcentration (or uptake from water only) factors (BCF) exceed cutoff values greater than 1000–5000 (Abelkop et al., 2015). Because empirical BCF data are available for such a limited number of environmental contaminants, computational methods, including quantitative structure activity relationships (QSARs), have historically been used to predict BCF values of organic chemicals. These models were initially derived for nonionizable organic contaminants, routinely depend on log K_{ow} to predict uptake, and fail to consider clearance rates and elimination. When pharmaceuticals are considered, physicochemical properties of therapeutics can routinely fall outside the applicability domain of historic bioaccumulation modeling tools (Brooks et al., 2009a).

It has been estimated that ~85% of pharmaceuticals are ionizable chemicals at physiologically and environmentally relevant pH (Valenti et al., 2009). When the presence of bases, which comprise the vast majority of human pharmaceuticals, are considered in surface waters, bioaccumulation and ecotoxicity are strongly influenced by pH influences on ionization state and thus bioavailability (Brooks et al., 2003; Nakamura et al., 2008; Valenti et al., 2009). Unfortunately, historical modeling approaches reliant on log K_{ow} values from nonionizable chemicals can mischaracterize bioaccumulation potential and subsequent adverse outcomes of ionizable pharmaceuticals. For example, Eq. (1) depicts a common equation (Veith and Kosian, 1983) used in QSAR models to estimate BCF values during bioaccumulation assessments:

$$\log \mathrm{BCF} = 0.79 \times \log K_{ow} - 0.40 \tag{1}$$

In addition to not considering ionization influences on bioaccumulation of organic compounds, these initial K_{ow}-driven BCF models predicted uptake and thus were not developed to consider elimination. Though BCF can be defined as $\mathrm{BCF} = C_{organism}/C_{water}$, it is also derived from the relationship between update (k_1) and elimination (k_2) kinetics. Thus, regulatory studies such as the OECD method 305 outline a technique to derive a kinetic BCF value that accounts for uptake and elimination. This appears important since the presence and/or functions of major cytochrome P450 isoenzymes (e.g., 3A4, 2D6, and 2C9) responsible for biotransformation of many pharmaceuticals in humans are absent or understudied in fish and other aquatic and terrestrial organisms (Connors et al. 2013).

Mechanistic bioaccumulation models, informed by robust empirical data for ionizable chemicals, are needed to develop the next generation of predictive bioaccumulation tools and to advance an understanding of bioaccumulation science (Brooks et al. 2012a). For example, Erickson et al. (2006a,b) and Nichols et al. (2015) developed and applied novel models to understand influence of pH on inhalational (i.e., gill) uptake of acids and bases, respectively, in fish. Spurred by initial reports of the antidepressants fluoxetine and sertraline and their primary active metabolites accumulating in fish from an effluent-dominated stream (Brooks et al., 2005), a number of studies have now reported accumulation of diverse pharmaceuticals in aquatic organisms (Daughton and Brooks, 2011; Brooks, 2014). Fortunately, ionizable base pharmaceuticals do not appear to biomagnify across trophic positions in aquatic systems (Du et al., 2014a, 2016). However, several studies report accumulation of pharmaceuticals in fish plasma at levels approaching or exceeding human therapeutic concentrations (Fick et al., 2010; Scott et al., 2016). As depicted in Eq. (2), determination and prediction of body burden (BB) represent a useful approach to understand such bioaccumulation dynamics of pharmaceuticals in wildlife:

$$BB = C_0 \times V_d \quad (2)$$

where BB is body burden, C_0 is plasma concentration at time 0, and V_d is apparent volume of distribution. Such efforts, which are paramount in clinical pharmacokinetics, would aid bioaccumulation assessments of environmental contaminants and could further support interpretation of ecotoxicological responses to ionizable chemicals in general and pharmaceuticals in particular during laboratory experiments and field monitoring studies.

Ecotoxicity during Prospective Studies of Chemicals

Numerous studies have been performed to summarize ecotoxicological responses to various pharmaceuticals in the environment. Instead of repeating these previous accounting efforts (e.g., Brausch et al. 2012), here we focus on challenges and opportunities associated with identifying adverse effects of pharmaceuticals in the environment. Descriptive ecotoxicology studies, similar to mammal-focused toxicology efforts, often are designed to identify whether an effect results from exposure and how much of an effect is observed. As depicted in Fig. 1, many ecotoxicological assessment frameworks employ a tiered structure where triaging occurs during identification of potentially problematic chemicals. As noted above, QSAR models are routinely used given limited to no empirical toxicity data for thousands of chemicals, costs associated with testing, and animal welfare concerns. During a PBT assessment, for example, QSAR models provide predictions of ecological and mammalian toxicity. In the case of ecotoxicity predictions, many QSARs have been historically developed based on relationships between lipophilicity (e.g., log K_{ow}) and ecotoxicological benchmark concentrations (e.g., LC_{50}) derived from the standardized ecotoxicity testing protocols described above. However, these first-tier modeling efforts are increasingly being supported by and new computational models are being derived from mechanistically focused

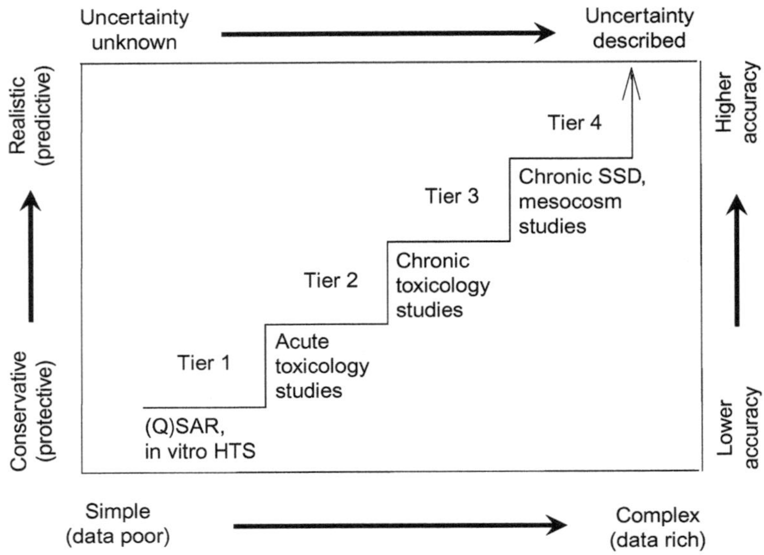

FIG. 1 Conceptual description of a tiered environmental assessment process for chemical contaminants.

high-throughput in vitro and zebra fish embryo screening efforts, such as included in the Tox 21 and ToxCast programs (www.epa.gov/ncct). Such next-generation computational approaches promise to advance mechanistic ecotoxicology assessments of pharmaceuticals and other environmental contaminants (McRobb et al., 2014). When concerns are identified in initial tiered screening, subsequent tiers examine acute ecotoxicity and then various chronic ecotoxicity responses. Thus, the first few tiers depicted in Fig. 1 are conceptually identical to tiered frameworks employed for human health assessments of chemicals.

Higher tiers during ecotoxicological assessment begin to examine adverse outcomes extending from ecological populations to communities. Specifically, mesocosm or microcosm studies, which include model artificial ecosystems such as streams or ponds, are typically performed in tier 4 to examine chemical influences on ecological communities and thus interactions (e.g., competition and predation) among various species and trophic levels (Fig. 1). For example, King et al. (2016) recently examined influences of environmentally realistic pulsed exposures of a common herbicide on the ecological structure and function of stream mesocosms. Perhaps the most well-known, but singular example for human pharmaceuticals, is an experiment in which 17a-ethinylestradiol (EE2) was added to a whole lake at low nanogram per liter concentrations (Kidd et al. 2007). Despite introduction of such low yet near environmentally relevant levels of a synthetic estrogen agonist, reproductive failure of a fish population (*Pimephales promelas*) was observed (Kidd et al. 2007). Mesocosm studies in general and whole-lake experiments in particular are expensive and thus are rarely performed. However, species sensitivity distributions (SSD; Posthuma et al., 2002) are more common. SSDs can be used as part of probabilistic hazard and risk assessments to predict how ecological communities are adversely affected by specific concentrations of a contaminant. SSDs include acute or chronic data from ecotoxicology studies with model species representing different trophic positions of an ecosystem (e.g., algae, plant, invertebrate, fish, amphibian). Thus, SSDs commonly rely on data produced by the previously described standardized experimental designs, model organisms, and toxicological endpoints employed for individual chemical and effluent and ambient toxicity testing.

As noted elsewhere in this volume, pharmaceuticals in developed countries are routinely detected in surface waters from $ng\,L^{-1}$ to low $\mu g\,L^{-1}$ concentrations (Monteiro and Boxall, 2010). Similarly, tissue concentrations in biota are typically observed from $ng\,kg^{-1}$ to low $\mu g\,kg^{-1}$ levels (Daughton and Brooks, 2011). Human medicines are present in surface waters at concentrations lower than thresholds for acute ecotoxicity (Brausch et al., 2012) with few exceptions, mostly from developing countries with less developed or absent treatment infrastructure capacity (Kookana et al., 2014) for waste streams from cities and manufacturing facilities (Larsson et al., 2007). Because pharmaceuticals are designed to provide therapeutic benefits by eliciting biological responses, we now appreciate that sublethal responses of aquatic and terrestrial organisms are more important to consider than acute mortality when mechanistically assessing hazards and risks of most pharmaceuticals in the environment (Brooks et al., 2003; Huggett et al., 2003). However, if the modes and mechanisms of action (MOA) of pharmaceuticals are not considered when selecting experimental designs, organisms, and endpoints for ecotoxicological study (Ankley et al., 2007), then chronic thresholds derived from standardized testing methods for other organic contaminants can yield incomplete at best, and misleading at worst, information (Berninger and Brooks, 2010).

Consider as a case study the ecotoxicological influences of EE2, a common medicine in contraceptives and hormone replacement therapies, on reproduction of common cladoceran models used for WET testing and prospective evaluations of pharmaceuticals and other contaminants. EE2 adversely affects reproduction of cladocera at low $mg\,L^{-1}$ or part per million levels (Clubbs and Brooks, 2007). However, when the same endpoint (reproduction) is examined in fish models, EE2 effectively impairs fish reproduction at much lower $ng\,L^{-1}$ (part per trillion) concentrations (Caldwell et al., 2008) because of sufficient conservation of estrogen receptor function across vertebrates (Gunnarsson et al., 2008). If a model organism is incorrectly selected for study, because it does not possess the biological target (e.g., estrogen receptors are absent in cladocera) for a pharmaceutical of interest (Clubbs and Brooks, 2007) and thus response variables measured are not mechanistically linked to chemical MOA (Ankley et al., 2007; Brain et al., 2008; Berninger and Brooks, 2010), then a six-order-of-magnitude error could be made during a regulatory determination. Unfortunately, cladoceran reproduction has been the most common and in many cases the only reproduction endpoint employed during prospective (e.g., new chemicals) and retrospective (e.g., whole effluent) ecotoxicological applications. Though for assessment of new pharmaceuticals with known estrogenicity, this is no longer the case in developed countries (see Chapter 13).

Misapplication of historical ecotoxicology tools, which are prescriptively applied during regulatory compliance programs, appears to lead to propagation of uncertainty from assumptions associated with common computational and biological models, largely due to diverging applicability domains of pharmaceuticals from traditional industrial contaminants. Fortunately, ecotoxicology studies with pharmaceuticals are increasingly examining sublethal endpoints that are plausibly linked to therapeutic or side effect MOA (Brooks, 2014). Herein, the adverse outcome pathway (AOP) conceptual framework (Ankley et al., 2010) has high utility to aid the design of experiments to test mechanistic hypotheses and to define ecotoxicity pathways extending from molecular initiation events to individual and population consequences, which represent the first and second anchors of an AOP, respectively. In fact, future ecotoxicology studies of pharmaceuticals in the environment would decidedly benefit from routinely employing AOPs (Brausch et al., 2012). In the next section, we examine some emerging ecotoxicology approaches to more robustly understand and support future assessment and management of pharmaceuticals in the environment.

EMERGING ECOTOXICOLOGY APPROACHES FOR STUDYING HEALTH-CARE IMPACTS

Biological Read-Across, Fish Plasma Modeling and Behavioral Ecotoxicology

Unlike the majority of environmental contaminants, pharmaceuticals receive extensive testing and evaluation during the drug development and approval process (Seiler, 2002; see Chapter 13). Identifying ways to inform ecotoxicological studies and environmental assessments from such preclinical pharmacology and toxicology information has received recent study. For example, the concept of biological read-across has emerged as a way to leverage available mammalian safety data to support ecotoxicological studies of pharmaceuticals (Brooks et al., 2009b; Berninger and Brooks, 2010). One interesting application of biological read-across is now known as the Huggett model (Huggett et al., 2003). Huggett et al. (2003) from the pharmaceutical industry recognized the evolutionary conservation of hundreds of drug targets across vertebrates and other species (Gunnarsson et al., 2008) and proposed a model to identify pharmaceuticals as priorities for study. They specifically proposed generation of an effect ratio of human therapeutic plasma concentrations (e.g., C_{max}) to fish plasma concentrations and recommended an application factor of 1000 to account for mammalian to fish and other common uncertainties in ecological and health risk assessments (Huggett et al., 2003). The Huggett model relied on a water-to-plasma partitioning relationships ($\log P_{plasma:water}$) in fish for nonionizable chemicals originally derived by Fitzsimmons et al. (2001), which was further modified by others (Daughton and Brooks, 2011; Berninger et al., 2011) using an equation more closely fitting most human pharmaceuticals, which more commonly are characterized by an apparent $\log K_{ow}$ ($\log D_{ow}$) of 3 or lower in surface waters:

$$\log P_{plasma:water} = \log\left[\left(10^{0.73 \log k_{ow}} \times 0.16\right) + 0.84)\right] \quad (3)$$

Instead of solely examining whole body concentrations of pharmaceuticals in organisms, Brooks et al. (2009a) recommend that residues of medicines also be examined in target tissues, including plasma. The Huggett et al. (2003) effect ratio approach has subsequently been used to support interpretation of bioaccumulation monitoring results in the field (Fick et al., 2010; Scott et al., 2016).

An initial test of the Huggett model was performed by Valenti et al. (2012) within an AOP context using sertraline, a selective serotonin reuptake inhibitor (SSRI). As depicted in Fig. 2, Valenti et al. (2012) predicted and then observed adult fathead minnow (*Pimephales promelas*) plasma concentrations from aqueous sertraline concentrations; these treatment levels were selected to reach and exceed human therapeutic (e.g., C_{max}) plasma values. When fish plasma concentrations exceeded human C_{max} values, binding to serotonin reuptake transporters in brain tissue of fathead minnows was significantly decreased (Valenti et al., 2012; Fig. 2), representing a molecular initiation event paralleling previous studies in rodents (Gould et al., 2007). Such observations were coupled with significant antianxiety behavioral responses (Fig. 2), which again were hypothesized to occur based on previous mammalian studies (Handley and McBlane, 1993) and evolutionary conservation of pharmacological targets among vertebrates (Gunnarsson et al., 2008).

It is important to note that these antianxiety responses observed by Valenti et al. (2012) were considered adverse outcomes because such physiological changes in fish behavior are expected to result in decreased competition for resources and predation risks. Subsequent studies by Margiotta-Casaluci et al. (2016) have examined and advanced such coupling of fish plasma observations within an adverse outcome context for synthetic glucocorticoids (2016) and for behavioral responses to the antidepressant fluoxetine (2014). Much like the case study presented above for EE2, similar observations

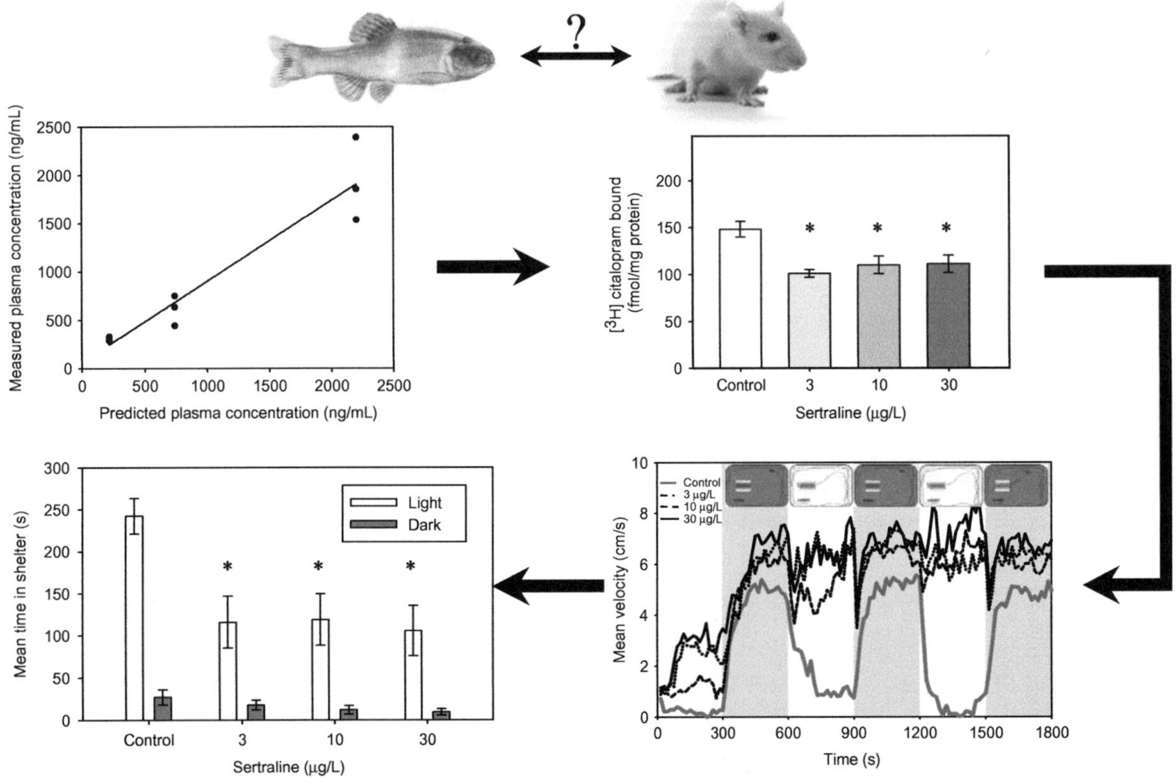

FIG. 2 Inhalational uptake, serotonin reuptake transporter binding, and antianxiety responses of adult male fathead minnows following sublethal exposure to the selective serotonin reuptake inhibitor sertraline. *Modified from Valenti, T.W., Gould, G.G., Berninger, J.P., Connors, K.A., Keele, N.B., Prosser, K.N., Brooks, B.-W. (2012). Human therapeutic plasma levels of the selective serotonin reuptake inhibitor (SSRI) sertraline decrease serotonin reuptake transporter binding and shelter-seeking behavior in adult male fathead minnows. Environ. Sci. Technol. 46, 2427–2435.*

extend to SSRIs in the environment. For example, such sublethal behavioral responses observed by Valenti et al. (2012) and Margiotta-Casaluci et al. (2014) to sertraline and fluoxetine, respectively, can be compared with standardized ecotoxicity response thresholds to each of these SSRIs (Brooks et al., 2003; Valenti et al., 2009). In these cases, ecologically important behavioral responses, which are plausibly linked to SSRI MOA, would be observed at lower concentrations than those standardized endpoints employed during retrospective WET testing of effluent discharges or prospective regulatory environmental assessments of pharmaceuticals.

Many active pharmaceutical ingredients are capable of altering behavior at sublethal concentrations. In fact, diverse groups of pharmaceuticals can impair fish behavior resulting in adverse outcomes at concentrations close to or below those that have been detected in the environment (Painter et al., 2009; Brodin et al., 2013; Dzieweczynski and Hebert, 2013). Because behaviors include actions and reactions necessary for organism survival and fitness that are regulated by biochemical, physiological, and ecological processes, aquatic contaminants that alter behavioral processes have important implications in threatening populations and thus ecosystem integrity (Little and Finger, 1990; Little and Adams, 2002; Tierney, 2011). As noted above, fish demonstrate clear relevance as aquatic models in understanding the behavioral consequences of pharmaceuticals due to the high conservation of drug targets between mammals and fish and because many pharmaceutical compounds, such as psychiatric medications, are designed specifically to alter behavior in humans (Brooks et al., 2003).

The growing popularity of zebrafish in neuropharmacology and psychiatric studies has led to observations of several behavioral phenotypes associated with various moods, psychological disorders, and social preferences that can be altered by exposure to various neuroactive compounds (Stewart et al., 2015; Kalueff et al., 2016). This understanding has the potential to be read across in ecotoxicology to test hypotheses and possibly predict effects of pharmaceuticals on natural fish populations. For example, it is well known in the biomedical sciences that particular larval and adult zebrafish behaviors such as wall hugging and light-dark preferences are associated with anxiety levels (Steenbergen et al., 2011; Schnörr et al., 2012; Maximino et al., 2014). These behaviors can be altered when fish are exposed to pharmaceuticals used to treat anxiety in humans. As noted above, it is also demonstrated in ecotoxicology studies that anxiety influences predator avoidance behaviors in fish (Valenti et al., 2012; Melvin and Wilson, 2013; Margiotta-Casaluci et al., 2014). Future research is needed to more quantitatively leverage data from psychiatric studies with zebrafish and other fish models to understand comparative effects of pharmaceuticals on ecologically relevant behaviors in natural fish populations and in turn to support diagnostic evaluation of behavioral responses elicited by other industrial chemicals.

Because behaviors are rapid and sensitive indicators of toxicity (Melvin and Wilson, 2013), behavioral endpoints can be used to quickly and effectively detect changes at the molecular and biochemical level to system changes in the individual organism (Kokel et al., 2010). Advances in tracking technology and computational capabilities now allow for behavioral effects of pharmaceuticals to be determined in HTS using early life stage (ELS) fish models (Noyes et al., 2015). Larval and embryonic fish models (FET and OECD methods) demonstrate the potential to be used to rapidly and sensitively screen compounds for behavioral responses using automated tracking software (Noyes et al., 2015; Reif et al., 2015). Due to their small size and fast development, zebrafish embryos as young as 24 h post fertilization (hpf) can be observed simultaneously in multichannel well plates (Reif et al., 2015). Using ELS models in behavioral ecotoxicology with the aid of digital tracking software has the potential to greatly reduce time and resources and significantly enhance accuracy of behavioral observations (Sloman and McNeil, 2012). Early life stage models, however, cannot be presumed to reflect behavioral responses representative of natural populations of fish. A key challenge for behavioral ecotoxicology is ensuring that laboratory models are representative of and have the ability to understand behavioral perturbations in the field. Data from pharmacological studies, however, can be used to elucidate how effects on known behavioral phenotypes in larval zebrafish may or may not translate to other fish species. Understanding the biochemical and physiological mechanisms underpinning these behavioral effects will provide the foundation to begin to translate behavioral effects in larval zebrafish to effects in other fish species and life stages (Parker, 2016). Further research in comparative behavioral physiology, pharmacology and toxicology promises to also help fill the gaps among larval zebrafish behavior patterns and behavioral ecotoxicology of pharmaceuticals and other chemicals in various fish species.

Ecotoxicology and Bioaccumulation Approaches to Study Medicines in the Environment

Building from behavioral, mechanistic, and bioaccumulation observations by Valenti et al. (2012), Brooks (2014) proposed the therapeutic hazard value (THV) as a way to translate surface water observations to concentrations potentially presenting therapeutic hazards to fish. The THV is described in Eq. (4):

$$\mathrm{THV}\ \left(\mu g\ \mathrm{L}^{-1}\right) = C_{\max} / \log P_{\text{plasma water}} \tag{4}$$

In this case, a "therapeutic hazard" represents responses elicited through mammalian MOAs for therapeutic or side effect activities and potentially similar molecular initiation events and pathways in fish. The THV concept appears useful as a diagnostic to identify chemicals for future study

(Brooks, 2014); to examine potential hazards of effluent discharges (Du et al., 2014b); to observe how surface water quality may change in time and space (Scott et al., 2016); and to support global chemical scanning of pharmaceutical hazards in effluents, surface waters, and various countries (Kristofco and Brooks, 2017; Saari et al., 2017). Building on earlier work by Brooks et al. (2012a) and Berninger et al. (2016), Fig. 3 illustrates how the THV concept can be extended to 1030 pharmaceuticals and specific groups of therapeutics,

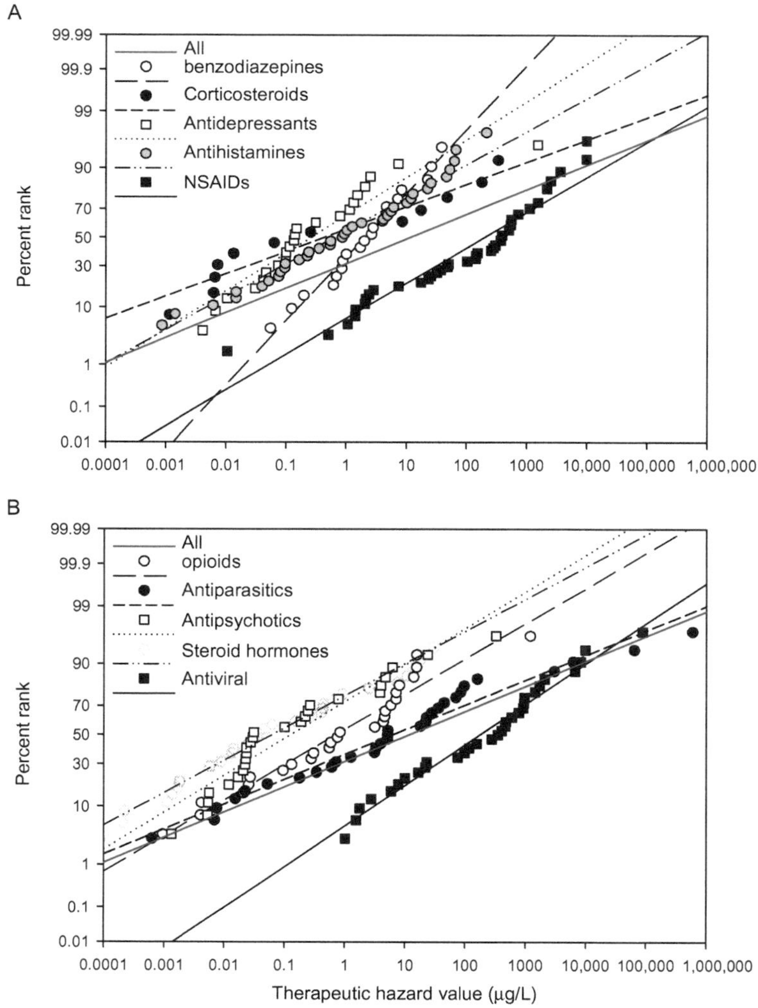

FIG. 3 Pharmaceutical probability distributions of therapeutic hazard values (THV) for 10 groups of pharmaceuticals. A distribution for 1030 individual therapeutics is plotted as a red reference line.

using a probabilistic hazard assessment approach (Berninger and Brooks, 2010). In fact, the likelihood of encountering a THV below 1 $\mu g L^{-1}$, a common historical trigger value to initiate performing environmental assessments of pharmaceuticals by regulatory agencies, would be expected 31.9% of the time for all pharmaceuticals (Table 1). Among groups of similar pharmaceutical classes, THV predictions relative to specific concentrations (e.g., 1 $\mu g L^{-1}$) vary. For example, the likelihood of encountering antidepressants, antihistamines, and steroid hormones with THV values lower than 1 $\mu g L^{-1}$ would be expected more than two-thirds of the time, whereas similar observations would be expected for antiviral medications and nonsteroidal antiinflammatory drugs less than 8% of the time (Table 1). Though the fish plasma modeling approach has much promise, it is important to recall some of the uncertainties of this ecotoxicological approach (Brooks, 2014), which represent important research needs identified during big question exercises (Boxall et al., 2012; Rudd et al., 2014). Future studies are needed to test the utility and limitations of the THV approach for additional pharmaceuticals and to define relationships among internal plasma and tissue doses of pharmaceuticals in fish and ecologically important adverse outcomes across life histories within and among species.

As noted above, Nichols et al. (2015) recently advanced bioconcentration efforts for ionizable chemicals by developing a predictive model for the influences of pH gradients on gill uptake using the antihistamine diphenhydramine as a model base. Nichols et al. (2015) also calculated the apparent V_d for diphenhydramine in the adult fathead minnow model. Interestingly, the empirically derived apparent V_d value of 3 L kg^{-1} for diphenhydramine in this common fish model at each of the experimental pH levels mirrored the apparent V_d in humans, which ranges from 3 to 8 L kg^{-1} for diphenhydramine (Madan et al., 2009). Though additional studies are necessary to determine whether apparent V_d values of pharmaceuticals in humans can be used to predict apparent V_d in fish models, such observations with the model base diphenhydramine have the potential to support development of more robust bioaccumulation predictions using biological read-across and comparative pharmacology. For example, in recent years, an increasing number of studies identify the presence of pharmaceutical bioaccumulation in aquatic organisms and terrestrial wildlife. A common practice is to determine chemical concentrations in whole tissues, but translating such observations to levels of concern remains elusive. From the body burden equation given in Eq. (2) and the THV equation in Eq. (4), it appears possible to derive a bioaccumulation hazard value (BHV):

$$\text{BHV}\ \left(\text{mg kg}^{-1}\right) = \left(\text{THV}^{*} \log P_{\text{plasma water}}\right)^{*} V_d \quad (5)$$

As observed in Fig. 4, the BHV concept, which is conceptually similar to a critical body residue approach, was extended to 574 pharmaceuticals and various groups of therapeutics. For example, the probability of exceeding the

TABLE 1 Equations for Regression Lines and Values Corresponding to the 1st, 5th, 10th, 25th, 50th, 75th, 90th, 95th, and 99th Centiles for Probabilistic Pharmaceutical Distributions of Therapeutic Hazard Values (THVs) for 10 Different Groups of Therapeutics

					THV Centile Value (μg/L)									Percent
Distribution	*n*	r^2	*a* (Slope)	*b* (Intercept)	*1%*	*5%*	*10%*	*25%*	*50%*	*75%*	*90%*	*95%*	*99%*	Below $1\ \mu g L^{-1}$
All pharmaceuticals	1030	0.98	0.45	−0.48	8.38E−05	0.0027	0.0170	0.3719	11.47	353.89	7748.70	49,134.05	1,570,686.67	31.9%
Antidepressants	22	0.88	0.70	0.41	0.0001	0.0011	0.0037	0.0278	0.26	2.40	17.83	59.19	562.34	66.3%
Antihistamines	37	0.97	0.61	0.12	9.34E−05	0.0012	0.0049	0.0485	0.62	8.02	79.80	315.74	4166.65	54.6%
Antiparasitics	31	0.97	0.45	−0.35	3.87E−05	0.0013	0.0084	0.1909	6.15	197.99	4506.02	29,241.45	976,249.77	36.0%
Antipsychotics	26	0.91	0.66	0.60	3.92E−05	0.0004	0.0015	0.0121	0.13	1.31	10.81	38.17	406.82	73.0%
Antivirals	31	0.98	0.73	−1.64	0.1134	0.9742	3.0661	20.8258	174.99	1470.45	9987.58	31,432.97	270,028.60	5.1%
Benzodiazepines	20	0.98	1.17	−0.35	0.0206	0.0787	0.1609	0.5313	2.00	7.55	24.95	50.99	194.95	36.5%
Corticosteroids	12	0.92	0.41	0.15	9.41E−07	4.26E−05	0.0003	0.0097	0.42	18.32	546.28	4167.52	188,464.46	56.9%
NSAIDs	38	0.93	0.65	−1.49	0.0508	0.5708	2.0727	17.8788	195.92	2146.98	18,519.55	67,246.63	755,461.91	7.7%
Opioids	26	0.95	0.64	0.09	0.0002	0.0019	0.0069	0.0620	0.71	8.13	72.91	271.03	3181.76	54.0%
Steroid hormones	34	0.98	0.58	0.70	6.28E−06	9.29E−05	0.0004	0.0043	0.06	0.89	9.82	41.29	610.67	75.9%

NSAID, nonsteroidal anti-inflammatory drugs.

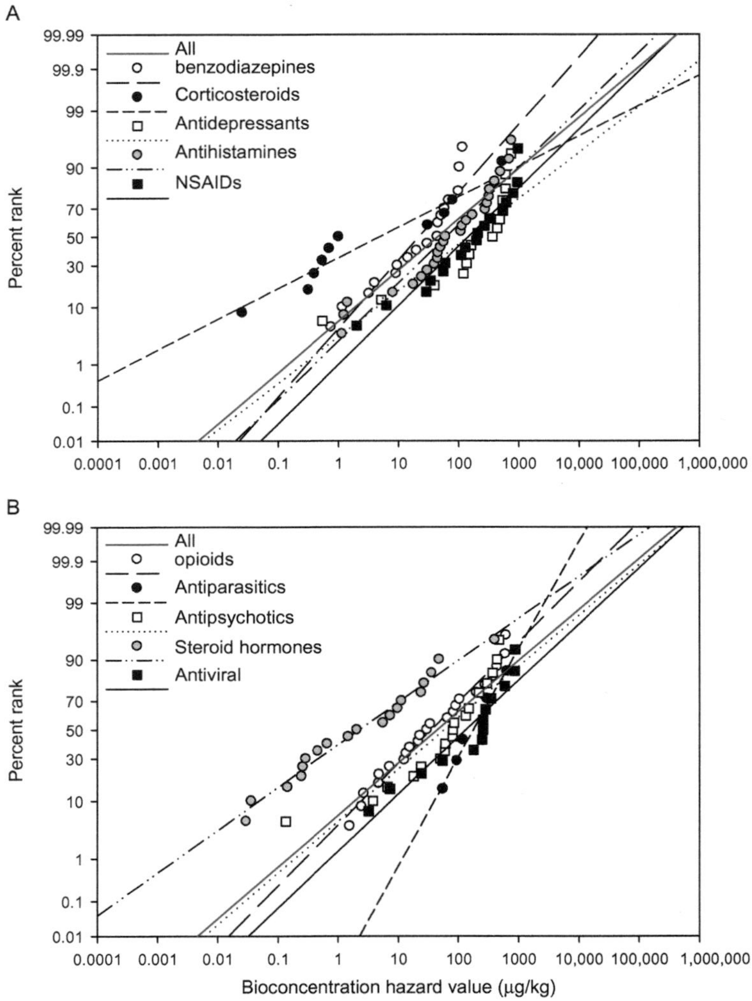

FIG. 4 Pharmaceutical probability distributions of bioaccumulation hazard values (BHV) for 10 groups of pharmaceuticals. A distribution of 574 individual therapeutics is plotted as a red reference line.

BHVs for all drugs observed in tissues above 1 μg kg^{-1} would be expected 94.2% of the time (Table 2). Here again, differences among groups of similar pharmaceutical classes vary (Fig. 4). Similar to the THV concept, the BHV may provide a useful tool for interpreting environmental monitoring results from bioaccumulation studies in the field, but future research is needed to specifically test its strengths and weaknesses.

TABLE 2 Equations for Regression Lines and Values Corresponding to the 1st, 5th, 10th, 25th, 50th, 75th, 90th, 95th, and 99th Centiles for Probabilistic Pharmaceutical Distributions of Bioaccumulation Hazard Values (BHVs) for 10 Different Groups of Therapeutics

					BHV Centile Value (μg/kg)									Percent
Distribution	*n*	r^2	a (Slope)	*b* (Intercept)	*1%*	*5%*	*10%*	*25%*	*50%*	*75%*	*90%*	*95%*	*99%*	Above 1 $\mu g\, kg^{-1}$
All pharmaceuticals	574	0.93	0.94	−1.55	0.1491	0.786	1.91	8.37	43.39	224.77	987.87	2396.01	12,626.58	94.2%
Antidepressants	15	0.74	0.84	−1.82	0.2534	1.627	4.39	23.00	144.91	913.17	4787.28	12,903.38	82,882.76	96.6%
Antihistamines	25	0.92	1.06	−1.90	0.3963	1.744	3.84	14.39	62.37	270.37	1012.15	2230.15	9815.56	97.2%
Antiparasitics	6	0.98	1.97	−4.42	11.6128	25.806	39.50	80.44	177.30	390.77	795.83	1218.12	2706.88	100.0%
Antipsychotics	19	0.83	0.93	−1.65	0.1879	1.007	2.47	11.00	57.95	305.28	1362.11	3333.62	17,867.43	95.0%
Antivirals	13	0.87	1.03	−2.19	0.7288	3.350	7.55	29.40	133.03	601.97	2342.47	5282.29	24,281.17	98.6%
Benzodiazepines	19	0.90	1.24	−1.67	0.2944	1.042	2.04	6.30	21.99	76.81	236.72	464.29	1642.74	95.1%
Corticosteroids	11	0.93	0.56	−0.38	0.0003	0.006	0.02	0.30	4.86	77.92	947.12	4222.85	69,726.08	64.6%
NSAIDs	18	0.92	1.07	−2.33	0.9993	4.302	9.37	34.38	145.80	618.31	2269.45	4941.76	21,273.55	99.0%
Opioids	23	0.98	1.10	−1.71	0.2761	1.144	2.44	8.65	35.31	144.15	511.22	1090.53	4516.96	95.6%
Steroid hormones	19	0.98	0.77	−0.27	0.0021	0.016	0.05	0.30	2.26	16.97	104.28	309.08	2372.63	60.7%

NSAID, nonsteroidal anti-inflammatory drugs.

ECOTOXICOLOGY OUTLOOK FOR MEDICINES: OPPORTUNITIES TO ADVANCE SUSTAINABILITY SCIENCE?

In addition to understanding ecological consequences of health care, ecotoxicological studies of pharmaceuticals appear to provide opportunities to advance sustainability science. As noted above, the presence of pharmaceuticals in the natural world simply reflects in increasingly urban planet and an urbanizing water cycle. How can occurrence of specific medicines in cities help advance an understanding of broader health and societal patterns? Much like a watershed includes a network of subwatersheds with smaller streams flowing to larger river systems and ultimately coastlines, waste streams in urban areas are collected by diverse sewerage systems from residential, commercial, and industrial users. In fact, different components of wastewater collection systems represent "sewersheds" because they are conceptually similar geospatial units to natural watersheds and thus can be characterized by differential urban activities and status. Over the last few years, increasing attention has been given to these urban systems where innovative determinations of human behavior (e.g., drugs of abuse patterns and locations detected in sewers) and potential health status and trends (e.g., excreted biomarkers associated with specific diseases) can be examined through an emerging area known as sewage epidemiology (Thomas et al., 2012; Baker et al., 2012; Ort et al., 2014; Jones et al., 2014). Sewage epidemiology, though in its infancy, promises to advance an understanding of urban metabolism, which is necessary to design and manage more sustainable cities and communities, one of the key Sustainable Development Goals (www.globalgoals.org).

Because pharmaceuticals receive more study than any other groups of chemicals in the environment, medicines represent useful model chemicals to advance an understanding of bioaccumulation and toxicity of health care and other organic chemicals, particularly by serving as "surgical scalpels" to comparatively understand structure and function relationships among species and habitats. As noted above, Brausch et al. (2012) proposed that future ecotoxicology studies of pharmaceuticals in the environment should employ AOP approaches (Ankley et al., 2010), which is being applied to other classes of environmental contaminants for both ecotoxicology and mammalian toxicology efforts. We concur. However, current regulatory ecotoxicology testing requirements depend on robust standardized experimental designs to concretely identify chemical thresholds resulting in adverse outcomes to the population level of biological organization or to individual organisms, as is the case for endangered and threatened species. Thus, future efforts are needed to advance the integration of toxicity pathways resulting from diverse molecular initiation events within a tiered framework. Herein, high-throughput testing strategies will be critical to characterize biological activities and advance ecotoxicology and its regulatory applications, but in vitro and embryonic fish model assay results must be linked to adverse outcomes at individual and population levels if commonly accepted ecosystem protection goals (e.g., protecting biodiversity and maintaining the physical, chemical, and biological integrity of surface waters) are to be realized.

Another timely Sustainable Development Goal focuses on responsible consumption and production (www.globalgoals.org). Herein, sustainable and green chemistry and engineering efforts present outstanding opportunities to stimulate innovation while supporting transition to more sustainable and circular economies. For example, the fourth principle of green chemistry aims to rationally design chemicals that maintain intended functions and are inherently less hazardous (Anastas and Warner, 1998). When recent efforts employed computational models that were originally developed design of less hazardous medicines to examine aquatic toxicity of industrial chemicals, two common physicochemical properties, now known as the "rule of two," were identified to significantly reduce acute and chronic aquatic toxicity (Voutchkova et al., 2011, Voutchkova-Kostal et al., 2012; Kostal et al., 2015), based on standardized experimental designs, model organisms, and endpoints. We then examined these observations using a probabilistic environmental hazard assessment approach, which identified how marked reduction of highly and moderately toxic chemicals in commerce could be achieved if these rule of two design guidelines were implemented (Connors et al., 2014). More recent work by our research team has identified additional design guidelines that may be useful for designing industrial chemicals with reduced likelihood of producing oxidative stress (Corrales et al., 2017).

Selection of less hazardous alternatives for chemicals in commerce coupled with sustainable molecular design of less hazardous substances would result in lower aggregate hazards across the commercial chemical universe space. Such efforts, though challenging, will particularly be needed to reduce ecological and human exposures in urban regions of developing countries where chemical use is increasing and being concentrated faster than environmental management systems and technologies are being implemented. Clearly, a number of timely ecotoxicology research needs exist to understand risks of health-care products and practices in the environment (Boxall et al., 2012; Rudd et al., 2014) and to leverage these lessons learned to support more sustainable chemical innovation, assessment, and management.

ACKNOWLEDGMENTS

BWB was supported by the US National Science Foundation (project no. CHE-1339637) with additional support from the US Environmental Protection Agency and the National Institute of Food and Agriculture, U.S. Department of Agriculture (project no. 20166900725093) during the preparation of this chapter. We thank Dr. RA Brain for preparing an earlier version of Fig. 1 and Drs. JP Berninger, RA Brain, TW Valenti, and LA Kristofco for useful discussions.

REFERENCES

Abelkop, A.D.K., Graham, J.D., Royer, T.V. (Eds.), 2015. Persistent, Bioaccumulative, and Toxic (PBT) Chemicals: Technical Aspects, Policies, and Practices. CRC Press, Boca Raton, FL.

Anastas, P.T., Warner, J.C., 1998. Green Chemistry: Theory and Practice. Oxford University Press, New York, NY, USA.

Ankley, G.T., Bennett, R.S., Erickson, R.J., Hoff, D.J., Hornugn, M.W., Johnson, R.D., Mount, D.R., Nichols, J.W., Russom, C.L., Schmieder, P.K., Serrano, J.A., Tietge, J.E., Villeneuve, D.L., 2010. Adverse outcome pathways: a conceptual framework to support ecotoxicology research and risk assessment. Environ. Toxicol. Chem. 29, 730–741.

Ankley, G.T., Brooks, B.W., Huggett, D.B., Sumpter, J.P., 2007. Repeating history: pharmaceuticals in the environment. Environ. Sci. Technol. 41, 8211–8217.

Baker, D.R., Očenášková, V., Kvicalova, M., Kasprzyk-Hordern, B., 2012. Drugs of abuse in wastewater and suspended particulate matter—further developments in sewage epidemiology. Environ. Int. 48:28–38.

Berninger, J.P., Brooks, B.W., 2010. Leveraging mammalian pharmaceutical toxicology and pharmacology data to predict chronic fish responses to pharmaceuticals. Toxicol. Lett. 193, 69–78.

Berninger, J.P., Du, B., Connors, K.A., Eytcheson, S.A., Kolkmeier, M.A., Prosser, K.N., Valenti, T.W., Chambliss, C.K., Brooks, B.W., 2011. Effects of the antihistamine diphenhydramine on selected aquatic organisms. Environ. Toxicol. Chem. 30, 2065–2072.

Berninger, J.P., LaLone, C.A., Villeneuve, D.L., Ankley, G.T., 2016. Prioritization of pharmaceuticals for potential environmental hazard through leveraging a large-scale mammalian pharmacological dataset. Environ. Toxicol. Chem. 35, 1007–1020.

Boxall, A.B.A., Rudd, M., Brooks, B.W., Caldwell, D., Choi, K., Hickmann, S., Innes, E., Ostapyk, K., Staveley, J., Verslycke, T., Ankley, G.T., Beazley, K., Belanger, S., Berninger, J.P., Carriquiriborde, P., Coors, A., DeLeo, P., Dyer, S., Ericson, J., Gagne, F., Giesy, J.P., Gouin, T., Hallstrom, L., Karlsson, M., Larsson, D.G.J., Lazorchak, J., Mastrocco, F., McLaughlin, A., McMaster, M., Meyerhoff, R., Moore, R., Parrott, J., Snape, J., Murray-Smith, R., Servos, M., Sibley, P.K., Straub, J.O., Szabo, N., Tetrault, G., Topp, E., Trudeau, V.L., van Der Kraak, G., 2012. Pharmaceuticals and personal care products in the environment: what are the big questions? Environ. Health Perspect. 120, 1221–1229.

Brain, R.A., Ramirez, A.J., Fulton, B.A., Chambliss, C.K., & Brooks, B.W., 2008. Herbicidal effects of sulfamethoxazole in *Lemna gibba*: using p-aminobenzoic acid as a biomarker of effect. Environ. Sci. Technol. 42, 8965–8970.

Brausch, J.M., Connors, K.A., Brooks, B.W., Rand, G.M., 2012. Human pharmaceuticals in the aquatic environment: a critical review of recent toxicological studies and considerations for toxicity testing. Rev. Environ. Contam. Toxicol. 218, 1–99.

Brodin, T., Fick, J., Jonsson, M., Klaminder, J., 2013. Dilute concentrations of a psychiatric drug alter behavior of fish from natural populations. Science 339, 814–815.

Brooks, B.W., 2014. Fish on Prozac (and Zoloft): ten years later. Aquat. Toxicol. 151, 61–67.

Brooks, B.W., Berninger, J.P., Kristofco, L.A., Ramirez, A.J., Stanley, J.K., Valenti, T.W., 2012a. Pharmaceuticals in the environment: lessons learned for reducing uncertainties in environmental risk assessment. Prog. Mol. Biol. Transl. Sci. 112, 231–258.

Brooks, B.W., Berninger, J.P., Ramirez, A.J., Huggett, D.B., 2012b. Perspectives on human pharmaceuticals in the environment. In: Brooks, B.W., Huggett, D.B. (Eds.), Human Pharmaceuticals in the Environment: Current and Future Perspectives. Springer, New York, pp. 1–16.

Brooks, B.W., Chambliss, C.K., Stanley, J.K., Ramirez, A., Banks, K.E., Johnson, R.D., et al., 2005. Determination of select antidepressants in fish from an effluent-dominated stream. Environ. Toxicol. Chem. 24, 464–469.

Brooks, B.W., Foran, C.M., Richards, S.M., Weston, J., Turner, P.K., Stanley, J.K., et al., 2003. Aquatic ecotoxicology of fluoxetine. Toxicol. Lett. 142, 169–183.

Brooks, B.W., Huggett, D.B., Boxall, A., 2009a. Pharmaceuticals and personal care products: research needs for the next decade. Environ. Toxicol. Chem. 28, 2469–2472.

Brooks, B.W., Huggett, D.B., Brain, R.A., Ankley, G.T., 2009b. Risk assessment considerations for veterinary medicines in aquatic systems. In: Henderson, K., Coats, J. (Eds.), Veterinary Pharmaceuticals in the Environment. American Chemical Society, Washington DC, pp. 205–223.

Brooks, B.W., Riley, T.M., Taylor, R.D., 2006. Water quality of effluent-dominated ecosystems: ecotoxicological, hydrological, and management considerations. Hydrobiologia 556, 365–379.

Caldwell, D.J., Mastrocco, F., Hutchinson, T.H., Länge, R., Heijerick, D., Janssen, C., et al., 2008. Derivation of an aquatic predicted no-effect concentration for the synthetic hormone, 17α-ethinyl estradiol. Environ. Sci. Technol. 42, 7046–7054.

Clubbs, R.L., Brooks, B.W., 2007. Daphnia magna responses to a vertebrate estrogen receptor agonist and an antagonist: a multigenerational study. Ecotoxicol. Environ. Saf. 67, 385–398.

Connors, K.A., Du, B., Fitzsimmons, P.N., Hoffman, A.D., Chambliss, C.K., Nichols, J.W., Brooks, B.W., 2013. Comparative pharmaceutical metabolism by rainbow trout (Oncorhynchus mykiss) liver S9 fractions. Environ. Toxicol. Chem. 32, 1810–1818.

Connors, K.A., Voutchkova-Kostal, A.M., Kostal, J., Anastas, P., Zimmerman, J.B., Brooks, B.W., 2014. Reducing aquatic toxicity: probabilistic hazard evaluation of sustainable molecular design guidelines. Environ. Toxicol. Chem. 33, 1894–1902.

Corrales, J., Kristofco, L.A., Steele, W.B., Saari, G.N., Kostal, J., Williams, E.S., Mills, M., Gallagher, E.P., Kavanagh, T.J., Simcox, N., Shen, L.Q., Melnikov, F., Zimmerman, J.B., Voutchkova-Kostal, A., Anastas, P., Brooks, B.W., 2017. Toward the design of less hazardous chemicals: exploring comparative oxidative stress in two common animal models. Chem. Res. Toxicol. 30, 893–904.

Daughton, C.G., Brooks, B.W., 2011. Active pharmaceuticals ingredients and aquatic organisms. In: Meador, J., Beyer, N. (Eds.), Environmental Contaminants in Wildlife: Interpreting Tissue Concentrations, second ed. Taylor and Francis, pp. 281–341.

Dickson, K.L., Waller, W.T., Kennedy, J.H., Ammann, L.P., 1992. Assessing the relationship between ambient toxicity and instream biological response. Environ. Toxicol. Chem. 11, 1307–1322.

Du, B., Haddad, S.P., Luek, A., Scott, C.S., Saari, G.N., Burket, S.R., Kelly, M., Broach, L., Rasmussen, J.B., Chambliss, C.K., Brooks, B.W., 2016. Bioaccumulation of ionizable pharmaceuticals in fish across habitats of a tidally-influenced urban bayou. Environ. Toxicol. Chem. 35, 966–974.

Du, B., Haddad, S.P., Luek, A., Scott, W.C., Saari, G.N., Kristofco, L.A., Connors, K.A., Rash, C., Rasmussen, J.B., Chambliss, C.K., Brooks, B.W., 2014a. Bioaccumulation and trophic dilution of human pharmaceuticals across trophic levels of an effluent-dependent wadeable stream. Philos. Trans. R. Soc. B: Biol. Sci. 369 (1656), 20140058. https://doi.org/10.1098/rstb.2014.0058.

Du, B., Price, A.E., Scott, W.C., Kristofco, L.A., Ramirez, A.J., Chambliss, C.K., Yelderman, J.C., Brooks, B.W., 2014b. Comparison of contaminants of emerging concern removal, discharge and water quality hazards among centralized and on-site wastewater treatment systems receiving common wastewater influent. Sci. Total Environ. 466–467, 976–984.

Dzieweczynski, T.L., Hebert, O.L., 2013. The effects of short-term exposure to an endocrine disrupter on behavioral consistency in male juvenile and adult siamese fighting fish. Arch. Environ. Contam. Toxicol. 64, 316–326.

Erickson, R.J., McKim, J.M., Lien, G.J., Hoffman, A.D., Batterman, S.L., 2006a. Uptake and elimination of ionizable organic chemicals at fish gills: I. Model formulation, parameterization, and behavior. Environ. Toxicol. Chem. 25, 1512–1521.

Erickson, R.J., McKim, J.M., Lien, G.J., Hoffman, A.D., Batterman, S.L., 2006b. Uptake and elimination of ionizable organic chemicals at fish gills: II. Observed and predicted effects of pH, alkalinity, and chemical properties. Environ. Toxicol. Chem. 25, 1522–1532.

Fick, J., Lindberg, R.H., Parkkonen, J., Arvidsson, B., Tysklind, M., Larsson, D.G.J., 2010. Therapeutic levels of levonorgestrel detected in blood plasma of fish: results from screening rainbow trout exposed to treated sewage effluents. Environ. Sci. Technol. 44, 2661–2666.

Fitzsimmons, P.N., Fernandez, J.D., Hoffman, A.D., Butterworth, B.C., Nichols, J.W., 2001. Branchial elimination of superhydrophobic organic compounds by rainbow trout (*Oncorhynchus mykiss*). Aquat. Toxicol. 55, 23–34.

Gaw, S., Brooks, B.W., 2016. Changing tides: adaptive monitoring, assessment and management of pharmaceutical hazards in the environment through time. Environ. Toxicol. Chem. 35, 1037–1042.

Gould, G.G., Brooks, B.W., Frazer, A., 2007. [^{3}H] Citalopram binding to serotonin transporter sites in minnow brains. Basic Clin. Pharmacol. Toxicol. 101, 203–210.

Grothe, D.R., Dickson, K.L., Reed-Judkins, D.K. (Eds.), 1996. Whole Effluent Toxicity Testing: An Evaluation of Methods and Prediction of Receiving System Impacts. SETAC, Pensacola, FL, USA.

Gunnarsson, L., Jauhiainen, A., Kristiansson, E., Nerman, O., Larsson, D.J., 2008. Evolutionary conservation of human drug targets in organisms used for environmental risk assessments. Environ. Sci. Technol. 42, 5807–5813.

Handley, S.L., McBlane, J.W., 1993. An assessment of the elevated X-maze for studying anxiety and anxiety-modulating drugs. J. Pharmacol. Toxicol. Methods 29, 129–138.

Huggett, D.B., Cook, J.C., Ericson, J.F., Williams, R.T., 2003. A theoretical model for utilizing mammalian pharmacology and safety data to prioritize potential impacts of human pharmaceuticals to fish. Hum. Ecol. Risk. Assess. 9, 1789–1799.

Jones, H.E., Hickman, M., Kasprzyk-Hordern, B., Welton, N.J., Baker, D.R., Ades, A., 2014. Illicit and pharmaceutical drug consumption estimated via wastewater analysis. Part B: placing back-calculations in a formal statistical framework. Sci. Total Environ. 487, 642–650.

Kalueff, A.V., Echevarria, D.J., Homechaudhuri, S., Stewart, A.M., Collier, A.D., Kaluyeva, A.A., Li, S., Liu, Y., Chen, P., Wang, J., 2016. Zebrafish neurobehavioral phenomics for aquatic neuropharmacology and toxicology research. Aquat. Toxicol. 170, 297–309.

Kidd, K.A., Blanchfield, P.J., Mills, K.H., Palace, V.P., Evans, R.E., Lazorchak, J.M., et al., 2007. Collapse of a fish population after exposure to a synthetic estrogen. Proc. Natl. Acad. Sci. U. S. A. 104, 8897–8901.

King, R.S., Brain, R.A., Back, J.A., Becker, C., Wright, M.V., Toteu Djomte, V., Scott, W.C., Virgil, S.R., Brooks, B.W., Hosmer, A.J., Chambliss, C.K., 2016. Effects of pulsed atrazine exposures on autotrophic community structure, biomass, and production in field-based stream mesocosms. Environ. Toxicol. Chem. 35, 660–675.

Kokel, D., Bryan, J., Laggner, C., White, R., Cheung, C.Y.J., Mateus, R., Healey, D., Kim, S., Werdich, A.A., Haggarty, S.J., MacRae, C.A., Shoichet, B., Peterson, R.T., 2010. Rapid behavior-based identification of neuroactive small molecules in the zebrafish. Nat. Chem. Biol. 6, 231–237.

Kookana, R.S., Williams, M., Boxall, A.B., Larsson, D.G., Gaw, S., Choi, K., Yamamoto, H., Thatikonda, S., Zhu, Y.G., Carriquiriborde, P., 2014. Potential ecological footprints of active pharmaceutical ingredients: an examination of risk factors in low-, middle- and high-income countries. Philos. Trans. R. Soc. Lond. Ser. B Biol. Sci. 369. https://doi.org/10.1098/rstb.2013.0586.

Kostal, J., Voutchkova-Kostal, A., Anastas, P.T., Zimmerman, J.B., 2015. Identifying and designing chemicals with minimal acute aquatic toxicity. Proc. Natl. Acad. Sci. U. S. A. 112, 6289–6294.

Kristiansson, E., Fick, J., Janzon, A., Grabic, R., Rutgersson, C., Weijdegård, B., et al., 2011. Pyrosequencing of antibiotic-contaminated river sediments reveals high levels of resistance and gene transfer elements. PLoS ONE 6, e17038.

Kristofco, L.A., Brooks, B.W., 2017. Global scanning of antihistamines in the environment: analysis of occurrence and hazards in aquatic systems. Sci. Total Environ. 592, 477–487.

La Point, T.W., Waller, W.T., 2000. Field assessments in conjunction with WET testing. Environ. Toxicol. Chem. 19, 14–24.

Larsson, D.J., de Pedro, C., Paxeus, N., 2007. Effluent from drug manufactures contains extremely high levels of pharmaceuticals. J. Hazard. Mater. 148, 751–755.

Little, E.E., Adams, S., 2002. Behavioural measures of environmental stressors in fish. In: Adams, S.M. (Ed.), Biological Indicators of Aquatic Ecosystem Stress. American Fisheries Society, pp. 431–472.

Little, E.E., Finger, S.E., 1990. Swimming behavior as an indicator of sublethal toxicity in fish. Environ. Toxicol. Chem. 9, 13–19.

Luthy, R.G., Sedlak, D.L., Plumlee, M.H., Austin, D., Resh, V.H., 2015. Wastewater-effluent-dominated streams as ecosystem-management tools in a drier climate. Front. Ecol. Environ. 13, 477–485.

Madan, A., O'Brien, Z., Wen, J., O'Brien, C., Farber, R.H., Beaton, G., Crowe, P., Oosterhuis, B., Garner, R.C., Lappin, G., Bozigian, H.P., 2009. A pharmacokinetic evaluation of five H1 antagonists after an oral and intravenous microdose to human subjects. Brit. J. Clin. Pharmacol. 67, 288–298.

Margiotta-Casaluci, L., Owen, S.F., Cumming, R.I., de Polo, A., Winter, M.J., Panter, G.H., Rand-Weaver, M., Sumpter, J.P., 2014. Quantitative cross species extrapolation between humans and fish: the case of the antidepressant fluoxetine. PLoS ONE 9, e110467.

Margiotta-Casaluci, L., Owen, S.F., Huerta, B., Rodríguez-Mozaz, S., Kugathas, S., Barceló, D., et al., 2016. Internal exposure dynamics drive the adverse outcome pathways of synthetic glucocorticoids in fish. Sci. Rep. 6, 21978.

Maximino, C., da Silva, A.W.B., Araújo, J., Lima, M.G., Miranda, V., Puty, B., Benzecry, R., Picanço-Diniz, D.L.W., Gouveia Jr., A., Oliveira, K.R.M., 2014. Fingerprinting of psychoactive drugs in zebrafish anxiety-like behaviors. PLoS ONE 9, e103943.

McRobb, F.M., Sahagún, V., Kufareva, I., Abagyan, R., 2014. In silico analysis of the conservation of human toxicity and endocrine disruption targets in aquatic species. Environ. Sci. Technol. 48, 1964–1972.

Melvin, S.D., Wilson, S.P., 2013. The utility of behavioral studies for aquatic toxicology testing: a meta-analysis. Chemosphere 93, 2217–2223.

Monteiro, S.C., Boxall, A.B., 2010. Occurrence and fate of human pharmaceuticals in the environment. Rev. Environ. Contam. Toxicol. 202, 53–154.

Nakamura, Y., Yamamoto, H., Sekizawa, J., Kondo, T., Hirai, N., Tatarazako, N., 2008. The effects of pH on fluoxetine in japanese medaka (*Oryzias latipes*): acute toxicity in fish larvae and bioaccumulation in juvenile fish. Chemosphere 70, 865–873.

Nichols, J.W., Du, B., Berninger, J.P., Connors, K.A., Chambliss, C.K., Erickson, R., Hoffman, A., Brooks, B.W., 2015. Observed and modeled effects of pH on bioconcentration of diphenhydramine, a weakly basic pharmaceutical, in fathead minnows. Environ. Toxicol. Chem. 34, 1425–1435.

Noyes, P.D., Haggard, D.E., Gonnerman, G.D., Tanguay, R.L., 2015. Advanced morphological-behavioral test platform reveals neurodevelopmental defects in embryonic zebrafish exposed to comprehensive suite of halogenated and organophosphate flame retardants. Toxicol. Sci. 145, 177–195.

Ort, C., Nuijs, A.L., Berset, J.D., Bijlsma, L., Castiglioni, S., Covaci, A., et al., 2014. Spatial differences and temporal changes in illicit drug use in Europe quantified by wastewater analysis. Addiction 109, 1338–1352.

Painter, M.M., Buerkley, M.A., Julius, M.L., Vajda, A.M., Norris, D.O., Barber, L.B., Furlong, E.T., Schultz, M.M., Schoenfuss, H.L., 2009. Antidepressants at environmentally relevant concentrations affect predator avoidance behavior of larval fathead minnows (*Pimephales promelas*). Environ. Toxicol. Chem. 28, 2677–2684.

Parker, M.O., 2016. Adult vertebrate behavioural aquatic toxicology: reliability and validity. Aquat. Toxicol. 170, 323–329.

Perkins, E.J., Ankley, G.T., Crofton, K.M., Garcia-Reyero, N., LaLone, C.A., Johnson, M.S., Tietge, J.E., Villeneuve, D.L., 2013. Current perspectives on the use of alternative species in human health and ecological hazard assessments. Environ. Health Perspect. 121, 1002–1010.

Posthuma, L., Suter II, G.W., Traas, T.P. (Eds.), 2002. Species Sensitivity Distributions in Ecotoxicology. CRC Press, Boca Raton, FL.

Reif, D.M., Truong, L., Mandrell, D., Marvel, S., Zhang, G., Tanguay, R.L., 2015. High-throughput characterization of chemical-associated embryonic behavioral changes predicts teratogenic outcomes. Arch. Toxicol. 90, 1459–1470.

Rovida, C., Hartung, T., 2009. Re-evaluation of animal numbers and costs for in vivo tests to accomplish REACH legislation requirements for chemicals—a report by the Transatlantic Think Tank for Toxicology (t4). Altern. Anim. Exp. 26, 187–208.

Rudd, M.A., Ankley, G.T., Boxall, A.B.A., Brooks, B.W., 2014. International scientists' research priorities for pharmaceuticals and personal care products in the environment. Integr. Environ. Assess. Manage 10, 576–587.

Saari, G.N., Scott, W.C., Brooks, B.W., 2017. Global scanning assessment of calcium channel blockers in the environment: Review and analysis of occurrence, ecotoxicology and hazards in aquatic systems. Chemosphere 189, 466–478.

Schnörr, S., Steenbergen, P., Richardson, M., Champagne, D., 2012. Measuring thigmotaxis in larval zebrafish. Behav. Brain Res. 228, 367–374.

Scott, W.C., Du, B., Haddad, S.P., Breed, C.S., Saari, G.N., Kelly, M., Broach, L., Chambliss, C.K., Brooks, B.W., 2016. Predicted and observed therapeutic dose exceedances of ionizable pharmaceuticals in fish plasma from urban coastal systems. Environ. Toxicol. Chem. 35, 983–995.

Seiler, J.P., 2002. Pharmacodynamic activity of drugs and ecotoxicology—can the two be connected? Toxicol. Lett. 131, 105–115.

Sloman, K., McNeil, P., 2012. Using physiology and behaviour to understand the responses of fish early life stages to toxicants. J. Fish Biol. 81, 2175–2198.

Steenbergen, P.J., Richardson, M.K., Champagne, D.L., 2011. Patterns of avoidance behaviours in the light/dark preference test in young juvenile zebrafish: a pharmacological study. Behav. Brain Res. 222, 15–25.

Stewart, A.M., Ullmann, J.F., Norton, W.H., Parker, M., Brennan, C., Gerlai, R., Kalueff, A.V., 2015. Molecular psychiatry of zebrafish. Mol. Psychiatry 20, 2–17.

Straub, J.O., Hutchinson, T.H., 2012. Environmental risk assessment for human pharmaceuticals: the current state of international regulations. In: Brooks, B.W., Huggett, D.B. (Eds.), Human Pharmaceuticals in the Environment: Current and Future Perspectives. Springer, New York, pp. 17–47.

Thomas, K.V., Bijlsma, L., Castiglioni, S., Covaci, A., Emke, E., Grabic, R., et al., 2012. Comparing illicit drug use in 19 European cities through sewage analysis. Sci. Total Environ. 432, 432–439.

Tierney, K.B., 2011. Behavioural assessments of neurotoxic effects and neurodegeneration in zebrafish. Biochim. Biophys. Acta (BBA) – Mol. Basis Dis. 1812, 381–389.

Valenti, T.W., Gould, G.G., Berninger, J.P., Connors, K.A., Keele, N.B., Prosser, K.N., Brooks, B.W., 2012. Human therapeutic plasma levels of the selective serotonin reuptake inhibitor (SSRI) sertraline decrease serotonin reuptake transporter binding and shelter-seeking behavior in adult male fathead minnows. Environ. Sci. Technol. 46, 2427–2435.

Valenti, T.W., Perez Hurtado, P., Chambliss, C.K., Brooks, B.W., 2009. Aquatic toxicity of sertraline to Pimephales promelas at environmentally relevant surface water pH. Environ. Toxicol. Chem. 28, 2685–2694.

Veith, G.D., Kosian, P., 1983. Estimating bioconcentration potential from octanol/water partition coefficients. In: MacKay, D., et al. (Eds.), Physical Behavior of PCBs in the Great Lakes. Ann Arbor Science, Ann Arbor, MI, pp. 269–282.

Voutchkova, A.M., Kostal, J., Steinfeld, J.B., Emerson, J.W., Brooks, B.W., Anastas, P., Zimmerman, J.M., 2011. Towards rational molecular design: derivation of property guidelines for reduced acute aquatic toxicity. Green Chem. 13, 2373–2379.

Voutchkova-Kostal, A.M., Kostal, J., Connors, K.A., Brooks, B.W., Anastas, P., Zimmerman, J.M., 2012. Towards rational molecular design for reduced chronic aquatic toxicity. Green Chem. 14, 1001–1008.

FURTHER READING

Brooks, B.W., Huggett, D.B. (Eds), 2012. *Human Pharmaceuticals in the Environment: Current and Future Perspectives*. Springer, New York. ISBN 978-1-4614-3419-1. 302 p.

Brown, J.N., Paxeus, N., Forlin, L., Larsson, D.G.J., 2007. Variations in bioconcentration of human pharmaceuticals from sewage effluents into fish blood plasma. Environ. Toxicol. Chem. 24, 267–274.

Coish, P., Brooks, B.W., Gallagher, E.P., Kavanagh, T.J., Voutchkova-Kostal, A., Zimmerman, J.B., Anastas, P.T., 2016. Current status and future challenges in the molecular design for reduced hazard. ACS Sustain. Chem. Eng. 4, 5900–5906.

LaLone, C.A., Berninger, J.P., Villeneuve, D.L., Ankley, G.T., 2014. Leveraging existing data for prioritization of the ecological risks of human and veterinary pharmaceuticals to aquatic organisms. Philos. Trans. R. Soc. Lond. Ser. B Biol. Sci. 369. https://doi.org/10.1098/rstb.2014.0022.

Ramirez AJ, Brain RA, Usenko S, Mottaleb MA, O'Donnell JG, Stahl LL, Wathen JB, Snyder BD, Pitt JL, Perez-Hurtado P. 2009. Occurrence of pharmaceuticals and personal care products in fish: results of a national pilot study in the United States. Environ. Toxicol. Chem. 28:2587–2597.

Chapter 5

Terrestrial Ecotoxicity

Laura J. Carter* and Chad A. Kinney†
**University of York, Heslington, United Kingdom*
†*Colorado State University-Pueblo, Pueblo, CO, United States*

INTRODUCTION OF HEALTH-CARE PRODUCTS INTO TERRESTRIAL ENVIRONMENTS

As discussed in Chapter 2, health-care products (HCPs) including prescription and over-the-counter pharmaceuticals (e.g., antibiotics, analgesics, lipid regulators, and antidepressants), disinfectants, and antifungal agents and a variety of other organic chemicals commonly enter the municipal waste stream (sewage system). Many HCPs are known to be recalcitrant to degradation during the treatment process. Modern wastewater treatment facilities (WWTFs) are not primarily designed to remove HCPs and related organic chemicals. WWTFs produce two end products, liquid treated wastewater effluent and semisolid sewage sludge. As stresses on available potable water resources continue to increase, alternative sources of water, particularly for nonpotable applications, are growing in demand. Increasingly, treated wastewater is being recycled (reclaimed wastewater) for irrigation of public and agricultural lands (USEPA, 2004). If the sewage sludge meets regulatory requirements (primarily for pathogens and metal/metalloid content), it can be classified as biosolids and land-applied, primarily on agricultural soil, as a source of plant nutrients and soil organic carbon (USEPA, 1999, 2015). As a result of incomplete removal during wastewater treatment, reclaimed wastewater and biosolids are well documented to contain a range of organic contaminants, including HCPs (Fig. 1). The typical concentration of most pharmaceuticals in reclaimed wastewater and biosolids is of the order of tens to hundreds of nanogram per liter and nanogram per gram, respectively. However, in the case of other HCPs such as the disinfectants triclosan and triclocarban, the concentration can reach thousands to tens of thousands of nanogram per liter and nanogram per gram in reclaimed water and biosolids, respectively (Chefetz et al., 2008; Kinney et al., 2006a,b; McClellan and Halden, 2010; Schultz and Furlong, 2008; USEPA, 2009) (Table 1). During wastewater treatment, many HCPs and other organic chemical contaminants, even those with

Environmental Contaminants, Vol. 11. https://doi.org/10.1016/B978-0-444-63857-1.00005-X

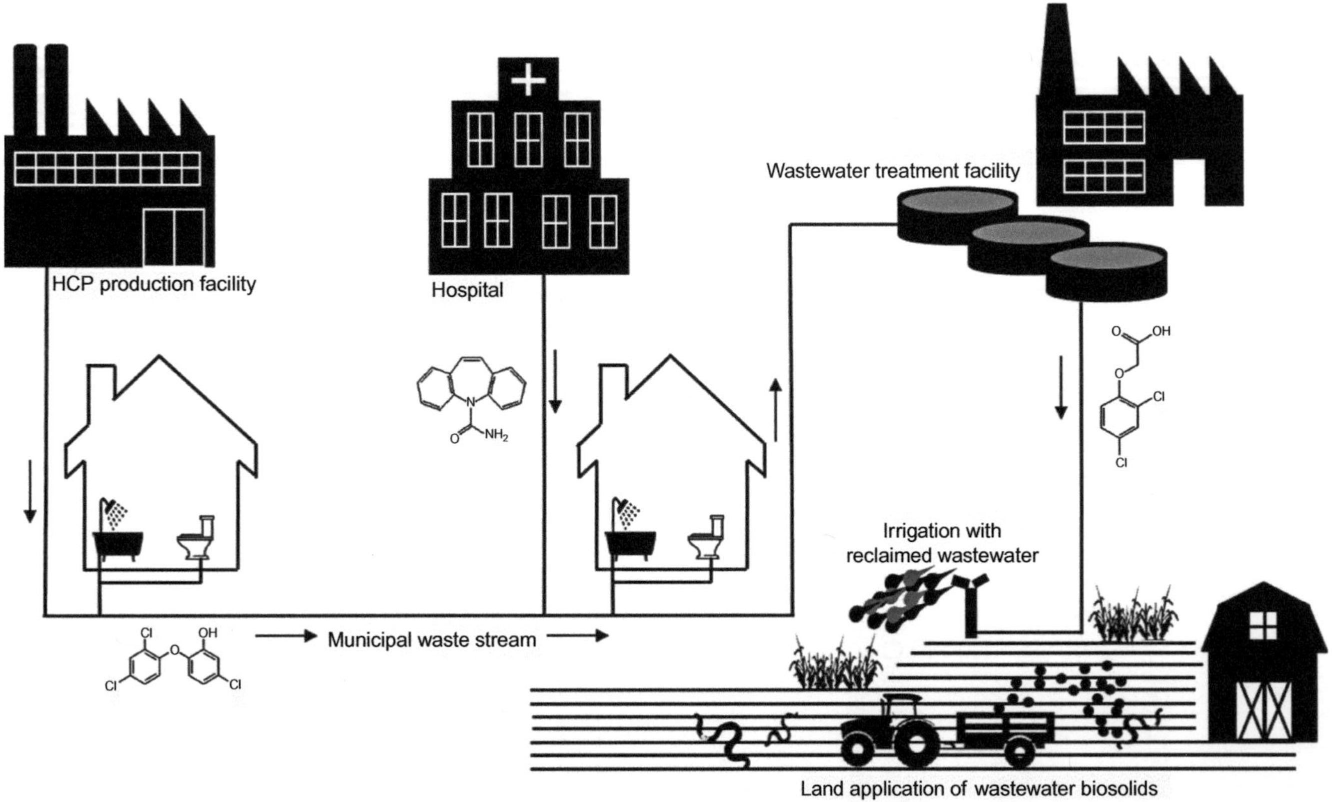

FIG. 1 Illustration highlighting some of the pathways that health-care products can enter the municipal waste stream and subsequently be released into the terrestrial environment.

TABLE 1 Select HCPs, Uses, Physicochemical Properties, and Representative Concentrations in Wastewater End Products

HCP	CAS	Use	Log K_{ow}	Water Solubility (mg/L)	pK_a	Conc. in Biosolids (ng/g)	Conc. in Reclaimed Water (ng/L)
Carbamazepine	298-46-4	Antiepileptic	2.45	112	7	99[a]	70[d]
Carboplatin	41575-94-4	Chemotherapy	−0.46	1.17×10^4	NA	NA	40[e]
Ciprofloxacin	85721-33-1	Antibiotic	0.28	3×10^4	6.07	2348[b]	36[f]
Diphenhydramine	58-73-1	Antihistamine	3.44	3.06	8.98	322[a]	40[d]
Fluoxetine	54910-89-3	Antidepressant	4.06	60.3	10.09	77[a]	3[d]
Salbutamol	18559-94-9	Bronchospasm/ asthma	0.64	1.41×10^4	9.3	271[a]	6[d]
Sertraline	79617-96-2	Antidepressant	5.29	3.52×10^3	8.9	104[b]	55[g]
Sulfamethoxazole	723-46-6	Antibiotic	0.89	610	1.85	155[a]	21[d]
Triclocarban	101-20-2	Disinfectant	4.90	2.37×10^{-2}	12.7	36,060[c]	2658[h]
Triclosan	3380-34-5	Disinfectant	4.76	10	7.9	12,640[c]	1715[h]
Trimethoprim	738-70-5	Antibiotic	0.91	4.0×10^2	3.23	26[a]	15[d]
Warfarin	81-81-2	Anticoagulant	2.70	17	4.50	65[a]	25[d]

[a]*Kinney et al. (2006a).*
[b]*McClellan and Halden (2010).*
[c]*Kinney et al. (2010).*
[d]*Kinney et al. (2006b).*
[e]*Lenz et al. (2007).*
[f]*Giger et al. (2003).*
[g]*Schultz and Furlong (2008).*
[h]*Kumar et al. (2010).*

moderate polarity, will preferentially partition into the organic carbon-rich semisolid sludge or biosolids phase as a result of hydrophobic partitioning (Xia et al., 2005). So while results of some studies document a reduction in HCP concentrations in the liquid phase during wastewater treatment (Metcalfe et al., 2003; Nakada et al., 2006; Yang and Metcalfe, 2006), this cannot necessarily be solely attributed to degradation of the HCPs alone but rather can likely be largely attributed to partitioning into the sludge phase during treatment.

The US Environmental Protection Agency (USEPA) estimates that 7.7 billion liters of wastewater are reused each day for a variety of applications and that reclaimed wastewater use is growing at a rate of about 15% each year. Some reclaimed wastewater is used for irrigation of agricultural and public lands. In Florida and California, about 11% and 6% of the total irrigation water is reclaimed wastewater, respectively, with much of it used for agricultural lands (USEPA, 2004). In arid regions of the world, such as Israel, agriculture relies on reclaimed wastewater as a major source of irrigation water to overcome water shortages. More than 50% of the treated wastewater in Israel is used for irrigation (about 1×10^9 m^3/year), accounting for about 50% of the total irrigation water. In Spain, about 17 % of the irrigation water is reclaimed wastewater. In the United States, approximately 8×10^6 kg of biosolids are produced on an annual basis. The EPA estimates that 50% of all biosolids produced are land-applied on an annual basis (USEPA, 2006). In Europe, it is estimated that about 37% of biosolids (2.4×10^9 kg annually) are land-applied on agricultural soils per year (Chang et al., 2002). However, in some countries, such as France where a reported 65% of sewage sludge is land-applied, use of biosolids is greater (Bonnin, 2001). Biosolid application rates on agricultural soils range from about 2.1 to 8.3 dry tonnes/ha in the United States. By comparison, the application rate of biosolids onto reclamation sites (e.g., mine sites or following wildfires), rangeland, and forestland can be considerably higher, as much as 41.3 dry tonnes/ha (USEPA, 2000).

The quantities of some HCPs entering the municipal waste stream and therefore the environment can be quite staggering. For instance, the disinfectants triclosan or triclocarban are used as antibacterial compounds in such a large number of soaps and a wide variety of other products that the combined discharge into the environment is greater than 1×10^6 kg/year in the United States primarily through the land application of biosolids and discharge of treated wastewater (Chu and Metcalfe, 2007; Heidler and Halden, 2007; Heidler et al., 2006; Perencevich et al., 2001). And as much as 28,000 kg of the antibiotic ciprofloxacin is land-applied in biosolids in the United States annually (McClellan and Halden, 2010).

REGULATORY AND TECHNICAL ASPECTS OF HEALTH-CARE PRODUCTS IN THE TERRESTRIAL ENVIRONMENT

In North America, Europe, and Asia, an environmental risk assessment (ERA) is required to evaluate (see Chapter 13) the impacts of pharmaceuticals, in particular, entering the environment as part of the drug approval process (EMEA, 2006; FDA, 1998). Results of various widely accepted test methodologies (laid down mainly by the Organization for Economic Cooperation and Development, OECD) form the basis of the risk assessment process. With regard to the assessment of the terrestrial environment, if the substance is not readily biodegradable and adsorption/desorption data indicate a high affinity for the pharmaceutical to bind to sewage sludge in the WWTF, an assessment of the drug substance in the terrestrial compartment is required. Specifically, if the adsorption coefficient in sludge (K_{OC}) is greater than 10,000 L/kg, a terrestrial risk assessment will be undertaken. Earthworms and plants are key terrestrial species included in a set of tests to investigate the potential toxicity to soil invertebrates and acute effects on higher plants. The terrestrial effect studies recommended in the Phase II Tier B of the EMEA ERA include terrestrial plants, growth test (OECD, 208 and USEPA OCSPP 850.4100), earthworm, acute toxicity tests (OECD 207), and Collembola reproduction test (ISO 11267) (ISO, 1999; OECD, 1984, 2006). Typically, these studies involve exposure of the test organism to a range of chemical concentrations to allow for a calculation of the dose at which an effect (end point, e.g., mortality) is observed in comparison with the control, also known as an effect concentration (EC_x) or lethal concentration (LC_x). In plant growth tests, the objective of the seedlings test is to measure the effect of test substances upon plants during early critical stages of their development. End points measured include visual assessment of seedling emergence, biomass (fresh or dry shoot weight or shoot height), and visual detrimental effects (chlorosis, mortality, and plant development abnormalities). For earthworms, a screening test on moist filter paper and exposure to the test substance in an artificial soil at a range of concentrations allow for the percentage of mortality after 7 and 14 days to be calculated.

SOIL INVERTEBRATE (EARTHWORM) TOXICITY

Earthworms represent an ideal sentinel organism for assessing soil contamination, as they are in contact with soil and soil solution, tend to migrate over only short distances, and are widely distributed in soils around the globe. Earthworms can comprise as much as 60%–80% of total soil biomass in some locations. Furthermore, earthworms are common primary consumers of organic matter in soil and, as such, may be exposed to organic and inorganic contaminants contained in the organic fractions within soil (Bouche, 1983, 1992; Hopkin, 1989). Earthworms can accumulate organic contaminants both by passive means through dermal absorption and actively through ingestion. The relative importance of dermal versus dietary exposure is dependent upon the individual soil contaminant, with the relative importance of dietary exposure generally increasing with increasing hydrophobicity of the contaminant (Jager et al., 2003; Ma et al., 1998).

While there is generally a paucity of available HCP toxicity data for terrestrial organisms, there are a variety of data available for earthworm exposure and bioaccumulation of HCPs and a limited amount of ecotoxicity data available for earthworms. Much of the data that are available are derived from controlled laboratory studies involving individual or select groups of HCPs in artificial soils. Survival is one of the most common end points studied, but measures of reproduction and behavior, such as aestivation (a curling behavior indicating stress), surfacing, or avoidance, have also been studied. In an ecotoxicity study of 18 pharmaceuticals including compounds from nonsteroidal anti-inflammatory drugs (NSAIDs), lipid regulators, β-blockers, and antibiotics classes of compounds, researchers found that a number of the pharmaceuticals studied were acutely toxic (lethality) to earthworms (*Eisenia fetida*, (Pino et al., 2015)). Fourteen-day exposure to some of the blood lipid regulators, the β-blockers, and the antibiotics did not result in any observed acute toxicity. However, the NSAIDs and some blood lipid-regulating pharmaceuticals resulted in toxicity to *E. fetida*. The greatest toxicity was reported for ibuprofen (median lethal concentration (LC_{50}) = 64.80 mg/kg) followed by diclofenac (LC_{50} = 90.49 mg/kg) and simvastatin (LC_{50} = 92.70 mg/kg). The LC_{50} concentrations reported for the pharmaceuticals in this study far exceed environmentally relevant concentrations reported for soils irrigated with reclaimed wastewater or amended with biosolids. Similar conclusions were reached in a study of the ecotoxicity of biosolids-borne triclocarban (disinfectant) in earthworms (*E. fetida*, (Snyder et al., 2011)). The LC_{50} for triclocarban was greater than 40 mg/kg for biosolid-amended soils, which far exceeds the concentration of triclocarban previously reported in soils amended with biosolids. In the same study, the researchers determined that triclocarban in biosolids, even at concentrations exceeding the native concentration, did not affect common measures of microbial activity in soil, soil respiration and ammonification.

While available research suggests that individual HCPs at environmentally relevant concentrations are likely not acutely toxic to some soil organisms, this does not address all concerns about the environmental release of HCPs. Scientists know that reclaimed wastewater and biosolids contain a complex mixture of HCPs and other organic and inorganic contaminants (Kinney et al., 2006a,b; McClellan and Halden, 2010; USEPA, 2009). A recent study of the ecotoxicity of biosolids containing a wide variety of HCPs and other organic chemicals, found biosolids to be acutely toxic to earthworms (*E. fetida*) even at environmental relevant application rates of biosolids to agricultural soils in the United States (Kinney et al., 2012). End points measured in the study included earthworm survival and measures of earthworm reproduction. Biosolids at application rates as low as 1% by mass in soil resulted in a significant reduction in adult earthworm survival and complete lethality at 3% and 4% biosolids in soil. Similar results have been reported in by others (Artuso et al., 2011). The production of juvenile earthworms and cocoons (measures of reproductive success) was also reduced with increasing exposure to biosolids from 1% up to 4% biosolids in soil (Kinney et al., 2012). It is unknown if the toxic effects observed in these studies were the result of synergistic interactions (where mixtures of chemicals interact to increases toxicity) of HCPs and other organic chemical contaminants or other components in the biosolids.

Springtails (*Folsomia candida*) are another standard test organism for soil contaminants. Springtails are a common small arthropod widely distributed in soils around the globe. Ecotoxicity studies into the effects of HCPs with springtails are even more limited than for earthworms. Climbazole, an antifungal agent used in some antidandruff shampoo, was found to have no effect on reproduction in springtails at concentrations as high as 1000 mg/kg soil dry mass (Richter et al., 2013). However, climbazole did disrupt dehydrogenase enzyme activity in the soil bacterium *Arthrobacter globiformis* with a median effective concentration (EC_{50}) of 456 mg/kg soil dry mass.

More abundant than toxicology studies involving HCPs and earthworms or other terrestrial organisms are laboratory and field studies that document the ability of earthworms to bioaccumulate HCPs and other organic chemicals available in the environment into their tissues (Carter et al., 2014a; Kinney et al., 2008, 2010, 2012; Markman et al., 2007). Given that earthworms occupy a low trophic level in terrestrial food webs, they might serve as the entry point for HCPs into terrestrial food webs and a route of exposure for higher trophic organisms to HCPs. For example, male European starlings (*Sturnus vulgaris*) exposed to select endocrine disrupting compounds with estrogenic activity displayed endocrine disruption. Male starlings exposed to xenobiotic estrogenic compounds at concentrations similar to that observed in earthworms collected from trickling filter beds at a WWTF resulted in significant enlargement of the high vocal center (HVC, portion of the brain controlling song production), increased song production and complexity, and a decrease in immune function (Markman et al., 2007, 2008). This demonstrates the need to consider ecotoxicity of HCPs beyond just soil dwelling organisms in terrestrial environments. Furthermore, if biomagnification of HCPs occurs in terrestrial food webs, higher trophic organisms may be at greater risk of toxic responses to HCPs than lower trophic organisms (this topic is discussed in more detail in the next chapter).

PLANT TOXICITY

As well as the uptake and accumulation of HCPs in soil invertebrates, research has demonstrated, through field trials and laboratory experiments, that plants also take up these chemicals. Chemicals present in the soil solution can cross the root membrane and be transported around the plant via mass flow. The accumulation of HCPs has been shown in a range of vegetable crops (e.g., lettuce, cucumber, radish, and tomato) and grasses and high volume production crops such as soybean and wheat (Carter et al., 2014b; Goldstein et al., 2014; Malchi et al., 2014; Wu et al., 2015).

In field experiments, plants are typically exposed to biosolids or reclaimed wastewater containing HCP residues. Experiments such as this provide insight into the plant uptake of contaminants under environmentally realistic conditions. By minimizing the influence of external factors, laboratory studies offer a means of investigating the complex interplay between the factors and processes, which influence the plant uptake of contaminants. To date, a majority of these studies have focussed on the uptake of contaminants into plants with the aim to improve our understanding of the health risks associated with the presence of contaminants in crops destined for human consumption. While current findings suggest that there is minimal risk to human health from ingesting crops containing chemical residues from HCPs (Prosser and Sibley, 2015), the uptake of HCPs can have a direct impact on plant growth and development. The potential human health risks of pharmaceutical residues in plants are discussed in Chapter 7.

When a plant is under toxic stress, visual symptoms are often easy to report and measure. End points consisting of effects on shoot length, plant biomass, and seedling emergence have been reported in the scientific literature in relation to exposure of contaminants. With regard to the ecotoxicity of HCPs in particular, one of the most well-documented classes of chemicals in terms of effects on plants is the antibiotics. Antibiotics are used widely in the treatment and prevention of bacterial infections. Some of the first observations of the impacts of pharmaceuticals on plant growth were reported for tetracycline antibiotics. Specifically, bean yields, plant heights, and top and root dry weights were decreased by the presence of antibiotics (Batchelder, 1982). Since then, our understanding of the phytotoxic effects of antibiotics has increased with experiments looking at effects on germination and early plant growth in a range of vegetable crops (e.g., Hillis et al., 2011; Liu et al., 2009b). The presence of antibiotic residues in soils can also affect soil microbial and enzyme activities ultimately interfering with the abundance and interactions between the different microbial species (Ding and He, 2010; Thiele-Bruhn, 2003). Soil microorganisms play an essential role in maintaining soil fertility and cycling nutrients, among other functions, and therefore, any disruption to these key soil processes can result in detrimental effect to the terrestrial environment. It however remains unclear whether the negative effects of antibiotics on plant growth result from the direct damage of the plant by the antibiotic active pharmaceutical ingredients or whether antimicrobial action on beneficial soil microorganisms is responsible for the damage by affecting the plant-microbe symbiosis (Chander et al., 2005; Grassi et al., 2013).

As well as the reported effects of antibiotics on plant growth, other classes of HCPs can induce phytotoxic responses in terrestrial plant species. Contaminants belonging to the group of NSAIDs (e.g., ibuprofen and diclofenac) have been reported to affect early-stage root development and root shoot in radish plants (Schmidt and Redshaw, 2015). In addition to the previously mentioned effects on dehydrogenase enzyme activity in the soil bacterium, a clear concentration-response curve for a series of end points including shoot length and shoot biomass has been observed in response to exposure to the fungicide climbazole (Richter et al., 2013). Even though effects on plant biomass yield were observed at soil concentrations ranging from 4.9 to 15.0 mg/kg (EC_{10}), the presence of climbazole in the soil did not affect the emergence of oat and turnip seedlings (up to 100 mg/kg) (Richter et al., 2013).

Similar effects were noted in plants exposed to the lipid-lowering drug gemfibrozil and tamoxifen, which is used as a hormone therapy treatment for breast cancer. However, exposure to these chemicals inhibited both seedling germination and root elongation in cultivated plant species and wild plant species (D'Abrosca et al., 2008). This study in particular highlights the notable impact on the different vegetal communities as evidenced by differing sensitivity among cultivated and wild species, probably due to differences in plant physiology. Differences in plant physiology may also be responsible for the enhanced cotyledon opening in lettuce, which was suppressed in radish plants after exposure to the NSAID, diclofenac (Schmidt and Redshaw, 2015). Collectively, the reported observations on the impact of HCPs clearly highlight differential toxicity with regard to the chemical and the plant species and the plant organ in question. However, we are able to draw a number of general conclusions, including that root material appears to be the more sensitive end point in comparison with the shoots for a variety of HCPs (Carter et al., 2015; Hillis et al., 2011), and while the uptake of HCPs may result in effects on plant biomass, this is not always accompanied by a reduction in seedling emergence that appears to be impacted at much higher concentrations.

As has been previously highlighted in this chapter, the discharge of triclosan, an antibacterial and antifungal agent found in consumer products (e.g., shampoo, cosmetics, and soap), into the environment is relatively high. However, a review of scientific literature reveals contrasting findings on the ecotoxicity of triclosan in plants. Elongation of roots was significantly affected at triclosan concentrations of 10,000 and 30,000 ng/g (dry weight) for rice and cucumber, respectively, while shoot length was significantly inhibited in both species at 30,000 ng/g dry weight (Liu et al., 2009a). These effect concentrations are comparable with measured concentrations of triclosan in biosolids (~10,500 ng/g) but greater than the concentrations of triclosan in biosolid-amended soil (~1000 ng/g) (Kinney et al., 2008). Similar effects were also observed in wheat seedlings after exposure to triclosan in a nutrient medium where shoot and root elongation was inhibited by up to 60% at the highest exposure concentration (250 mg/L) (An et al., 2009). It has been suggested that these changes may be a result of the mode of action of triclosan in humans, namely, how it selectively targets fatty acid synthesis and disrupts cell membranes of bacteria (Escalada et al., 2005; Herklotz et al., 2010). However, it can be argued that exposure to triclosan in a nutrient medium, or concentrations in excess of 10 mg/kg, is not environmentally realistic. When an environmentally relevant scenario is incorporated into the study design, consisting of exposure via biosolid-derived triclosan, no consistent triclosan-dependent trends in seedling emergence or plant growth parameters were observed in a range of plants (radish, carrot, soybean spring wheat, and corn) (Prosser et al., 2014).

Nevertheless, the lack of triclosan ecotoxicity described above is not representative of all exposures that utilize biosolid-derived HCPs in context of environmentally relevant scenarios. The application of biosolids containing a multitude of HCPs such as personal care products, pharmaceuticals, and other wastewater indicator compounds at agronomically relevant levels has been found to affect lettuce seedling emergence. A higher percentage incorporation of biosolids resulted in an obvious impairment to both lettuce emergence and seedling fitness. Overall, seedling emergence after 14 days was reduced to 27%–73% of the controls for biosolid incorporations of up to 4% (Kinney et al., 2012).

An effect on biomass, as has been discussed, is typically the end result or the sum of individual responses manifesting in a larger impact. This would suggest that contaminants accumulating in the plant at lower concentrations can induce sublethal effects on plant growth and development processes. The uptake of the antiepileptic and antihypertensive drugs, for example, can be attributed to an effect on plant nutrient and hormone homeostasis (Carter et al., 2015). Specifically, the concentrations of the macronutrients potassium and phosphorus in the zucchini leaf were measured in excess of the controls. If there are changes in ion flow across the cell membrane resulting in this increase in macronutrient composition of the leaf, this will change cell membrane potential. Membrane potentials ultimately enable sugars to be transported from source organs (e.g., mature leaves) to sink organs (e.g., young leaves and fruit), and disturbing this could explain why the plants treated with the pharmaceuticals had reduced biomass and eventually went necrotic at the highest concentrations, displaying symptoms akin to sugar deficiency (Fig. 2). Interestingly, while effects on biomass and nutrient content were observed at high concentrations (>8 mg/kg), plant hormone production was affected at environmentally relevant concentrations (0.005 mg/kg) when no effects on biomass ensued. Plant hormones including jasmonates and abscisic acid are responsible for regulating how a plant responds to additional abiotic or biotic stress. On a large scale, impacts on the production and metabolism of plant hormones such as this pose a threat to wider ecosystem health and sustainable crop production (Carter et al., 2015).

A number of other nonstandard, sublethal end points have been described in relation to exposure of a small number of HCPs. In wheat, triclosan accumulation along with the uptake of the synthetic musk fragrance, galaxolide, caused damage to a suite of biochemical responses including the accumulation of chlorophyll *a*, synthesis of soluble protein, and activity of antioxidant

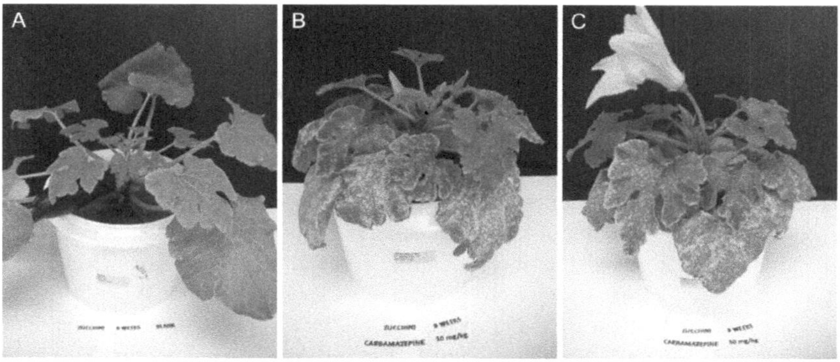

FIG. 2 Carbamazepine uptake by zucchini plants induces necrosis in leaves, increasing in intensity with increasing treatment concentration; (A) control and (B) and (C) exposure to 10 mg/kg carbamazepine. Burnt leaf edges and dead spots were noted in treatments >2 mg/kg. *(Photograph credit Laura J. Carter.)*

enzymes at 0.2–3.0 mg/L (An et al., 2009). Published findings also reveal exposure to the bioactive estrogen, ethinyl estradiol, commonly found in formulations of the oral contraceptive pill that causes a reduction of photosynthetic pigments in lettuce (50% of both chlorophylls and carotenoids). In addition, the synthesis of total sugars, phenols, and flavonoids in lettuce decreased in the presence of ethinyl estradiol (D'Abrosca et al., 2008). Effects on end points such as this (e.g., nutrient or sugar content) may have implications on the consistent high-quality production of vegetable crops, where certain quality standards need to be met in order to be suitable for human consumption. While it is important to consider the potential for acute, lethal effects on plants, it is also important to investigate sublethal impacts on important homeostasis processes, which may ultimately result in unknown implications in the future.

OUTLOOK AND FUTURE RESEARCH NEEDS

Beyond the lack of existing data on the ecotoxicity of HCPs in terrestrial organisms, significant challenges and opportunities exist for scientists when considering the release of HCPs into the natural environment. Our current understanding of the terrestrial ecotoxicological risks of HCPs primarily originate from short-term, acute laboratory studies evaluating mortality or effects on biomass in comparison to controls. These studies are often not environmentally realistic having been carried out in artificial soil or in a nutrient medium, and typically, the exposure concentrations are elevated in comparison with what we would expect in the environment. There is a real need to further our understanding of the chronic effects of exposure as long-term uptake may have a significant impact on the flower formation, fruit maturation, or seed setting, for example, in plants. Long-term effects may have deleterious impacts on both soil invertebrate and plant species fitness through transgenerational effects. Existing laboratory studies have also typically focused on only one or a select group of HCPs. However, in the real world, nontarget terrestrial organisms are exposed under noncontrolled conditions and are likely exposed to a complex mixture of HCPs and other contaminants. A mixture of contaminants may elicit additive or synergistic effects on exposed organisms, which may not be observed when the organism is exposed to a single contaminant at environmentally relevant concentrations (Markman et al., 2008).

As previously suggested, the impact of HCPs on higher trophic terrestrial organisms is needed, and this is discussed more in the next chapter. This may require new innovations in sampling protocols and end points. This might include protocols geared toward noninvasive sampling approaches (hair, feathers, fecal matter, etc.), the use of existing invasive species, or the use of carcasses (Taggart et al., 2015). Historically, emphasis has been placed on analysis of parent HCPs and a limited number of common degradation or transformation products. Scientists are just now beginning to appreciate the potential importance of HCP transformation products that may have similar or different biological activity as the parent compound. To fully understand the fate, behavior, and impacts of HCPs in nontarget organisms, transformation products must be considered. Transformation products may result from biological activities such as human metabolism, biological activity in WWTFs, or even metabolism within nontarget organism exposed to HCPs in the natural environment (Evgenidou et al., 2015; Grossberger et al., 2014; Malchi et al., 2014; Pycke et al., 2014). Likewise, abiotic processes such as disinfection as part of the wastewater treatment process or photodegradation in the environment can contribute to the production of transformation products (Bedner and MacCrehan, 2006; Young et al., 2014). We do know that exposure to and toxicity of HCPs may not be limited to the parent compound and can include transformation products (Carter et al., 2014a; Pino et al., 2015). Our historic focus on parent HCPs likely means that we have underestimated the total impact of HCPs in the waste stream and the environment. One of the biggest challenges to overcoming this deficiency is the identification of previously unknown transformation products, especially those that occur at low concentration or in complex matrices.

New advances in analytical chemistry and scientific instrumentation are fuelling the ability to measure HCPs in complex matrices in support of ERAs. This includes improving our ability to detect the presence of and identity of transformation products and monitor the fate and behavior of HCPs in the natural environment. Given the low environmental concentration of many HCPs and the likelihood of chronic exposure to HCPs, traditional toxicological end points focused on survival and reproduction may not have the sensitivity required to detect subtle effects. New novel approaches to address this gap are being developed and applied in risk assessment studies. This includes new "omics" approaches (e.g., genomics and proteomics) to monitor biochemical alterations and behavioral studies (e.g., feeding, avoidance, and movement) to identify manifestations of biochemical or physiological changes occurring as a result of exposure to HCPs. These new tools coupled with existing approaches will be needed to more fully understand the impact that HCPs have on the terrestrial environment.

The combination of an aging demographic and the growing global population will undoubtedly lead to an increase in the use of HCPs. Persistent chemicals from these products may build up in the environment and present a potential threat to all inhabiting species. To date, limited research findings suggest that some HCPs are not acutely toxic to the terrestrial organisms tested. However, in many cases, nontarget organisms will experience chronic exposure to low concentration, and this exposure may occur in the presence of other natural stressors such as food shortages, drought, or disease. To better understand the risks posed by HCPs and other contaminants to nontarget terrestrial organisms, scientists will need to account for chronic exposure to mixtures of contaminants (and transformation products) under field conditions using relevant environmental monitoring. Further research is therefore required to fully understand the potential environmental issues associated with the release of HCPs into the terrestrial environment.

REFERENCES

An, J., Zhou, Q.X., Sun, Y.B., Xu, Z.Q., 2009. Ecotoxicological effects of typical personal care products on seed germination and seedling development of wheat (*Triticum aestivum* L.). Chemosphere 76 (10), 1428–1434. https://doi.org/10.1016/j.chemosphere.2009.06.004.

Artuso, N., Kennedy, T.F., Connery, J., Grant, J., Schmidt, O., 2011. Assessment of biosolids in earthworm choice tests with different species and soils. Global Nest J. 13 (3), 255–265.

Batchelder, A.R., 1982. Chlortetracycline and oxytetracycline efffects on plant-growth and development in soil systems. J. Environ. Qual. 11 (4), 675–678.

Bedner, M., MacCrehan, W.A., 2006. Transformation of acetaminophen by chlorination produces the toxicants 1,4-benzoquinone and *N*-acetyl-*p*-benzoquinone imine. Environ. Sci. Technol. 40 (2), 516–522. https://doi.org/10.1021/es0509073.

Bonnin, C., 2001. In: Organic Pollutants and Sludge—The French Experience. Research The Sludge Directive. Paper Presented at the A Conference on Sewage Sludge, Belgium.

Bouche, A., 1983. The establishment of earthworm communities. In: Satchell, J.E. (Ed.), Earthworm Ecology: From Darwin to Vermiculture. Chapman and Hall, London, pp. 431–438.

Bouche, A., 1992. Earthworm species and ecotoxicological studies. In: Greig-Smith, P.W., Becker, H., Edwards, P.J., Heimbach, F. (Eds.), Ecotoxicology of Earthworms. Intercept, Andover, UK, pp. 22–35.

Carter, L.J., Garman, C.D., Ryan, J., Dowle, A., Bergstrom, E., Thomas-Oates, J., Boxall, A.B.A., 2014a. Fate and uptake of pharmaceuticals in soil-earthworm systems. Environ. Sci. Technol. 48 (10), 5955–5963. https://doi.org/10.1021/es500567w.

Carter, L.J., Harris, E., Williams, M., Ryan, J.J., Kookana, R.S., Boxall, A.B.A., 2014b. Fate and uptake of pharmaceuticals in soil-plant systems. J. Agric. Food Chem. 62 (4), 816–825. https://doi.org/10.1021/jf404282y.

Carter, L.J., Williams, M., Boettcher, C., Kookana, R.S., 2015. Uptake of pharmaceuticals influences plant development and affects nutrient and hormone homeostases. Environ. Sci. Technol. 49 (20), 12509–12518. https://doi.org/10.1021/acs.est.5b03468.

Chander, Y., Kumar, K., Goyal, S.M., Gupta, S.C., 2005. Antibacterial activity of soil-bound antibiotics. J. Environ. Qual. 34 (6), 1952–1957. https://doi.org/10.2134/jeq2005.0017.

Chang, A. C., Pan, G., Page, A. L., Asano, T., 2002. Developing human health-related chemical guidelines for reclaimed water and sewage sludge applications in agriculture (retrieved from Geneva).

Chefetz, B., Mualem, T., Ben-Ari, J., 2008. Sorption and mobility of pharmaceutical compounds in soil irrigated with reclaimed wastewater. Chemosphere 73 (8), 1335–1343. https://doi.org/10.1016/j.chemosphere.2008.06.070.

Chu, S., Metcalfe, C.D., 2007. Simultaneous determination of triclocarban and triclosan in municipal biosolids by liquid chromatography tandem mass spectrometry. J. Chromatogr. A 1164 (1-2), 212–218. https://doi.org/10.1016/j.chroma.2007.07.024.

D'Abrosca, B., Fiorentino, A., Izzo, A., Cefarelli, G., Pascarella, M.T., Uzzo, P., Monaco, P., 2008. Phytotoxicity evaluation of five pharmaceutical pollutants detected in surface water on germination and growth of cultivated and spontaneous plants. J. Environ. Sci. Health Part A—Toxic/Hazard. Subst. Environ. Eng. 43 (3), 285–294. https://doi.org/10.1080/10934520701792803.

Ding, C., He, J., 2010. Effect of antibiotics in the environment on microbial populations. Appl. Microbiol. Biotechnol. 87 (3), 925–941. https://doi.org/10.1007/s00253-010-2649-5.

EMEA, 2006. Guideline on the environmental risk assessment of medicinal products for human use. EMEA, London, UK.

Escalada, M.G., Harwood, J.L., Maillard, J.Y., Ochs, D., 2005. Triclosan inhibition of fatty acid synthesis and its effect on growth of *Escherichia coli* and *Pseudomonas aeruginosa*. J. Antimicrob. Chemother. 55 (6), 879–882. https://doi.org/10.1093/jac/dki123.

Evgenidou, E.N., Konstantinou, I.K., Lambropoulou, D.A., 2015. Occurrence and removal of transformation products of PPCPs and illicit drugs in wastewaters: a review. Sci. Total Environ. 505, 905–926.

FDA, 1998. Guidance for Industry Environmental Assessment of Human Drug and Biologics Applications. U.S. Department of Health and Human Services, Food and Drug Administration.

Giger, W., Alder, A.C., Golet, E.M., Kohler, H.P.E., McArdell, C.S., Molnar, E., Siegrist, H., Suter, M.J.F., 2003. Occurrence and fate of antibiotics as trace contaminants in wastewaters, sewage sludges, and surface waters. Chimia 57 (9), 485–491. https://doi.org/10.2533/000942903777679064.

Goldstein, M., Shenker, M., Chefetz, B., 2014. Insights into the uptake processes of wastewater-borne pharmaceuticals by vegetables. Environ. Sci. Technol. 48 (10), 5593–5600. https://doi.org/10.1021/es5008615.

Grassi, M., Rizzo, L., Farina, A., 2013. Endocrine disruptors compounds, pharmaceuticals and personal care products in urban wastewater: implications for agricultural reuse and their removal by adsorption process. Environ. Sci. Pollut. Res. 20 (6), 3616–3628. https://doi.org/10.1007/s11356-013-1636-7.

Grossberger, A., Hadar, Y., Borch, T., Chefetz, B., 2014. Biodegradability of pharmaceutical compounds in agricultural soils irrigated with treated wastewater. Environ. Pollut. 185, 168–177. https://doi.org/10.1016/j.envpol.2013.10.038.

Heidler, J., Halden, R.U., 2007. Mass balance assessment of triclosan removal during conventional sewage treatment. Chemosphere 66 (2), 362–369. https://doi.org/10.1016/j.chemosphere.2006.04.066.

Heidler, J., Sapkota, A., Halden, R.U., 2006. Partitioning, persistence, and accumulation in digested sludge of the topical antiseptic triclocarban during wastewater treatment. Environ. Sci. Technol. 40 (11), 3634–3639. https://doi.org/10.1021/es052245n.

Herklotz, P.A., Gurung, P., Heuvel, B.V., Kinney, C.A., 2010. Uptake of human pharmaceuticals by plants grown under hydroponic conditions. Chemosphere 78 (11), 1416–1421. https://doi.org/10.1016/j.chemosphere.2009.12.048.

Hillis, D.G., Fletcher, J., Solomon, K.R., Sibley, P.K., 2011. effects of ten antibiotics on seed germination and root elongation in three plant species. Arch. Environ. Contam. Toxicol. 60 (2), 220–232. https://doi.org/10.1007/s00244-010-9624-0.

Hopkin, S.P., 1989. Ecophysiology of Metals in Terrestrial Invertebrates. Springer, Netherlands.

ISO, 1999. Soil Quality—Inhibition of Reproduction of Collembola (*Folsomia candida*) by Soil Contaminants, second ed. Geneva, Switzerland.

Jager, T., Fleuren, R., Hogendoorn, E.A., De Korte, G., 2003. Elucidating the routes of exposure for organic chemicals in the earthworm, *Eisenia andrei* (Oligochaeta). Environ. Sci. Technol. 37 (15), 3399–3404. https://doi.org/10.1021/es0340578.

Kinney, C.A., Campbell, B.R., Thompson, R., Furlong, E.T., Kolpin, D.W., Burkhardt, M.R., Zaugg, S.D., Werner, S.L., Hay, A.G., 2012. Earthworm bioassays and seedling emergence for monitoring toxicity, aging and bioaccumulation of anthropogenic waste indicator compounds in biosolids-amended soil. Sci. Total Environ. 433, 507–515. https://doi.org/10.1016/j.scitotenv.2012.06.097.

Kinney, C.A., Furlong, E.T., Kolpin, D.W., Burkhardt, M.R., Zaugg, S.D., Werner, S.L., Bossio, J.P., Benotti, M.J., 2008. Bioaccumulation of pharmaceuticals and other anthropogenic waste indicators in earthworms from agricultural soil amended with biosolid or swine manure. Environ. Sci. Technol. 42 (6), 1863–1870. https://doi.org/10.1021/es702304c.

Kinney, C.A., Furlong, E.T., Kolpin, D.W., Zaugg, S.D., Burkhardt, M.R., Bossio, J.P., Werner, S.L., 2010. Earthworms: diagnostic indicators of wastewater derived anthropogenic organic contaminants in terrestrial environments. In: Halden, R.U. (Ed.), Contaminants of Emerging Concern in the Environment: Ecological and Human Health Considerations, vol. 1048. American Chemical Society, Washington, pp. 297–317.

Kinney, C.A., Furlong, E.T., Werner, S.L., Cahill, J.D., 2006a. Presence and distribution of wastewater-derived pharmaceuticals in soil irrigated with reclaimed water. Environ. Toxicol. Chem. 25 (2), 317–326. https://doi.org/10.1897/05-187r.1.

Kinney, C.A., Furlong, E.T., Zaugg, S.D., Burkhardt, M.R., Werner, S.L., Cahill, J.D., Jorgensen, G.R., 2006b. Survey of organic wastewater contaminants in biosolids destined for land application. Environ. Sci. Technol. 40 (23), 7207–7215. https://doi.org/10.1021/es0603406.

Kumar, K.S., Priya, S.M., Peck, A.M., Sajwan, K.S., 2010. Mass loadings of triclosan and triclocarbon from four wastewater treatment plants to three rivers and landfill in Savannah, Georgia, USA. Arch. Environ. Contam. Toxicol. 58 (2), 275–285. https://doi.org/10.1007/s00244-009-9383-y.

Lenz, K., Mahnik, S.N., Weissenbacher, N., Mader, R.M., Krenn, P., Hann, S., Koellensperger, G., Uhl, M., Knasmueller, S., Ferk, F., Bursch, W., Fuerhacker, M., 2007. Monitoring, removal and risk assessment of cytostatic drugs in hospital wastewater. Water Sci. Technol. 56 (12), 141–149. https://doi.org/10.2166/wst.2007.828.

Liu, F., Ying, G.-G., Yang, L.-H., Zhou, Q.-X., 2009a. Terrestrial ecotoxicological effects of the antimicrobial agent triclosan. Ecotoxicol. Environ. Safety 72 (1), 86–92. https://doi.org/10.1016/j.ecoenv.2008.06.009.

Liu, F., Ying, G.G., Tao, R., Jian-Liang, Z., Yang, J.F., Zhao, L.F., 2009b. Effects of six selected antibiotics on plant growth and soil microbial and enzymatic activities. Environ. Pollut. 157 (5), 1636–1642. https://doi.org/10.1016/j.envpol.2008.12.021.

Ma, W.C., van Kleunen, A., Immerzeel, J., de Maagd, P.G.J., 1998. Bioaccumulation of polycyclic aromatic hydrocarbons by earthworms: assessment of equilibrium partitioning theory in in situ studies and water experiments. Environ. Toxicol. Chem. 17 (9), 1730–1737. https://doi.org/10.1897/1551-5028(1998)017<1730:bopahb>2.3.co;2.

Malchi, T., Maor, Y., Tadmor, G., Shenker, M., Chefetz, B., 2014. Irrigation of root vegetables with treated wastewater: evaluating uptake of pharmaceuticals and the associated human health risks. Environ. Sci. Technol. 48 (16), 9325–9333. https://doi.org/10.1021/es5017894.

Markman, S., Guschina, I.A., Barnsley, S., Buchanan, K.L., Pascoe, D., Muller, C.T., 2007. Endocrine disrupting chemicals accumulate in earthworms exposed to sewage effluent. Chemosphere 70 (1), 119–125. https://doi.org/10.1016/j.chemosphere.2007.06.045.

Markman, S., Leitner, S., Catchpole, C., Barnsley, S., Muller, C.T., Pascoe, D., Buchanan, K.L., 2008. Pollutants increase song complexity and the volume of the brain area HVC in a songbird. PLoS One 3 (2). https://doi.org/10.1371/journal.pone.0001674.

McClellan, K., Halden, R.U., 2010. Pharmaceuticals and personal care products in archived US biosolids from the 2001 EPA national sewage sludge survey. Water Res. 44 (2), 658–668. https://doi.org/10.1016/j.watres.2009.12.032.

Metcalfe, C.D., Koenig, B.G., Bennie, D.T., Servos, M., Ternes, T.A., Hirsch, R., 2003. Occurrence of neutral and acidic drugs in the effluents of Canadian sewage treatment plants. Environ. Toxicol. Chem. 22 (12), 2872–2880. https://doi.org/10.1897/02-469.

Nakada, N., Tanishima, T., Shinohara, H., Kiri, K., Takada, H., 2006. Pharmaceutical chemicals and endocrine disrupters in municipal wastewater in Tokyo and their removal during activated sludge treatment. Water Res. 40 (17), 3297–3303. https://doi.org/10.1016/j.watres.2006.06.039.

OECD, 1984. Test No. 207: Earthworm, Acute Toxicity Tests. OECD Publishing.

OECD, 2006. Test No. 208: Terrestrial Plant Test: Seedling Emergence and Seedling Growth Test. OECD Publishing.

Perencevich, E.N., Wong, M.T., Harris, A.D., 2001. National and regional assessment of the antibacterial soap market: a step toward determining the impact of prevalent antibacterial soaps. Am. J. Infect. Control 29 (5), 281–283. https://doi.org/10.1067/mic.2001.115469.

Pino, M.R., Val, J., Mainar, A.M., Zuriaga, E., Espanol, C., Langa, E., 2015. Acute toxicological effects on the earthworm *Eisenia fetida* of 18 common pharmaceuticals in artificial soil. Sci. Total Environ. 518, 225–237. https://doi.org/10.1016/j.scitotenv.2015.02.080.

Prosser, R.S., Lissemore, L., Solomon, K.R., Sibley, P.K., 2014. Toxicity of biosolids-derived triclosan and triclocarban to six crop species. Environ. Toxicol. Chem. 33 (8), 1840–1848. https://doi.org/10.1002/etc.2624.

Prosser, R.S., Sibley, P.K., 2015. Human health risk assessment of pharmaceuticals and personal care products in plant tissue due to biosolids and manure amendments, and wastewater irrigation. Environ. Int. 75, 223–233. https://doi.org/10.1016/j.envint.2014.11.020.

Pycke, B.F.G., Roll, I.B., Brownawell, B.J., Kinney, C.A., Furlong, E.T., Kolpin, D.W., Halden, R.U., 2014. Transformation products and human metabolites of triclocarban and triclosan in sewage sludge across the United States. Environ. Sci. Technol. 48 (14), 7881–7890. https://doi.org/10.1021/es5006362.

Richter, E., Wick, A., Ternes, T.A., Coors, A., 2013. Ecotoxicity of climbazole, a fungicide contained in antidandruff shampoo. Environ. Toxicol. Chem. 32 (12), 2816–2825. https://doi.org/10.1002/etc.2367.

Schmidt, W., Redshaw, C.H., 2015. Evaluation of biological endpoints in crop plants after exposure to non-steroidal anti-inflammatory drugs (NSAIDs): implications for phytotoxicological assessment of novel contaminants. Ecotoxicol. Environ. Saf. 112, 212–222. https://doi.org/10.1016/j.ecoenv.2014.11.008.

Schultz, M.M., Furlong, E.T., 2008. Trace analysis of antidepressant pharmaceuticals and their select degradates in aquatic matrixes by LC/ESI/MS/MS. Anal. Chem. 80 (5), 1756–1762. https://doi.org/10.1021/ac702154e.

Snyder, E.H., O'Connor, G.A., McAvoy, D.C., 2011. Toxicity and bioaccumulation of biosolids-borne triclocarban (TCC) in terrestrial organisms. Chemosphere 82 (3), 460–467. https://doi.org/10.1016/j.chemosphere.2010.09.054.

Taggart, M.A., Richards, N., Kinney, C.A., 2015. Impacts of Pharmaceuticals on Terrestrial Wildlife. In: Hester, R.E., Harrison, R.M. (Eds.), Pharmaceuticals in the Environment. Royal Society of Chemistry.

Thiele-Bruhn, S., 2003. Pharmaceutical antibiotic compounds, in soils—a review (vol. 166, p. 145, 2003). J. Plant Nutri. Soil Sci.—Z. Pflanzenernahr. Bodend. 166 (4), 546. https://doi.org/10.1002/jpln.200390066.

USEPA, 1999. U.S. Environmental Protection Agency: Biosolids Generation, Use, and Disposal in the United States. EPA530-R-99-009.

USEPA, 2000. U.S. Environmental Protection Agency: Biosolids Technology Fact Sheet: Land Application of Biosolids. EPA 832-F-00-064.

USEPA, 2004. U.S. Environmental Protection Agency: Guidelines for Water Reuse. EPA/625/R-04/108.

USEPA, 2006. U.S. Environmental Protection Agency: Emerging Technologies for Biosolids Management. EPA 832-R-06-005.

USEPA, 2009. U.S. Environmental Protection Agency: Targeted National Sewage Sludge Survey. EPA-822-R-08-016.

USEPA, 2015. U.S. Environmental Protection Agency: Sewage Sludge (Biosolids) frequently asked questions. Retrieved from: http://www.epa.gov/biosolids/frequently-asked-questions-about-biosolids.

Wu, X.Q., Dodgen, L.K., Conkle, J.L., Gan, J., 2015. Plant uptake of pharmaceutical and personal care products from recycled water and biosolids: a review. Sci. Total Environ. 536, 655–666. https://doi.org/10.1016/j.scitotenv.2015.07.129.

Xia, K., Bhandari, A., Das, K., Pillar, G., 2005. Occurrence and fate of pharmaceuticals and personal care products (PPCPs) in biosolids. J. Environ. Qual. 34 (1), 91–104.

Yang, J.J., Metcalfe, C.D., 2006. Fate of synthetic musks in a domestic wastewater treatment plant and in an agricultural field amended with biosolids. Sci. Total Environ. 363 (1–3), 149–165. https://doi.org/10.1016/j.scitotenv.2005.06.022.

Young, R.B., Chefetz, B., Liu, A., Desyaterik, Y., Borch, T., 2014. Direct photodegradation of lamotrigine (an antiepileptic) in simulated sunlight—pH influenced rates and products. Environ. Sci. Process. Impacts 16 (4), 848–857. https://doi.org/10.1039/c3em00581j.

Chapter 6

Environmental Contaminants of Health-Care Origin: Exposure and Potential Effects in Wildlife

Thomas G. Bean* **and Barnett A. Rattner**[†]
University of Maryland, College Park, MD, United States
[†]*US Geological Survey, Patuxent Wildlife Research Center, Beltsville, MD, United States*

INTRODUCTION

Concerns over anthropogenic effects on the environment, particularly on wildlife species abundance and diversity, have grown since the publication of Rachel Carson's "Silent Spring" in 1962 (Carson, 1962). Much of the focus of "Silent Spring" was on the consequences of indiscriminate pesticide use on bird populations. With the birth of ecotoxicology in 1969 (Rattner, 2009), scientists' understanding of how use (and misuse) of chemicals have adversely affected wildlife (from molecular to population level) has been greatly improved but is far from complete. Contaminants, both legacy (often persistent) and contemporary (often more ephemeral) compounds, are present in the environment, and some may represent significant stressors alongside other imposed pressures faced by wildlife (e.g., habitat destruction, climate change, and urbanization). Pharmaceuticals are an emerging group of contaminants that have gained attention as a result of a few cases of population-level effects on fish and wildlife (Daughton and Ternes, 1999). As discussed in the previous chapters, active pharmaceutical ingredients (APIs) and some personal care products (PCPs; e.g., triclosan and triclocarban antimicrobials) contain one or more chemicals designed to affect specific biochemical, cellular, physiological, and behavioral processes within an organism. Evolution has often conserved protein/DNA targets across species (Gunnarson et al., 2008), and thus, APIs and PCPs can evoke effects in nontarget organisms. However, evidence of APIs directly affecting nontarget wildlife (i.e., amphibians, reptiles, birds and mammals) is scarce (Arnold et al., 2014; Taggart et al., 2015). Furthermore, biomarkers and bioindicators

Environmental Contaminants, Vol. 11. https://doi.org/10.1016/B978-0-444-63857-1.00006-1

of exposure at environmentally relevant concentrations for wildlife species have yet to emerge (Boxall et al., 2012; Arnold et al., 2013). This chapter examines significant developments and issues related to the risks of contaminants of health-care origin to wildlife. An overview of potential API exposure pathways for wildlife is provided followed by a discussion of exposure (level at which contaminant is encountered), hazard (inherent ability to induce adverse biological response, i.e., toxicity) and risk (actual likelihood of toxicological consequence for biota) which highlights future research priorities.

SOURCES OF HEALTH-CARE CONTAMINANTS AND EXPOSURE PATHWAYS FOR WILDLIFE

As discussed in Chapter 2, aided by advances in analytical chemistry (e.g., high-resolution mass spectrometry) (Taggart et al., 2015), APIs and PCPs from a variety of therapeutic classes have been detected in a range of matrices from aquatic (sediment, surface water and groundwater, invertebrates, fish, and mammals) and terrestrial environments (sludge-amended soils, plants, invertebrates and vertebrates) (Daughton and Ternes, 1999; Daughton and Brooks, 2011; Monteiro and Boxall, 2010; Carter et al., 2014a,b; Meredith-Williams et al., 2012; Bean et al., 2014; Monteiro and Boxall, 2009; McClellan and Halden, 2010; Pycke et al., 2014; Walters et al., 2010; Markman et al., 2007; Brooks et al., 2005; Lazarus et al., 2015a). Fig. 1 portrays API and PCP exposure pathways for aquatic and terrestrial wildlife.

Aquatic Environment

The key route of exposure involves ingestion of invertebrates and fish, which themselves have attained a body burden of pharmaceuticals directly from water (bioconcentration), sediments (dermal or ingestion) or diet (Fig. 1). An understanding of physico-chemical properties of different groups of APIs (e.g., persistence and potential for biomagnification) would be a useful starting point when selecting target species for risk assessment. For pharmaceuticals that are persistent or bioaccumulative, it is expected that top predators (e.g., piscivorous birds and mammals) will be exposed to greatest concentrations and, thus, sentinels for monitoring APIs and PCPs (Lazarus et al., 2015a; Richards et al., 2011). Nevertheless, wildlife feeding at lower trophic levels may also be exposed to APIs via diet (e.g., wading birds feeding on sediment-dwelling invertebrates; aerial insectivores, reptiles, and amphibians feeding on emerging invertebrates). Dermal exposure of amphibians (and perhaps even reptiles and birds) to water and sediments containing APIs are plausible routes of pharmaceutical exposure, but largely unexplored (also see Stankevičiūtė et al., 2015). In high-income countries, approximately 30 % of effluents enter surface waters untreated via misconnections, pipe failures,

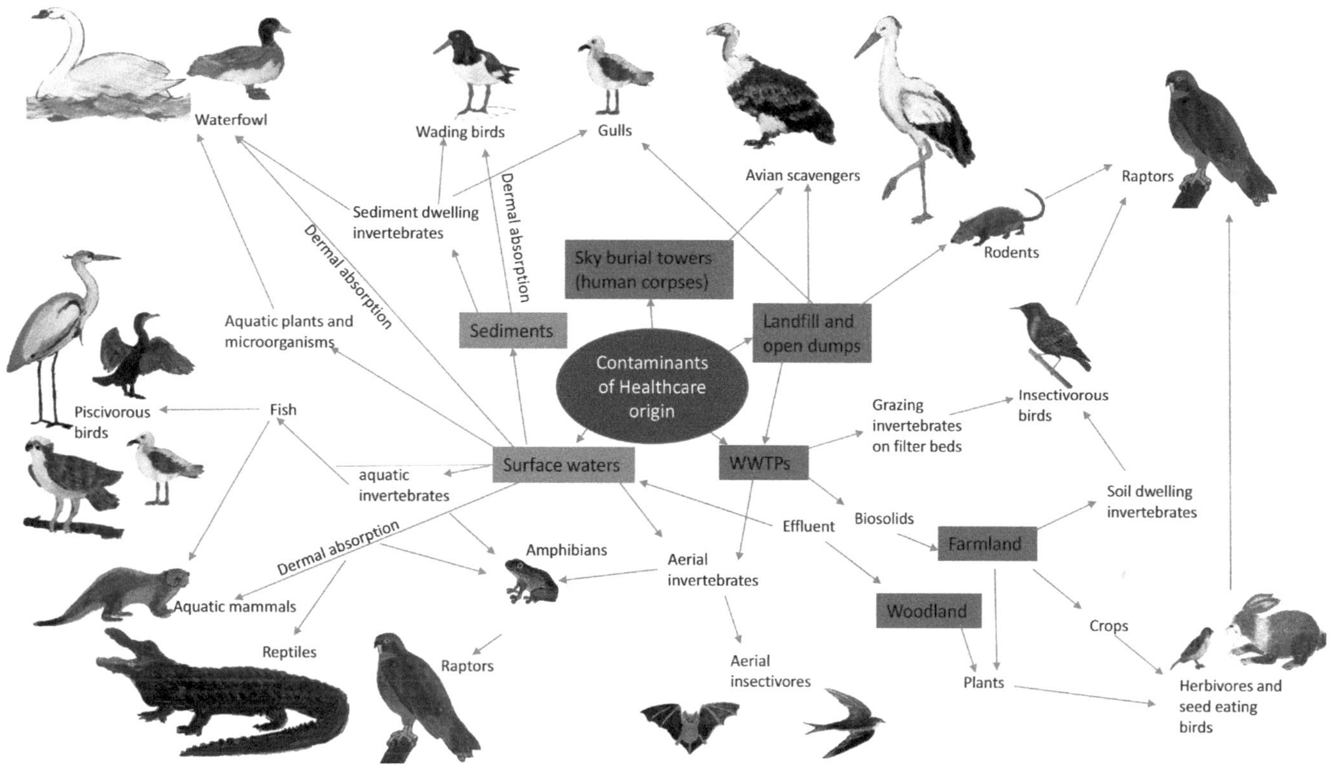

FIG. 1 Pathways for uptake of pharmaceuticals into terrestrial and aquatic wildlife species.

combined sewer overflows and septic tanks (Sato et al., 2013; Hoornweg and Bhada-Tata, 2012). In low-income countries that lack sanitation systems, as much as 92 % of wastewater is released untreated (Sato et al., 2013; Hoornweg and Bhada-Tata, 2012), enhancing the likelihood of wildlife exposure to APIs and PCPs.

Terrestrial Environment

The four habitats where wildlife could be exposed to APIs and PCPs include wastewater treatment plants (WWTPs), farmland where biosolids (a by-product of wastewater treatment) are used as fertilizer or to amend soil, fields and woodland that have been irrigated with "medicated" effluent from WWTPs, and landfills or open dump sites containing API waste. In the case of WWTPs, insectivorous and omnivorous birds (e.g., starlings, *Sturnus vulgaris*; crows and rooks, *Corvus* spp.; pied wagtail, *Motacilla alba*) foraging on the invertebrates in sludge may be exposed to APIs via diet and possibly dermal contact with effluent and biosolids (Bean et al., 2014; Markman et al., 2011). Aerial insectivores (e.g., swallows and martins (Hirundinidae), swifts (Apodidae), and bats (e.g., *Pipistrellus* spp.)) that frequent WWTPs may also be exposed to APIs and PCPs via diet (Fig. 1) (Gough et al., 2003; Park et al., 2009; Park and Cristinacce, 2006).

Biosolids previously dumped in oceans now constitute approximately 60 % of the 7 million dry tons of sewage sludge spread on farmland in the United States (McClellan and Halden, 2010; Park and Cristinacce, 2006; Gielen et al., 2009; USEPA, 2016a). Notably, inputs of veterinary medicines to fields via animal dung-based manures greatly exceed that of human APIs contained in sewage sludge (Sabourin et al., 2009). The application of biosolids to farmland may also expose birds foraging for invertebrates or birds and small mammals eating the crops that have incorporated pharmaceuticals by their roots into seeds, fruit, and leaves (Carter et al., 2014b; Washburn and Begier, 2011; Carter et al., 2015). A greater diversity of species (waterfowl and other game birds, passerines and other aerial insectivores, gulls and wading birds) could be exposed to APIs from agricultural lands than WWTPs, which attract foraging specialists (Gough et al., 2003). This is likely, as agricultural fields represent a much greater area on the landscape than WWTPs.

Effluents from WWTPs are extensively used to irrigate crops and wooded fields (Dalkmann et al., 2012; Jordan et al., 1997). This could expose seed-eating birds and small mammals to APIs. Exposure could occur by direct dermal contact during irrigation, or by ingestion of APIs accumulated into plant tissues (Carter et al., 2014b, 2015).

Landfills and open dump sites are potential sources of APIs due to inappropriate disposal of unused product. This practice is known to occur even in high-income countries (Tong et al., 2011). It is probable that disposal in household waste or dumping will be similar and even more common in low-income countries (Kookana et al., 2014). Scavenging wildlife including gulls (Laridae), corvids, storks (Ciconiidae), vultures (Accipitridae and Cathartidae) and mammals such as raccoons (*Procyon lotor*), foxes (*Vulpes* spp.), and rodents may be exposed to pharmaceuticals that enter solution and contaminate food scraps. To date, this exposure route has not really been investigated; however, there is evidence of mortality of wildlife from veterinary-use pharmaceuticals and extensive use of landfills by some species. For example, mortality of bald and golden eagles (*Haliaeetus leucocephalus* and *Aquila chrysaetos*) scavenging on carcasses of companion animals that had been euthanized with pentobarbital before being dumped has been observed in North America (Langelier, 1993; Russell and Franson, 2014). The previously migratory white stork (*Ciconia ciconia*) populations of the Iberian Peninsula are now year-round residents feeding at landfills (Gilbert et al., 2016). Turning a migratory species into a year-round residents (feeding on rubbish) could have consequences at the population level and even for ecosystems, although API exposure of these birds has not been investigated.

Exposure of top predators to APIs in the terrestrial environment remains unexplored. It is plausible that biomagnification could occur (e.g., a raptor predates a small passerine which had accumulated APIs via ingestion of invertebrates from a biosolid-amended field). This exposure route remains purely speculative, and biomagnification is typically associated with contaminants that are more lipophilic and persistent (e.g., compounds such as polychlorinated biphenyls) than pharmaceuticals. To date, there are no data on the body burden of pharmaceuticals in free-ranging vertebrate prey. Nevertheless, the potential for API exposure, toxicity, and even lethality in organisms occupying higher trophic levels is an area of concern. This is best highlighted by the dramatic collapse of Asian vulture populations (the 1990s to mid-2000s, >99% decline) due to veterinary use of the nonsteroidal anti-inflammatory drug (NSAID) diclofenac (Oaks et al., 2004; Oaks and Watson, 2011). Some other NSAIDs (e.g., ketoprofen and ibuprofen) that are used in veterinary and human medicine have been shown to be toxic to vultures and other species of bird (Oaks et al., 2004; Rattner et al., 2008; Swan et al., 2006a; Naidoo et al., 2010a,b, 2011; Cuthbert et al., 2007; Klein et al., 1994).

At present, many of the aforementioned exposure pathways remain hypothetical. Exposure of wildlife to environmentally relevant levels of pharmaceuticals may or may not be a cause for concern. Pharmaceuticals are biologically active molecules designed to bind to specific receptors, alter metabolism, or even cause the death of cells and microbes. Many of the target receptors have been evolutionarily conserved in non-target wildlife (Gunnarson et al., 2008), giving potential for mode-of-action related effects. At present, we can only speculate (based on the assumption of similar activity in humans, laboratory mammals and wildlife) on potential effects of APIs and

PCPs in wildlife. The following points illustrate this for broad groups of pharmaceuticals categorized by their therapeutic use:

- Psychoactive drugs could alter foraging, energetics and reproductive behavior resulting in decreased survival, productivity and increased predation risk.
- Weight-loss medication and stimulant laxatives could alter nutrient absorption causing malnutrition, impair thermoregulation and immune function, and even result in mortality from starvation.
- Synthetic hormones could alter reproductive behavior, hormone synthesis, fertility, productivity and parental care, with potential effects at the population and ecosystem level.
- Analgesics and NSAIDs could cause organ damage (e.g., liver and kidney structural and functional changes leading to mortality).
- Antimicrobials may lead to buildup of resistant microbes, which are then transported throughout the environment by highly mobile species (e.g., migratory birds) with implications for human populations.
- Cytotoxic and cytostatic drugs could have effects on immune function, resulting in greater sensitivity to, and transmission of wildlife diseases. Individual mortality, population-level mortality, and even the need for humans to cull populations may be exacerbated by these drugs.

It is often difficult to predict effects in wildlife given the diversity of physiologies among species. Additionally, free-ranging animals have a multitude of other anthropogenic and environmental stressors to contend with that are not replicated in controlled laboratory exposures. These additional stressors may also increase sensitivity to chemical contaminants such as APIs; alternatively, the resilience acquired by surviving alongside humans may make wildlife species more tolerant than captive animals. Exposure to multiple stressors not only could affect sensitivity to APIs, but also could drive exposure pathways. For example, anthropogenically induced climate change, urbanization, loss of habitat, and intensification of agriculture may force some wildlife species to seek food sources that could be contaminated with APIs (e.g., see Markman et al., 2011; Gough et al., 2003; Gilbert et al., 2016).

CURRENT KNOWLEDGE OF EXPOSURE, HAZARD AND RISKS OF CONTAMINANTS OF HEALTH-CARE ORIGIN

A summary of studies into the exposure, hazard, and risk of human-use APIs for wildlife is presented in Table 1 (data for fish are summarized in Chapters 5 and 7). While a literature search revealed over 100 scientific publications, only 46 studies involving APIs and PCPs with biological activity (e.g., triclosan and triclocarban) and wildlife were germane (controlled laboratory and field studies investigating exposure and effects; 30 compounds from 15 API or PCP classes). To illustrate hazard, well-characterized examples of

TABLE 1 Summary of Studies on Exposure, Hazard, and Risk for APIs and PCPs in Wildlife

Class	Species	Type of Study	Compound(s) and Therapeutic Class(es)	Exposure Route, Concentration, Duration	Internal Concentrations or Effects	Ref.
Mammals	Marine mammals: pygmy sperm whale (*Kogia breviceps*), clymene dolphin (*Stenella clymene*), bottlenose dolphin (*Tursiops truncatus*), rough-toothed dolphin (*Steno bredanensis*), striped dolphin (*Stenella coeruleoalba*), sea otters (*Enhydra lutris nereis* and *E. lutris kenyoni*), polar bear (*Ursus maritimus*)	Field monitoring in samples from US coastal waters	Methyl paraben (MeP) and metabolite p-hydroxybenzoic acid (4-HB) (antimicrobial contained in some APIs and PCPs)	Dermal from water and bioaccumulation from prey	MeP, 4.57–95.2 ng/g liver 4-HB, 359–19,600 ng/g liver Toxicity remains to be evaluated, hypothetical link with poor body condition, even mortality	Xue et al. (2015)
	Eurasian otter (*Lutra lutra*)	Field monitoring in samples from UK rivers	Diclofenac and ibuprofen NSAID	Bioconcentration from water and bioaccumulation from prey	Detection, but not quantification in hair samples (5/28 diclofenac and 2/28 ibuprofen)	Richards et al. (2011)
	Bat (*Pipistrellus pipistrellus*)	Hypothetical exposure	17α-Ethinylestradiol (synthetic estrogen)	Hypothetical ingestion of aerial invertebrates at WWTPs (28.8 ng/foraging bout)	Effects not speculated	Park et al. (2009) and Park and Cristinacce (2006)

Continued

TABLE 1 Summary of Studies on Exposure, Hazard, and Risk for APIs and PCPs in Wildlife—Cont'd

Class	Species	Type of Study	Compound(s) and Therapeutic Class(es)	Exposure Route, Concentration, Duration	Internal Concentrations or Effects	Ref.
Scavenging birds	Old World *Gyps* vultures: Oriental white-backed vulture (*Gyps bengalensis*), Indian vulture (*Gyps indicus*), slender-billed vulture (*Gyps tenuirostris*)	Field investigation of mass mortality in Indian subcontinent	Diclofenac (NSAID)	Oral, scavenging on treated livestock Field exposure	Median lethal dose 0.1–0.2 mg/kg BW Visceral gout, renal necrosis, and mortality >99% population declines	Oaks et al. (2004), Oaks and Watson (2011), Swan et al. (2006a), Green et al. (2007), and Cuthbert et al. (2014)
	Old World *Gyps* vultures: oriental white-backed vulture (*Gyps bengalensis*), Egyptian vulture (*Neophron percnopterus*) and griffon vulture, (*Gyps fulvus*)	Field monitoring in India			Diclofenac in blood, liver, and kidney samples from vultures killed or injured by kites <LOD-0.42 mg/kg liver <LOD-1.44 mg/kg kidney <LOD-0.17 mg/L blood	Muralidharan and Dhananjayan (2010)
	Gyps vultures and feral dogs (*Canis familiaris*)	Economic appraisal of benefit of vultures to humans and ecosystems			Increased incidence of dog bites and rabies in humans, estimated health cost 1992–2006 ~Rs. 998–1095 billion 1992–2003, vulture population declined by ~10M, while feral dog population increased to 3.7–7.3M	Markandya et al. (2008)

	Pied crow (*Corvus albus*)	Laboratory toxicity test, pharmacokinetics	Diclofenac (NSAID)	Single oral dose of 0.8–10 mg/kg BW	No toxic effects, rapid elimination	Naidoo et al. (2011)
	New World vulture Turkey vulture (*Cathartes aura*)	Laboratory toxicity test, pharmacokinetics	Diclofenac (NSAID)	Single oral dose 0.08–25 mg/kg BW	No adverse effects, plasma half-life 6 h	Rattner et al. (2008)
	79 species including species of vultures, raptors, storks, cranes, and owls	Survey of veterinary use of NSAIDs in wild caught birds held in captivity	Aspirin, carprofen, diclofenac, flunixin, ibuprofen, ketoprofen, meloxicam, phenylbutazone (NSAIDs)	Veterinary treatment in zoos and wildlife rehabilitation centers	Mortality of Marabou stork (*Leptoptilus crumeniferus*) following treatment with flunixin NSAIDs other than diclofenac may be toxic to avian scavengers/predators; meloxicam appears safe	Cuthbert et al. (2007)
	Cape griffon vulture (*Gyps coprotheres*) and African white-backed vultures (*Gyps africanus*)	Laboratory pharmaco- and toxicokinetics	Ketoprofen (NSAID)	Single oral dose, 0.5–5 mg/kg BW	Mortality at 1.5–5 mg/kg BW Environmentally relevant dose estimate 1.54 mg/kg BW 5 mg/kg BW resulted in death of 7 of 11 birds within ~40 h eliminated by zero-order kinetics	Naidoo et al. (2010a) and Naidoo et al. (2010b)
	Cape griffon vulture (*Gyps coprotheres*)			Spiked cattle tissues containing 0.5–1 mg/kg BW	No mortality	

Continued

TABLE 1 Summary of Studies on Exposure, Hazard, and Risk for APIs and PCPs in Wildlife—Cont'd

Class	Species	Type of Study	Compound(s) and Therapeutic Class(es)	Exposure Route, Concentration, Duration	Internal Concentrations or Effects	Ref.
	African white-backed vulture (*Gyps africanus*), Oriental white-backed vulture (*Gyps bengalensis*), Indian vulture (*Gyps indicus*)	Laboratory toxicity test	Meloxicam (NSAID)	Oral gavage (liquid) 0.5–2 mg/kg BW gavage 0.03–1.98 mg/kg BW cattle tissue	No obvious clinical effects or mortality Elimination half-life 0.5–2.4 h Meloxicam has low toxicity to vultures	Swan et al. (2006b)
	Eurasian griffon vulture, (*Gyps fulvus*)	Forensic study of field exposure of a single vulture in Spain	Flunixin (NSAID)	Oral scavenging on contaminated livestock	2.7 mg/kg liver and 6.5 mg/kg kidney	Zorrilla et al. (2015)
	Cape griffon vulture (*Gyps coprotheres*)	Laboratory toxicity and pharmacokinetic study	Carprofen, flunixin, phenylbutazone (NSAIDs)	Single oral dose 11.5 mg carprofen/kg BW, 1 mg flunixin/kg, 1.7 mg phenylbutazone/kg	All three NSAIDs were less toxic to vultures than diclofenac but further investigation of safety required No mortality or lesions but inhibition of uric acid excretion observed	Fourie et al. (2015)
	Bald and golden eagles (*Haliaeetus leucocephalus, Aquila chrysaetos*)	Field exposure	Sodium pentobarbital (barbiturate)	Oral scavenging on euthanized companion animals	Mortality of 62 bald eagles and 3 golden eagles, intoxicated birds sedated, drowsy or comatose, varying degrees of consciousness, slow heart, and respiratory rate	Langelier (1993), Russell and Franson (2014), and Thomas (1999)

Fish-eating birds	Osprey (*Pandion haliaetus*)	Field study in Chesapeake and Delaware Bays, the United States	Diltiazem (calcium channel blocker)	Oral ingestion of fish	0.56–8.63 ng/mL plasma detected in all 69 samples 22 other APIs not detected	Lazarus et al. (2015a)
			Paracetamol, diclofenac, and 17 APIs not detected (analgesic and NSAID)	Oral ingestion of fish from Delaware River and Bay	Adverse effects unlikely, exposure order of magnitude below effects levels Paracetamol, 1.42–3.95 ng/mL detected in 22 of 29 samples Diclofenac 2.33, 3.73 ng/mL detected in 2 of 29 samples	Bean et al. (2018)
		Field study in Chesapeake Bay, the United States	Methoxytriclosan (degradant of antimicrobial PCP)	Oral ingestion and transfer into eggs	0.4–7.4 ng/g egg WW detected in 17 of 75 samples Toxicity yet to be quantified	Lazarus et al. (2015b) and Lazarus et al. (2016)
	Common cormorant (*Phalacrocorax carbo*), grey heron (*Ardea cinerea*), black kite (*Milvus migrans*)	Specimen bank samples from Japan	14 APIs and 3 PCPs NSAIDs, hypertensive agents, antidepressants, and antimicrobials	Oral ingestion of fish Oral	<LOD-2.9 ng triclosan/g, <LOD-0.56 ng triclocarban/g, and <LOD-4.7 ng DEET/g detected in the liver and kidney of fish-eating birds (no detects in non-fish-eating birds) No APIs detected in birds but detected in fish	Tanoue et al. (2014)

Continued

TABLE 1 Summary of Studies on Exposure, Hazard, and Risk for APIs and PCPs in Wildlife—Cont'd

Class	Species	Type of Study	Compound(s) and Therapeutic Class(es)	Exposure Route, Concentration, Duration	Internal Concentrations or Effects	Ref.
Raptors	Owls (*Strix uralensis* spp.), northern goshawk (*Accipiter gentilis*), mountain hawk eagle (*Spizaetus nipalensis orientalis*)	Specimen bank samples from Japan	14 APIs and 3 PCPs NSAIDs, hypertensive agents, antidepressants and antimicrobials	Oral	No APIs or PCPs detected	
Waterfowl	Mute swan (*Cygnus olor*)	Field study	Screened for over 5000 pharmaceuticals	Oral	No APIs detected including stimulants, benzodiazepines, hypnotics, neuroleptics, barbiturates, antidepressants, and opioids	Stankevičiūtė et al. (2015)
Passerine birds	Starling (*Sturnus vulgaris*)	Laboratory chronic toxicity test, environmentally relevant exposure	Fluoxetine (antidepressant)	Oral, spiked wax worms fed at 0.13 mg/kg BW for 22 weeks	Effects on foraging behavior and change in corticosterone levels in response to body mass loss Brain residues 3.05 ng/g (1000 times below human therapeutic concentrations)	Bean et al. (2014) and Bean et al. (2017)

	Starling (*Sturnus vulgaris*)		17β-Estradiol (E2) (estrogen and other non-API endocrine-disrupting chemicals, EDCs)	Oral, spiked mealworms E2 200 ng or mixture 200 ng E2, 520 ng dioctyl phthalate, 80 ng bisphenol A, and 120 ng dibutyl phthalate	Brain area controlling male song complexity enlarged Enhanced male signaling but immunosuppressed	Markman et al. (2008)
	Starling nestlings 1–15 d old (*Sturnus vulgaris*)		17β-Estradiol (E2); (estrogen); and other non-API EDCs dibutyl phthalate, dioctyl phthalate, and bisphenol A	Oral gavage solution day 1–5, hand-fed spiked mealworm day 6–15 Days 1–5, 125 ng E2 Days 6–10, 250 ng E2 days 11–15, 225 ng E2	Reduced growth and immune function in nestlings Growth and immunologic effects could affect recruitment to adult population	Markman et al. (2011)
	Red-winged blackbird (*Agelaius phoeniceus*)	Laboratory toxicity test	Aspirin (NSAID)	LD50 estimate	All >100 mg/kg BW	Schafer et al. (1983)
			Zolamine (Antihistamine)			
			Oxazepam (Sedative)			

Continued

TABLE 1 Summary of Studies on Exposure, Hazard, and Risk for APIs and PCPs in Wildlife—Cont'd

Class	Species	Type of Study	Compound(s) and Therapeutic Class(es)	Exposure Route, Concentration, Duration	Internal Concentrations or Effects	Ref.
	Red-winged blackbird (*Agelaius phoeniceus*) Starling (*Sturnus vulgaris*)		Caffeine (stimulant)	LD50 estimate	316 mg/kg BW 500 mg/kg BW	Schafer et al. (1983)
	Red-winged blackbird, starling, brown-headed cowbird *Molothrus ater*, American crow (*Corvus brachyrhynchos*), white-crowned sparrow (*Zonotrichia leucophrys*)		Nicotine sulfate (stimulant)		75 mg/kg BW, >100 mg/kg BW, 31.6 mg/kg BW, 42.2 mg/kg BW, 56.2 mg/kg BW	
Reptiles	American alligator (*Alligator mississippiensis*)	Laboratory endocrine disruption study	17α-Ethinylestradiol (synthetic estrogen)	In ovo exposure 0.1–10 mg/kg	Intersex males	Matter et al. (1998)
		Pharmacokinetic study	Tetracycline (antibiotic)	Single oral dose via gavage, 50 mg/kg BW	Maximum concentration in plasma, 460–520 ng/mL Time to maximum concentration, 9–10.7 h	Rivera et al. (2012)

Amphibians	Western clawed frog (*Xenopus tropicalis*)	Laboratory test	17α-Ethinylestradiol (synthetic estrogen)	Bioconcentration from water 1.8 ng/L (environmentally relevant exposure)	Impaired spermatogenesis	Gyllenhammar et al. (2009a)
	African clawed frog (*Xenopus laevis*)		Ethyldihydro-testosterone (xeno-androgen)	Bioconcentration, 30 ng/L (environmentally relevant)	Enhanced calling in males	Hoffmann and Kloas (2012a)
	Western clawed frog (*Xenopus tropicalis*)		Levonorgestrel (progestin)	Bioconcentration, 156 ng/L (Above environmentally relevant exposure)	Females were sterile at sexual maturity	Kvarnryd et al. (2011)
	Female Western clawed frog (*Xenopus tropicalis*)			Bioconcentration, 1.3 ng/L (similar to environmentally relevant exposure)	Reproductive and courtship effects	Safholm et al. (2012)
	African clawed frog (*Xenopus laevis*)			Bioconcentration, 30 ng/L (above environmentally relevant exposure)	Enhanced calling in males	Hoffmann and Kloas (2012b)
	Western clawed frog (*Xenopus tropicalis*)		Fluoxetine (antidepressant)	Bioconcentration, 0.6 and 9.6 mg/L, (hatching metamorphosis; 8 weeks) (above environmentally relevant exposure)	At 9.6 mg/L, hypothalamic levels of serotonin metabolite, 5-hydroxyindoleacetic acid were elevated No effect on time to metamorphosis, thyroid histology, gonadal sex differentiation, or aromatase activity in the brain and the gonads at either concentration	Berg et al. (2013)

Continued

TABLE 1 Summary of Studies on Exposure, Hazard, and Risk for APIs and PCPs in Wildlife—Cont'd

Class	Species	Type of Study	Compound(s) and Therapeutic Class(es)	Exposure Route, Concentration, Duration	Internal Concentrations or Effects	Ref.
	African clawed frog tadpoles (*Xenopus laevis*)		Fluoxetine and sertraline (antidepressants)	Bioconcentration 10 μg/L fluoxetine 70d plus 0.1, 1.0, or 10 μg sertraline/L for 70 d; fluoxetine ~20 times greater than environmentally relevant but sertraline was at environmentally relevant concentration	Fluoxetine and sertraline reduced body weight at metamorphosis. Sertraline affected larval development at environmentally relevant concentrations (0.1–1 μg/L)	Conners et al. (2009)
	Leopard frogs (*Rana pipiens*)		Flavone (aromatase inhibitor)	Bioconcentration, 10 μg/L (above environmentally relevant)	Intersex	MacKenzie et al. (2003)
	Wood frogs (*Rana sylvatica tadpoles*)		Flavone (aromatase inhibitor)	Bioconcentration, 10 and 100 μg/L (above environmentally relevant)	No intersex effect	

	Japanese wrinkled frog (*Rana rugosa*)		17β-Estradiol (estrogen), androsterone (androgen); flutamide, and cyproterone acetate (androgen receptor antagonists) 4-OHA (aromatase inhibitor)	Bioconcentration, E2 1 μM, all others 0.5 μM Exposed from day 17 to day 40 post fertilization	Long-term treatment with aromatase inhibitor led to the development of testes from ovarian structure; short-term treatment did not Expression of aromatase and androgen receptor Development of testes or intersex gonads in females	Ohtani et al. (2003)
	Western clawed frog (*Xenopus tropicalis*)		Fadrozole (aromatase inhibitor)	Bioconcentration 1–64 μg/L, 24 h post fertilization until metamorphosis	No effects on time to metamorphosis, body mass, or body length; sex ratio male biased ≥1 μg/L; complete masculinization ≥16 μg/L	Olmstead et al. (2009)
			Clotrimazole (aromatase inhibitor)	Bioconcentration from water 7 weeks exposure 2, 14, and 129 μg/L (above environmentally relevant)	Above 14 μg/L modulation of aromatase activity in the gonads and the brain was observed during the development No effects on gonadal differentiation	Gyllenhammar et al. (2009b)
			Norethindrone (NET) and progesterone (P) (progestogens)	Bioconcentration from water 1 ng/L (NET), 10 ng/L (P) (similar to environmentally relevant)	Inhibited vitellogenesis	Safholm et al. (2014)

veterinary-use APIs are also included (Langelier, 1993; Russell and Franson, 2014; Oaks et al., 2004; Oaks and Watson, 2011; Rattner et al., 2008; Swan et al., 2006a; Naidoo et al., 2010a,b, 2011; Cuthbert et al., 2007; Klein et al., 1994; Thomas, 1999; Green et al., 2007). It is noteworthy that some APIs have been used as vertebrate pest control agents. For example, there are extensive literatures on warfarin and related compounds (coumarin-based therapeutic preventing thromboembolism that at greater doses evoke coagulopathy and hemorrhage) used as an anticoagulant rodenticide (Rattner et al., 2014), and paracetamol (analgesic causing glutathione depletion and liver necrosis in some species) used to control some reptiles, most notably the brown tree snake (*Boiga irregularis*) (Savarie et al., 2001). However, concentrations of vertebrate pest control agents used in baits exceed environmental levels of health-care APIs by orders of magnitude, and are not discussed.

To confirm that an API has adversely affected a wildlife species requires both the detection of residues (in plasma, fur/feathers, or feces for a living organism and tissues from a carcass) and a downstream response (e.g., biomarker) that can be directly associated with the mechanism-of-action of the API (therapeutic or toxic effect). For example, general biomarkers of toxicity commonly investigated in laboratory and field contaminant studies in wildlife (e.g., enzyme induction/inhibition, oxidative stress, blood biochemistry, immune function, thyroid hormones, glucocorticoids, and androgens) can be affected by a multitude of organic and inorganic contaminants (Huggett et al., 1989). Using these criteria, none of the studies show firm evidence that any human-use API represents a genuine risk to wildlife at the individual or population level. Existing evidence on exposure and hazard (Table 1) is far from complete, with only a handful of APIs and species investigated. The data collated in Table 1 highlight the focus on amphibians in the laboratory and birds in the field, while information for reptiles and wild mammals is lacking.

Mammals

Surprisingly, for wild mammals, only three germane studies could be found. Specifically, parabens and metabolites (preservative having antimicrobial activity) have been detected in marine mammals, and were hypothesized to affect body condition and even cause mortality (Xue et al., 2015). Secondly, diclofenac and ibuprofen have been detected using screening methods (but not quantified in terms of concentration) in hair samples of the Eurasian otter (*Lutra lutra*) (Richards et al., 2011). And finally, a study of foraging locations and diet for bats, suggested potential exposure for some endocrine-disrupting compounds including 17α-ethinylestradiol (Park et al., 2009; Park and Cristinacce, 2006). Clearly, data are lacking, with incomplete knowledge of exposure (pathways and concentration) and hazard, and seemingly little or no environmental risk to mammals.

Birds

Perhaps the most notable, and the only case of pharmaceuticals causing population-level effects, has been the Asian vulture crisis (Oaks et al., 2004; Oaks and Watson, 2011). Veterinary-use of the NSAID diclofenac in cattle resulted in >99% decline in *Gyps* vulture populations on the Indian subcontinent in the 1990s (Green et al., 2007; Cuthbert et al., 2014). Although exposure to vultures was from veterinary use of diclofenac, this drug and many other NSAIDs are licensed for use as human medications (e.g., see discussion of scavengers feeding on human corpses). The route of exposure and hazard are well-documented, and risk up to the population level has been studied and modelled in great detail (reviewed in Oaks and Watson, 2011). Regulatory and policy responses related to the use of diclofenac have been undertaken, and a safe alternative, meloxicam, has been identified and is in use (Cuthbert et al., 2014; Swan et al., 2006b). Low levels of diclofenac have also recently been detected in 2 of 29 plasma samples of nestling osprey (*Pandion haliaetus*) collected in Delaware Bay (Bean et al., 2018). Without knowledge of the timing of diclofenac exposure and elimination and potential toxicity in ospreys, it is difficult to interpret the consequences of such low-level exposure. However, based on *Gyps* vulture plasma data (70–170 ng/mL in vultures injured or dying from collisions with kites, and not direct mortalities caused by diclofenac intoxication) (Muralidharan and Dhananjayan, 2010), it is likely the risk for ospreys is minimal (2.33 or 3.73 ng/mL). Numerous NSAIDs (flunixin, carprofen, ibuprofen, phenylbutazone and ketoprofen) have now been investigated in birds (Table 1) (Naidoo et al., 2010a,b; Cuthbert et al., 2007; Klein et al., 1994; Zorrilla et al., 2015; Fourie et al., 2015). Clearly, differential sensitivity among avian species is the hallmark of many NSAIDs, making interspecific prediction difficult (Rattner et al., 2008).

A second example of veterinary-use drugs causing mortality in wild birds involves eagles in the United States and Canada consuming carcasses of domestic animals euthanized with barbiturates. In 1988, a single cattle carcass resulted in the poisoning of 29 bald eagles in British Columbia (Langelier, 1993). As of 2013, there were 33 cases of bald eagle poisoning and 3 cases of golden eagle poisoning from the barbiturate pentobarbital in the United States (Russell and Franson, 2014).

Although no cases of mortality or even detection of high concentrations of API and PCPs that originated from human use has been reported to date, several studies have documented low-level exposure in piscivorous birds. The calcium channel blocker diltiazem was detected at 0.54–8.63 ng/mL in all 69 nestling osprey samples from the Chesapeake Bay of the United States; 22 other APIs were not detected in osprey plasma, although 8 compounds were detected in fish and 18 in water (Lazarus et al., 2015a). In nestling osprey plasma samples collected in nearby Delaware Bay, diltiazem was not detected in 2015, but the analgesic paracetamol was present in 22 of

29 samples (1.42–3.95 ng/mL) with greater concentrations in the more populated regions of Delaware Bay (Bean et al., 2018). However, Stankevičiūtė and coworkers did not detect any APIs in plasma from mute swans in Lithuania, although it is not clear how robust the analysis was as only general therapeutic classes are mentioned (MS scan of 5000 compounds) and no analyte list was provided (Stankevičiūtė et al., 2015).

The antimicrobial PCP triclosan has been detected in the liver and kidney of cormorants, herons, and black kites (5 of 11 piscivorous birds, but 0 of 6 non-fish-eating predatory birds) in Japan (Tanoue et al., 2014). Its degradant, methoxytriclosan, was found in osprey eggs (17 of 65) collected in Chesapeake Bay (Lazarus et al., 2015b; Lazarus et al., 2016). While some antibiotics, antihistamines, analgesics, and psychoactive drugs have been detected in fish, they have not been detected in fish-eating birds (Lazarus et al., 2015a; Bean et al., 2018; Tanoue et al., 2014). The hazard posed to piscivorous birds by low-level exposures to APIs and PCPs is unknown.

For passerine birds feeding directly on WWTP invertebrates, there is evidence of deficits in fitness-related aspects of behavior and physiology. Exposure of starlings to 17β-estradiol and other non-API endocrine-disrupting compounds (EDCs) enhanced adult male advertising by enlarging the cerebral high vocal center of the brain (Markman et al., 2008). The exposed males were immunosuppressed (evidenced by humoral response to phytohemagglutinin), but paradoxically falsely advertising their quality to females that could have potential population-level consequences (Markman et al., 2008). Similarly, exposure of nestlings resulted in impaired growth and immunosuppression (Markman et al., 2011). Beyond 17β-estradiol, data collected by Bean and coworkers indicates exposure of passerines to fluoxetine is likely for birds feeding on invertebrates at WWTPs in the United Kingdom (Bean et al., 2014, 2017). In a chronic exposure study, Bean and coworkers (Bean et al., 2014) presented evidence that fluoxetine affects starling foraging behavior at environmentally relevant levels. However, analysis could not detect fluoxetine in plasma above the LOD, while levels in the brain were 1000 times lower than human therapeutic concentrations (Bean et al., 2017). These data suggest that behavioral effects are perhaps unlikely to be contaminant related unless starlings are dramatically more sensitive to fluoxetine than humans. Studies investigating exposure or assessing hazard of uptake from fields spread with sewage sludge have yet to be conducted.

Reptiles and Amphibians

For reptiles, only two reports in American alligator could be found: one reproductive study with 17α-ethinylestradiol (Matter et al., 1998) and one pharmacokinetic study with tetracycline (presumably veterinary application) (Rivera et al., 2012), but neither of these addressed environmental exposure. A significant number of laboratory exposure studies on APIs with endocrine-disrupting activity have been conducted in amphibians (Table 1 and reviewed in Safholm et al., 2014) with the overwhelming conclusion that this group of compounds can cause intersex conditions at environmentally relevant concentrations. This body of work seems to have been triggered by the discovery of intersex condition in fish which has been noted by many (Brion et al., 2004; Desbrow et al., 1998; Gross-Sorokin et al., 2006; Jobling et al., 2006; Kidd et al., 2007; Kidd et al., 2014; Blazer et al., 2014). As background, male fish have been studied downstream of WWTP effluent discharges containing EDCs (e.g., synthetic estrogens and progestins) and natural estrogens (e.g., estrone, estradiol, and estriol) but also non-API EDCs such as PCBs and PBDEs. Intersex "male" fish are found to have elevated concentrations of vitellogenin and the presence of eggs in their testes, but without controlled laboratory exposure studies, it is difficult to pinpoint which EDCs in the mixture are responsible for observed intersex (Blazer et al., 2014).

If endocrine disruption in free-ranging amphibians and reptiles was affected, as has been observed in laboratory studies simulating environmental exposures, then there could be effects at the individual and potentially population level as noted in fish (Kidd et al., 2007; Kidd et al., 2014). Both 17α-ethinylestradiol and 17β-estradiol (European Commission, 2016) have been proposed under the European Union Water Framework Directive to be added to the existing list of priority compounds (European Commission, 2011). However, to date, they have not been listed. Environmentalists have suggested this is due to the cost of upgrading WWTPs (£26–30 billion in the United Kingdom alone) (Gilbert, 2012). The USEPA has provided guidance documents to health-care facilities and the public on appropriate disposal and treatment of pharmaceutically contaminated waste (USEPA, 2016b), but at present, no legislative requirement is included to monitor APIs in WWTP effluents as safety levels have not been established (USEPA, 2016c).

APPROACHES FOR CHARACTERIZING EXPOSURE, HAZARD AND RISK POSED BY CONTAMINANTS FROM HEALTH CARE TO WILDLIFE

Ecological risk assessment (ERA) integrates both exposure and hazard to quantify risk. With over 4,000 APIs licensed globally (Boxall et al., 2012) and species of wildlife exceeding 30,000 (i.e., mammals 5,488, birds 10,000+, amphibians 6,000+, and reptiles 10,000+) (IUCN, 2016), data gaps for exposure, hazard, and ultimately risk cannot feasibly be assessed using in vivo approaches.

In silico approaches play a major role in prioritizing in vivo research effort, but the utility of theoretical models depends upon the availability of robust empirical data (Fig. 2). One approach is the use of read-across, for which there is interest in filling the hazard component by reading-across to established effect concentrations for model mammals and humans (e.g., see Rand-Weaver et al., 2013). The aim of such in silico/in vitro modeling

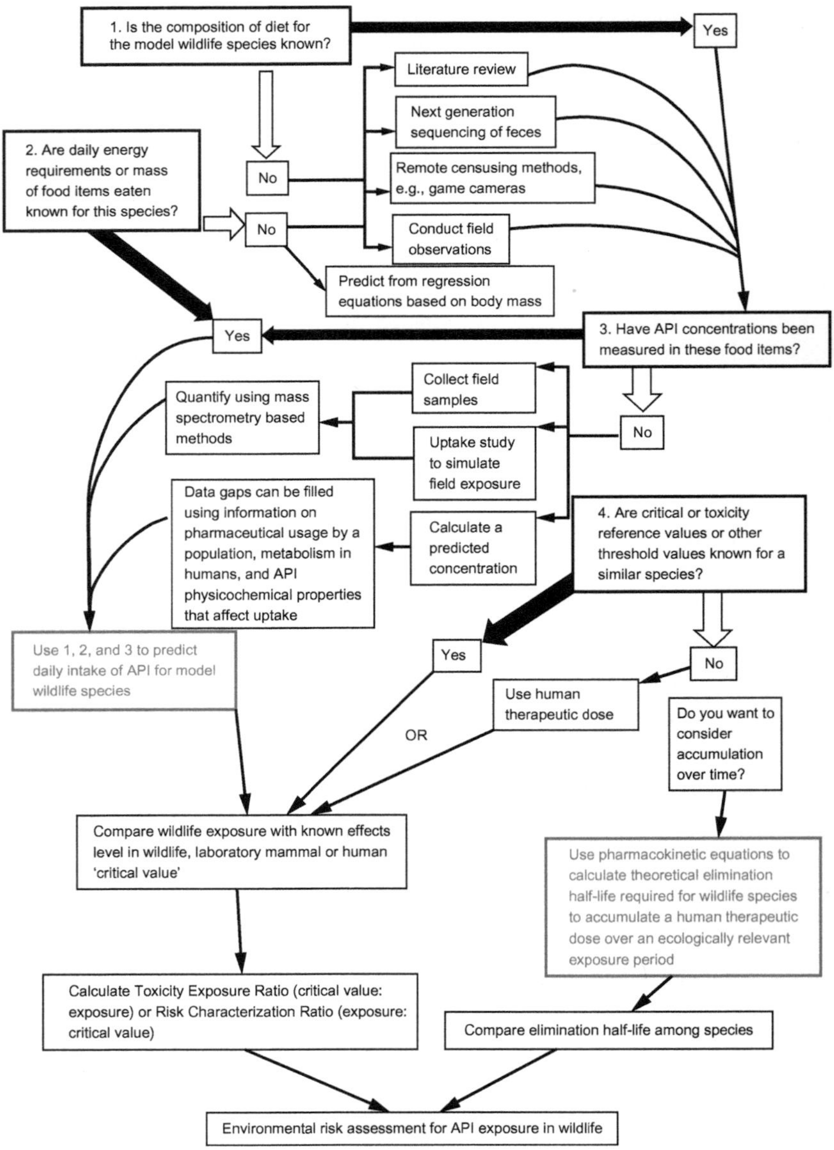

FIG. 2 Schematic illustrating approaches to conducting wildlife exposure, effects, and risk assessment for APIs and ways to fill data gaps.

approaches is to gain an estimate of exposure and then assess the likelihood of effects for a particular wildlife species or surrogate (e.g., mallard, *Anas platyrhynchos*, or bobwhite, *Colinus virginianus*, for birds). Usually, there is a focus on exposure via ingestion for terrestrial species and bioconcentration for amphibians.

Exposure Assessment

In the context of this chapter, the aim of exposure assessments is to estimate the level at which wildlife will encounter chemical contaminants. API concentration in food items can be calculated from field measurements, laboratory data simulating uptake into food items, or using in silico models that include estimates of human API use. In addition to API concentration in food items, knowledge of wildlife diet composition and foraging range are also required. These data can be generated by a combination of literature review, field observations and remote tracking, and in silico modeling.

The European Medicines Agency's (EMA) registration process, which is discussed in more detail in Chapter 13 by Murray Smith, specifically examines uptake of pharmaceuticals by invertebrates and fish but has no such mandate for wildlife species (European Medicines Agency, 2016). However, there are schemes for modelling uptake of pesticides by wildlife that could be adapted and applied to pharmaceuticals (EFSA (European Food Safety Authority), 2009.; Crocker, 2005) Using a worst-case dietary exposure model as a starting point (European Commission, 2002), it is possible to conduct a "Tier 1" exposure assessment that assumes 100 % of an indicator species' diet is derived from a highly contaminated source. The contaminant concentration in the food item can be based on field measurements, laboratory data, or in silico models. If risk is indicated using the Tier 1 exposure assessment, then the calculation of exposure is refined using a generic focal species or multiple species found at the site (EFSA (European Food Safety Authority), 2009). Refined exposure models have the capability to improve accuracy of risk assessments by specifically considering life history of a species. Adjustments can be made to parameters such as the proportion of diet that is contaminated while also taking into account different levels of contamination in different food items (Carter et al., 2016). These parameters can be defined using information on species foraging behavior from the literature, radio tracking efforts, and field observations (EFSA (European Food Safety Authority), 2009). Further refinements can be made by considering in vivo metabolism, avoidance of unpalatable contaminated food, spatial and temporal differences in food availability, and contamination as well as internal exposure (i.e., proportion of ingested contaminant that is absorbed), and even integration of multiple exposure routes (e.g., dermal in addition to oral). It is likely that there will be many data gaps when trying to collate species-specific data to model an exposure scenario, and these should be acknowledged as limitations when presenting the model.

The above exposure assessments have been designed for exposures involving acutely toxic pesticides. Given the potential for pharmaceuticals to initiate effects at low levels (i.e., evolutionary conservation of receptors), chronic exposure assessments are perhaps more relevant here. An approach to chronic exposure assessment that enables consideration of accumulation of multiple doses over time has been used by Lazarus and coworkers

(Lazarus et al., 2015a). This approach is based on pharmacokinetic equations (see Bardal et al., 2011) that assume elimination by first-order kinetics. Pharmacokinetic data for APIs are rarely available for wildlife or even closely related species. To some extent, it may be possible to substitute pharmacokinetic parameters defined in humans and laboratory mammals for missing values in wildlife. However, when doing such read-across extrapolations, it is important to acknowledge the influence of inter-specific differences in metabolic rate, and metabolic pathways (Bean et al., 2017, 2018). For example, it is possible to gain an idea of the likelihood of accumulating a critical cumulative body burden over time by determining how short the elimination half-life could be and yet still exceed the critical body burden for effects (e.g., Lazarus and coworkers did this for a fish-osprey uptake route) (Lazarus et al., 2015a). For example, wildlife with greater basal metabolic rates than humans might be expected to have more rapid compound loss rates than humans. If the theoretical elimination half-life required for wildlife was shorter than the known API elimination half-life for humans, then it would not be unreasonable to assume that internal concentrations of APIs approaching effects thresholds could be accumulated by wildlife within an ecologically relevant time frame.

Hazard Assessment: Empirical Data and the Use of Read-Across From Preclinical and Clinical Drug Trials

Ideally, in vivo dose hazard assessments would be conducted in wildlife species to assess effects. The doses used would cover a range of exposures (i) at, (ii) below, and (iii) exceeding environmentally relevant concentrations. Effects (or the absence of effects) at concentrations below and exceeding those considered environmentally relevant aid in interpretation of findings. For example, diclofenac has a lethal oral dose of 0.1–0.2 mg/kg BW for *Gyps* vultures (Swan et al., 2006a). When assessing the safety of diclofenac to a species of New World vultures, the turkey vulture (*Cathartes aura*), Rattner and coworkers assessed effects of concentrations ranging from 0.08 to 25 mg/kg BW (Rattner et al., 2008). No overt signs of toxicity, renal necrosis, visceral gout, or elevated uric acid levels in plasma were found at the highest dose. Thus, the study was able to conclude that diclofenac was over 100 times less toxic for turkey vultures than *Gyps* vultures; had the top dose in the turkey vulture trial merely been the lethal dose for *Gyps* vultures (Swan et al., 2006a), much less would be known about the sensitivity of New World vultures to diclofenac. Thus, in vivo effects studies should not only seek to produce dose-response curves that cover environmentally relevant or previously defined toxicity reference values but also ideally establish an upper threshold for no-observed-adverse-effects-levels.

Although in vivo ecotoxicology studies are ideal for defining toxicity thresholds, they are challenging to conduct in terrestrial vertebrate species (i.e., expensive, time-consuming, ethical considerations related to animal use, robust design). To fill the knowledge gaps on the hazards of APIs for wildlife, there is interest in using data (toxicity thresholds and pharmacokinetic measures including absorption, distribution, metabolism, and excretion, viz., ADME) collected in preclinical (laboratory mammals) and clinical (humans) drug trials to "readacross" to wildlife species (Rand-Weaver et al., 2013). The read-across approach is based on the hypothesis that therapeutic effects for an API should only occur at equivalent internal concentrations in different organisms (i.e., the human therapeutic dose, HTD) (Rand-Weaver et al., 2013). Although there has been some success in using the read-across approach for fish (e.g., see Margiotta-Casaluci et al., 2014; Owen et al., 2007, 2009), the approach has yet to be thoroughly evaluated for wildlife. Given the great diversity in physiology and life-history among wildlife, it is plausible that APIs have different pharmacokinetics and potencies in such biota compared with laboratory mammals and humans. Thus, studies to evaluate pharmacokinetics and read-across suitability for wildlife should be the priority (Toutain et al., 2010). Collection of ADME for interspecific read-across has been used for predicting toxicity of veterinary-use NSAIDs for scavenging birds (e.g., see Rattner et al., 2008; Naidoo et al., 2010a,b, 2011; Zorrilla et al., 2015).

Risk

Once daily exposure and toxicity reference values are known or predicted, then it is possible to conduct an ERA by relating exposure to effects (hazard). A toxicity-exposure ratio (TER) is generated by dividing a toxic effect threshold (hazard) by the exposure level, while the risk characterization ratio (RCR) is exposure divided by hazard (Shore et al., 2014); efforts are underway to replace this deterministic quotient with probability-based estimates of risk. Fig. 2 illustrates approaches for conducting an ERA for APIs in wildlife and data needs for conducting ERA in wildlife. For the API hazard component, the "critical value" for toxicity is likely to be unknown for wildlife (see Table 1). Additionally, data from a closely related surrogate species may not be available, leaving only the HTD as the critical value. The differences in physiologies between species make the HTD an uncertain benchmark. The ERA provides a means of quantifying risk (i.e., the likelihood that the expected exposure of known toxicity causes an adverse response). The ERA could also be used to prioritize research efforts (e.g., highlight species and APIs of concern).

FUTURE PRIORITIES

Exposure Assessment

An understanding of API exposure levels of wildlife would help assessors prioritize compounds, exposure pathways and species in hazard and risk assessment. It would be beneficial to develop and validate high-throughput in silico models for wildlife exposure assessment, similar to those currently used by regulators to predict human exposure for thousands of chemicals (USEPA, 2016d). A similar approach could be developed for wildlife. Far-field models, such as USEtox and RAIDAR, are used for estimating outdoor environmental exposure. Near-field models, such as the Stochastic Human Exposure and Dose Simulation (SHEDS) model, are used to incorporate multiple exposure routes, scenarios, and pathways into the exposure assessment. Additionally, SHEDS also facilitates adjustments for differences in the kinetics of chemicals in subpopulations (age and ethnic group) to provide a holistic approach for human exposure assessment to domestic products (USEPA, 2016d).

Exposure assessments are challenging to conduct in wildlife. Determination of API concentrations in wildlife food items would, in some cases, require obtaining collection permits and permissions to access private land. In order to quantify exposure, it is necessary to determine internal dose (i.e., bioavailability, including bioaccessibility for different compounds, species, and food items, currently assumed to be 100 %). Nevertheless, there is a strong analytical chemistry foundation for APIs and PCPs. This means method development, even for challenging matrices and compounds, should be achievable, enabling quantification down to trace levels (ng/g or even pg/g) (Taggart et al., 2015; USEPA, 2016e; Chu and Metcalfe, 2007; Ramirez et al., 2009). It may even be possible to build on existing wildlife exposure and mortality incident monitoring programs (e.g., currently in place in some regions of North America and Western Europe) (Shore et al., 2014). Shore and coworkers (2014) recommend expanding these schemes (in terms of analytes, species, and spatial scale) to improve API exposure assessments for wildlife. At present, when an intoxicated animal is observed or a carcass is collected (typically a high trophic level predator or scavenger), the analytic focus is on persistent, bioaccumulative, and toxic chemicals, but not APIs (Shore et al., 2014). It would be beneficial to also screen tissue samples (e.g., using Fourier transform ion cyclotron resonance mass spectrometry) for a wider range of compounds, including some APIs. Screening tissues collected from carcasses of predators and scavengers for APIs would further our knowledge of exposure to relatively bioaccumulative/nonpolar APIs. However, to gain an understanding of exposure to less persistent APIs, it would be necessary to include organisms feeding at lower trophic levels (granivorous and insectivorous species) (Shore et al., 2014). In contrast to collection of tissues from free-ranging wildlife, use of carcasses and other remains might enhance our understanding of API exposure. Ideally, screening should use models that consider both API use and physicochemical properties (Bean et al., 2017).

Hazard Assessments

To date, hazard assessments in the laboratory have focused on bioconcentration of synthetic hormones from the aquatic environment into early-life-stage amphibians and oral exposure of avian scavengers to NSAIDs (Table 1). Wildlife exposure studies of priority compounds to establish dose-response relationships are an important but challenging area for future research. Working with animals requires important but lengthy ethical and legislative considerations to ensure animal use is merited and welfare standards are met. For protected species, non-lethal sampling methods are likely to be required; this means quantifying API residues in plasma, whole blood, fur, or feathers rather than more toxicologically relevant tissue.

Exploration of alternative methods that seek to refine test procedures, reduce the numbers of test subjects, and ultimately replace animals with in vitro methods is a priority. At present, our knowledge of API exposure, pharmacokinetics, and potencies in wildlife is limited. Additionally, there is likely to be considerable variability between exposure and internal concentrations, and ultimately physiological and behavioral responses. Thus, it would also be beneficial to use in vivo data from controlled exposure studies to validate in silico and in vitro methods for the assessment of the following:

(1) Protein binding and active transport
(2) Metabolic pathway, CYP450 expression
(3) Receptor expression

These laboratory alternative methods could be used to rapidly and inexpensively (compared with in vivo studies) highlight APIs, exposure pathways and even species for detailed ERA.

Exposure and Risk of APIs to Wildlife in Developing Countries

Pharmaceuticals are used globally. With the exception of NSAIDs and vultures in South Asia, most of the research on API exposure of wildlife has focused on developed countries in North America and Europe (Table 1) (Kookana et al., 2014). In areas of the world where sanitation, environmental regulations, and waste treatment technologies are less well developed (Liu and Wong, 2013; der Beek et al., 2016), hotspots of API and PCP contamination likely exist. Exposure to APIs in these geographic areas could represent greater risk for wildlife than has been suggested in developed countries to date. We suggest that future research should seek to provide an understanding of API exposure levels and pathways for wildlife in Asia, Africa, and South America.

Novel Ideas and Thinking Outside of the Box

Birds scavenging on landfill sites (e.g., white storks in Portugal (see Gilbert et al., 2016) and gulls in other areas) may encounter inappropriately disposed unused APIs or even via APIs contained in incontinence pads from nursing homes. In the United Kingdom, regulators do not consider incontinence pads a biohazard requiring incineration even though patients from care homes are often heavily medicated (HSE, 2014). On a similar theme, scavenging birds in India have been observed feeding on human corpses released into the River Ganges (BBC, 2016) and at Parsi Sky Burial sites (open burial grounds where the intention is for spirit removal by scavengers) (Karkaria, 2016).

Do Wildlife Act as Vectors for the Transfer of Antimicrobial Resistance?

Development of antimicrobial resistance is one of the most important issues related to API use and contamination of the environment (Boxall et al., 2012). This is discussed in Chapter 8 by Pruden. While accumulation of an HTD of an antibiotic in wildlife may require a species to ingest quantities of food exceeding their normal dietary intake, the development of resistant microbes in the gastrointestinal tract and their transfer to the environment does not (Bonnedahl and Jarhult, 2014). In fact, low-level chronic API exposure typical of wildlife (including antibiotics) provides the ideal exposure regimen for antimicrobial resistance. Many wildlife species, particularly migratory birds, are highly mobile and thus could act as vectors for long-range transfer of resistant microbes (e.g., excreted via feces or transferred to a predator/scavenger via ingestion of a carcass) (Bonnedahl and Jarhult, 2014; Greig et al., 2015). The extent to which resistant microbes may flourish in the gastrointestinal tract of birds is unknown (e.g., birds have a body temperature of 42°C, around 5°C higher than mammals) (Bean et al., 2016). However, there is evidence that birds can transfer resistant microbes to humans and vice versa (Bonnedahl and Jarhult, 2014).

CONCLUSIONS

Our understanding of the risk of health-care contaminants to wildlife is limited. To some extent, exposure assessments in the field and hazard assessments in the laboratory have been conducted, but the only real examples of free-ranging wildlife being adversely affected by APIs involve veterinary rather than human therapeutics. Exposure to human-use APIs is certainly plausible for both terrestrial and aquatic wildlife. However, the limited evidence collected to date, particularly for birds, suggests risks are likely to be minimal (Bean et al., 2014; Lazarus et al., 2015a; Stankevičiūtė et al., 2015; Bean et al., 2018; Tanoue et al., 2014). In the context of the multitude of stressors encountered by free-ranging wildlife (e.g., climate change, habitat destruction, urbanization, intensive agriculture, disease, predators, population control measure, and environmental disasters), there are no "real-world" data to suggest human health-care contaminants' rank among these. However, it would certainly be premature to conclude that APIs from health care pose no risk to wildlife. Interspecific differences in sensitivity could result in unpredictable effects for a given species.

ACKNOWLEDGMENTS

We thank Drs. Bryan Brooks and W. Nelson Beyer for their comments on a draft of this manuscript. Any use of trade, firm, or product names is for descriptive purposes only and does not imply endorsement by the US Government.

REFERENCES

Arnold, K., Boxall, A., Brown, A., Cuthbert, R., Gaw, S., Hutchinson, T., Jobling, S., Madden, J., Metcalfe, C., Naidoo, V., Shore, R., Smits, J., Taggart, M., Thompson, H., 2013. Assessing the exposure risk and impacts of pharmaceuticals in the environment on individuals and ecosystems. Biol. Lett. 9, 20130492.

Arnold, K.E., Brown, A.R., Ankley, G.T., Sumpter, J.P., 2014. Medicating the environment: assessing risks of pharmaceuticals to wildlife and ecosystems. Philos. Trans. R. Soc. B 369. https://doi.org/10.1098/rstb.2013.0569.

Bardal, S., Eaechter, J., Martin, D., 2011. Applied Pharmacology. Elsevier Saunders, St. Louis, MO, p. 480.

BBC, 2016. More than 100 Bodies Recovered From India's Ganges. http://www.bbc.com/news/world-asia-india-30808745 (accessed 7 April 2016).

Bean, T.G., Boxall, A.B.A., Lane, J., Herborn, K.A., Pietravalle, S., Arnold, K.E., 2014. Behavioural and physiological responses of birds to environmentally relevant concentrations of an antidepressant. Philos. Trans. R. Soc. 369. https://doi.org/10.1098/rstb.2013.0575.

Bean, T., Arnold, K., Lane, J., Pietravalle, S., Boxall, A., 2016. An in vitro method for determining the bioaccessibility of pharmaceuticals in wildlife. Environ. Toxicol. Chem. (online early access, Published online 19 February 2016). https://doi.org/10.1002/etc.3406.

Bean, T.G., Arnold, K.E., Lane, J.M., Bergström, E., Thomas-Oates, J., Rattner, B.A., Boxall, A.B.A., 2017. Predictive framework for estimating exposure of birds to pharmaceuticals. Environ. Toxicol. Chem. https://doi.org/10.1002/etc.3771.

Bean, T.G., Rattner, B.A., Lazarus, R.S., Day, D.D., Burket, S.R., Brooks, B.W., Haddad, S.P., Bowerman, W.W., 2018. Pharmaceuticals in water, fish and osprey nestlings in Delaware River and Bay. Environ Pollut. Online only at this time. https://doi.org/10.1016/j.envpol.2017.09.083.

Berg, C., Backstrom, T., Winberg, S., Lindberg, R., Brandt, I., 2013. Developmental exposure to fluoxetine modulates the serotonin system in hypothalamus. PLoS ONE 8, e55053.

Blazer, V.S., Iwanowicz, D.D., Walsh, H.L., Sperry, A.J., Iwanowicz, L.R., Alvarez, D.A., Brightbill, R.A., Smith, G., Foreman, W.T., Manning, R., 2014. Reproductive health indicators of fishes from Pennsylvania watersheds: association with chemicals of emerging concern. Environ. Monit. Assess. 186, 6471–6491.

Bonnedahl, J., Jarhult, J.D., 2014. Antibiotic resistance in wild birds. Ups. J. Med. Sci. 119, 113–116.

Boxall, A., Rudd, M., Brooks, B., Caldwell, D., Choi, K., Hickmann, S., Innes, E., Ostapyk, K., Staveley, J., Verslycke, T., Ankley, G., Beazley, K., Belanger, S., Berninger, J., Carriquiriborde, P., Coors, A., DeLeo, P., Dyer, S., Ericson, J., Gagne, F., Giesy, J., Gouin, T., Hallstrom, L., Karlsson, M., Larsson, D., Lazorchak, J., Mastrocco, F., McLaughlin, A., McMaster, M., Meyerhoff, R., Moore, R., Parrott, J., Snape, J., Murray-Smith, R., Servos, M., Sibley, P., Straub, J., Szabo, N., Topp, E., Tetreault, G., Trudeau, V., Van Der Kraak, G., 2012. Pharmaceuticals and personal care products in the environment: what are the big questions? Environ. Health Perspect. 120, 1221–1229.

Brion, F., Tyler, C., Palazzi, X., Laillet, B., Porcher, J., Garric, J., Flammarion, P., 2004. Impacts of 17β-estradiol, including environmentally relevant concentrations, on reproduction after exposure during embryo-larval-, juvenile- and adult-life stages in zebrafish (*Danio rerio*). Aquat. Toxicol. 68, 193–217.

Brooks, B.W., Chambliss, C.K., Stanley, J.K., Ramirez, A., Banks, K.E., Johnson, R.D., Lewis, R.J., 2005. Determination of select antidepressants in fish from an effluent-dominated stream. Environ. Toxicol. Chem. 24, 464–469.

Carson, R., 1962. Silent Spring. Houghton Mifflin, Boston.

Carter, L., Garman, C., Ryan, J., Dowle, A., Bergstrom, E., Thomas-Oates, J., Boxall, A., 2014a. Fate and uptake of pharmaceuticals in soil-earthworm systems. Environ. Sci. Technol. 48, 5955–5963.

Carter, L.J., Harris, E., Williams, M., Ryan, J.J., Kookana, R.S., Boxall, A.B.A., 2014b. Fate and uptake of pharmaceuticals in soil-plant systems. J. Agric. Food Chem. 62, 816–825.

Carter, L.J., Williams, M., Bottcher, C., Kookana, R.S., 2015. Uptake of pharmaceuticals influences plant development and affects nutrient and hormone homeostases. Environ. Sci. Technol. 49, 12509–12518.

Carter, L., Bean, T., Boxall, A., 2016. In: Evaluating the risk of pharmaceuticals in a terrestrial food web. 26th SETAC Europe Meeting, Nantes, France.

Chu, S., Metcalfe, C.D., 2007. Analysis of paroxetine, fluoxetine and norfluoxetine in fish tissues using pressurized liquid extraction, mixed mode solid phase extraction cleanup and liquid chromatography-tandem mass spectrometry. J. Chromatogr. A 1163, 112–118.

Conners, D.E., Rogers, E.D., Armbrust, K.L., Kwon, J.W., Black, M.C., 2009. Growth and development of tadpoles (*Xenopus laevis*) exposed to selective serotonin reuptake inhibitors, fluoxetine and sertraline, throughout metamorphosis. Environ. Toxicol. Chem. 28, 2671–2676.

Crocker, D., 2005. Estimating the exposure of birds and mammals to pesticides in long-term risk assessments. Ecotoxicology 14, 833–852.

Cuthbert, R., Parry-Jones, J., Green, R.E., Pain, D.J., 2007. NSAIDs and scavenging birds: potential impacts beyond Asia's critically endangered vultures. Biol. Lett. 3, 90–93.

Cuthbert, R., Taggart, M., Prakash, V., Chakraborty, S., Deori, P., Galligan, T., Kulkarni, M., Ranade, S., Saini, M., Sharma, A., Shringarpure, R., Green, R., 2014. Avian scavengers and the threat from veterinary pharmaceuticals. Philos. Trans. R. Soc. B. https://doi.org/10.1098/rstb.2013.0574.

Dalkmann, P., Broszat, M., Siebe, C., Willaschek, E., Sakinc, T., Huebner, J., Amelung, W., Grohmann, E., Siemens, J., 2012. Accumulation of pharmaceuticals, Enterococcus, and resistance genes in soils irrigated with wastewater for zero to 100 years in Central Mexico. PLoS ONE 7, e45397.

Daughton, C.G., Brooks, B.W., 2011. Active pharmaceutical ingredients and aquatic organisms. In: Beyer, W.N., Meador, J. (Eds.), Environmental Contaminants in Biota: Interpreting Tissue Concentrations, second ed. CRC Press, London, pp. 1–86.

Daughton, C.G., Ternes, T.A., 1999. Pharmaceuticals and personal care products in the environment: agents of subtle change? Environ. Health Perspect. 107, 907–937.

der Beek, T.A., Weber, F.A., Bergmann, A., Hickmann, S., Ebert, I., Hein, A., Kuster, A., 2016. Pharmaceuticals in the environment-global occurrences and perspectives. Environ. Toxicol. Chem. 35 (4), 823–835.

Desbrow, C., Routledge, E.J., Brighty, G.C., Sumpter, J.P., Waldock, M., 1998. Identification of estrogenic chemicals in STW effluent. 1. Chemical fractionation and in vitro biological screening. Environ. Sci. Technol. 32, 1549–1558.

EFSA (European Food Safety Authority), 2009. Guidance document on risk assessment for birds and mammals on request from EFSA. EFSA J. 7 (12), 1438.

European Commission, 2002. Guidance Document on Risk Assessment for Birds and Mammals under Council Directive 91/414/EEC, p. 74.

European Commission, 2011. Proposal for a Directive of the European Parliament and Council Amending Directives 2000/EC and 2008/105/EC as Regards Priority Substances in the Field of Water Policy. European Commission, Brussels, pp. 1–35.

European Commission, 2016. EC Press Release: Environment and Water: Proposal to Reduce Water Pollution Risks. http://europa.eu/rapid/press-release_IP-12-88_en.html (accessed 19 April 2016).

European Medicines Agency, 2016. EMEA Committee for Medicinal Products for Human Use (CHMP). http://www.ema.europa.eu/ema/index.jsp?curl=pages/about_us/general/general_content_000094.jsp (accessed 9 August 2016).

Fourie, T., Cromarty, D., Duncan, N., Wolter, K., Naidoo, V., 2015. The safety and pharmacokinetics of carprofen, flunixin and phenylbutazone in the Cape vulture (*Gyps coprotheres*) following oral exposure. PLoS ONE 10, e0141419.

Gielen, G., van den Heuvel, M.R., Clinton, P.W., Greenfield, L.G., 2009. Factors impacting on pharmaceutical leaching following sewage application to land. Chemosphere 74, 537–542.

Gilbert, N., 2012. Drug-pollution law all washed up: EU initiative to clean up waterways faces tough opposition. Nature 491, 503–504.

Gilbert, N.I., Correia, R.A., Silva, J.P., Pacheco, C., Catry, I., Atkinson, P.W., Gill, J.A., Franco, A.M.A., 2016. Are white storks addicted to junk food? Impacts of landfill use on the movement and behaviour of resident white storks (*Ciconia ciconia*) from a partially migratory population. Move. Ecol. 4, 1–13.

Gough, S.J., Gillings, S., Vickery, J.A., 2003. The Value and Management of Wastewater Treatment Works for Breeding and Wintering Birds in Lowland Eastern England. British Trust for Ornithology, 50 p.

Green, R.E., Taggart, M.A., Senacha, K.R., Raghavan, B., Pain, D.J., Jhala, Y., Cuthbert, R., 2007. Rate of decline of the oriental white-backed vulture population in India estimated from a survey of diclofenac residues in carcasses of ungulates. PLoS ONE 2, e686.

Greig, J., Rajic, A., Young, I., Mascarenhas, M., Waddell, L., LeJeune, J., 2015. A scoping review of the role of wildlife in the transmission of bacterial pathogens and antimicrobial resistance to the food chain. Zoonoses Public Health 62, 269–284.

Gross-Sorokin, M.Y., Roast, S.D., Brighty, G.C., 2006. Assessment of feminization of male fish in English rivers by the Environment Agency of England and Wales. Environ. Health Perspect. 114, 147–151.

Gunnarson, L., Jauhainen, A., Kristiansson, E., Nerman, O., Joakim Larsson, D.G., 2008. Evolutionary conservation of human drug targets in organisms used for environmental risk assessments. Environ. Sci. Technol. 42, 5807–5813.

Gyllenhammar, I., Holm, L., Eklund, R., Berg, C., 2009a. Reproductive toxicity in *Xenopus tropicalis* after developmental exposure to environmental concentrations of ethynylestradiol. Aquat. Toxicol. 91, 171–178.

Gyllenhammar, I., Eriksson, H., Soderqvist, A., Lindberg, R.H., Fick, J., Berg, C., 2009b. Clotrimazole exposure modulates aromatase activity in gonads and brain during gonadal differentiation in *Xenopus tropicalis* frogs. Aquat. Toxicol. 91, 102–109.

Hoffmann, F., Kloas, W., 2012a. The synthetic progestogen, levonorgestrel, but not natural progesterone, affects male mate calling behavior of *Xenopus laevis*. Gen. Comp. Endocrinol. 176, 385–390.

Hoffmann, F., Kloas, W., 2012b. Effects of environmentally relevant concentrations of the xeno-androgen, methyldihydrotestosterone, on male and female mating behavior in *Xenopus laevis*. Chemosphere 87, 1246–1253.

Hoornweg, D., Bhada-Tata, P., 2012. What a waste: a global review of solid waste management. World Bank, Washington, DC, pp. 1–99.

HSE, 2014. Managing offensive/hygiene waste safely. In: Health and Safety Executive (UK), pp. 1–10. www.hse.gov.uk/pubns/waste22.pdf (accessed 10 April 2016).

Huggett, R.J., Kimerle, R.A., Mehrle Jr. P.M., Bergman, H.L. (Eds.), 1989. Biomarkers: biochemical, physiological, and histological markers of anthropogenic stress. Proceedings of the Eighth Pellston Workshop Keystone Colorado, July 23–28. Chelsea, MI, Lewis Publishers.

IUCN, 2016. IUCN Red List. http://www.iucnredlist.org/ (accessed 20 April 2016).

Jobling, S., Williams, R., Johnson, A., Taylor, A., Gross-Sorokin, M., Nolan, M., Tyler, C.R., van Aerle, R., Santos, E., Brighty, G., 2006. Predicted exposures to steroid estrogens in UK rivers correlate with widespread sexual disruption in wild fish populations. Environ. Health Perspect. 114, 32–39.

Jordan, M.J., Nadelhoffer, K.J., Fry, B., 1997. Nitrogen cycling in forest and grass ecosystems irrigated with N-15-enriched wastewater. Ecol. Appl. 7, 864–881.

Karkaria, B., 2016. Death in the City: How a Lack of Vultures Threatens Mumbai's 'Towers of Silence'. http://www.theguardian.com/cities/2015/jan/26/death-city-lack-vultures-threatens-mumbai-towers-of-silence (accessed 7 April 2016).

Kidd, K.A., Blanchfield, P.J., Mills, K.H., Palace, V.P., Evans, R.E., Lazorchak, J.M., Flick, R.W., 2007. Collapse of a fish population after exposure to a synthetic estrogen. Proc. Natl. Acad. Sci. U. S. A. 104, 8897–8901.

Kidd, K.A., Paterson, M.J., rennie, M.D., Podemski, C.L., Findlay, D.L., Blanchfield, P.J., Liber, K., 2014. Direct and indirect responses of a freshwater food web to a potent synthetic oestrogen. Philos. Trans. R. Soc. B 369. https://doi.org/10.1098/rstb.2013.0578.

Klein, P.N., Charmatz, K., Langenberg, J., 1994. In: The effect of flunixin meglumine (Banamine®) on the renal function in northern bobwhite (*Collinus virginianus*): an avian model. Proceedings, Annual Conference of the American Association of Zoo Veterinarians and association of Reptilian and Amphibian Veterinarians , Pittsburgh, PA, October 22–27, pp. 129–131.

Kookana, R.S., William, M., Boxall, A.B.A., Joakim Larsson, D.G., Gaw, S., Choi, K., Yamamoto, H., Tatikonda, S., Zhu, Y.-G., Carriquiriborde, P., 2014. Potential ecological footprints of active pharmaceutical ingredients: an examination of risk factors in low-, middle-, and high-income countries. Philos. Trans. R. Soc. B 369. https://doi.org/10.1098/rstb.2013.0586.

Kvarnryd, M., Grabic, R., Brandt, I., Berg, C., 2011. Early life progestin exposure causes arrested oocyte development, oviductal agenesis and sterility in adult *Xenopus tropicalis* frogs. Aquat. Toxicol. 103, 18–24.

Langelier, K.M., 1993. Barbiturate poisoning in twenty-nine bald eagles. In: Redig, P.T., Cooper, J.E., Remple, D., Hunter, D.B. (Eds.), Raptor Biomedicine. University of Minnesota Press, Minneapolis, MN, pp. 231–233.

Lazarus, R.S., Rattner, B.A., Brooks, B.W., Du, B., McGowan, P.C., Blazer, V.S., Ottinger, M.A., 2015a. Exposure and food web transfer of pharmaceuticals in ospreys (*Pandion haliaetus*): predictive model and empirical data. Integr. Environ. Assess. Manag. 11, 118–129.

Lazarus, R.S., Rattner, B.A., McGowan, P.C., Hale, R.C., Schultz, S.L., Karouna-Renier, N.K., Ottinger, M.A., 2015b. Decadal re-evaluation of contaminant exposure and productivity of ospreys (*Pandion haliaetus*) nesting in Chesapeake Bay regions of concern. Environ. Pollut. 205, 278–290.

Lazarus, R., Rattner, B., McGowan, P., Hale, R., Karouna-Renier, N., Erickson, R., Ottinger, M., 2016. Chesapeake Bay fish-osprey (*Pandion haliaetus*) food chain: evaluation of contaminant exposure and genetic damage. Environ. Toxicol. Chem. 16, 1560–1575.

Liu, J.L., Wong, M.H., 2013. Pharmaceuticals and personal care products (PPCPs): a review on environmental contamination in China. Environ. Int. 59, 208–224.

MacKenzie, C.A., Berrill, M., Metcalfe, C., Pauli, B.D., 2003. Gonadal differentiation in frogs exposed to estrogenic and antiestrogenic compounds. Environ. Toxicol. Chem. 22, 2466–2475.

Margiotta-Casaluci, L., Owen, S.F., Cumming, R.I., de Polo, A., Winter, M.J., Panter, G.H., Rand-Weaver, M., Sumpter, J.P., 2014. Quantitative cross-species extrapolation between humans and fish: the case of the anti-depressant fluoxetine. PLoS ONE 9, e110467.

Markandya, A., Taylor, T., Longo, A., Murty, M.N., Murty, S., Dhavala, K., 2008. Counting the cost of vulture decline – an appraisal of the human health and other benefits of vultures in India. Ecol. Econ. 67, 194–204.

Markman, S., Guschina, I.A., Barnsley, S., Buchanan, K.L., Pascoe, D., Muller, C.T., 2007. Endocrine disrupting chemicals accumulate in earthworms exposed to sewage effluent. Chemosphere 70, 119–125.

Markman, S., Leitner, S., Catchpole, C., Barnsley, S., Müller, C.T., Pascoe, D., Buchanan, K.L., 2008. Pollutants increase song complexity and the volume of the brain area HVC in a songbird. PLoS ONE 3, e1674.

Markman, S., Muller, C.T., Pascoe, D., Dawson, A., Buchanan, K.L., 2011. Pollutants affect development in nestling starlings *Sturnus vulgaris*. J. Appl. Ecol. 48 (2), 391–397.

Matter, J., Crain, D., Sills-McMurry, C., Pickford, D., Rainwater, T., Reynolds, K., Rooney, A., Dickerson, R., Guillette Jr. L., 1998. Effects of endocrine disrupting chemicals in reptiles: alligators. In: Kendall, R., Dickerson, R., Giesy, J., Suk, W. (Eds.), Principles and Processes for Evaluating Endocrine Disruption in Wildlife. SETAC Press, Pensacola, FL, pp. 267–289.

McClellan, K., Halden, R.U., 2010. Pharmaceuticals and personal care products in archived US biosolids from the 2001 EPA national sewage sludge survey. Water Res. 44, 658–668.

Meredith-Williams, M., Carter, L., Fussell, R., Raffaelli, D., Ashauer, R., Boxall, A., 2012. Uptake and depuration of pharmaceuticals in aquatic invertebrates. Environ. Pollut. 165, 250–258.

Monteiro, S., Boxall, A., 2009. factors affecting the degradation of pharmaceuticals in agricultural soils. Environ. Toxicol. Chem. 28, 2546–2554.

Monteiro, S.C., Boxall, A.B.A., 2010. Occurrence and fate of human pharmaceuticals in the environment. Rev. Environ. Contam. Toxicol. 202, 53–154.

Muralidharan, S., Dhananjayan, V., 2010. Diclofenac residues in blood plasma and tissues of vultures collected from Ahmedabad, India. Bull. Environ. Contam. Toxicol. 85, 377–380.

Naidoo, V., Venter, L., Wolter, K., Taggart, M., Cuthbert, R., 2010a. The toxicokinetics of ketoprofen in *Gyps coprotheres*: toxicity due to zero-order metabolism. Arch. Toxicol. 84, 761–766.

Naidoo, V., Wolter, K., Cromarty, D., Diekmann, M., Duncan, N., Meharg, A., Taggart, M., Venter, L., Cuthbert, R., 2010b. Toxicity of non-steroidal anti-inflammatory drugs to *Gyps* vultures: a new threat from ketoprofen. Biol. Lett. 6, 339–341.

Naidoo, V., Mompati, K., Duncan, N., Taggart, M., 2011. The pied crow (*Corvus albus*) is insensitive to diclofenac at concentrations present in carrion. J. Wildl. Dis. 47, 936–944.

Oaks, J.L., Watson, R.T., 2011. South Asian vultures in crisis: environmental contamination with a pharmaceutical. In: Elliot, J., Bishop, C., Morrisey, C. (Eds.), Wildlife Ecotoxicology: Forensic Approaches. Springer, New York, pp. 413–442.

Oaks, J.L., Gilbert, M., Virani, M.Z., Watson, R.T., Meteyer, C.U., Rideout, B.A., Shivaprasad, H.L., Ahmed, S., Chaudry, M.J.I., Arshad, M., Mamood, S., Ali, A., Khan, A.A., 2004. Diclofenac residues as the cause of vulture population decline in Pakistan. Lett. Nat. 427, 630–633.

Ohtani, H., Miura, I., Ichikawa, Y., 2003. Role of aromatase and androgen receptor expression in gonadal sex differentiation of ZW/ZZ-type frogs, *Rana rugosa*. Compar. Biochem. Physiol. C – Toxicol Pharmacol. 134, 215–225.

Olmstead, A.W., Kosian, P.A., Korte, J.J., Holcombe, G.W., Woodis, K.K., Degitz, S.J., 2009. Sex reversal of the amphibian, *Xenopus tropicalis*, following larval exposure to an aromatase inhibitor. Aquat. Toxicol. 91, 143–150.

Owen, S., Giltrow, E., Huggett, D., Hutchinson, T., Saye, J., Winter, M., Sumpter, J., 2007. Comparative physiology, pharmacology and toxicology of beta-blockers: mammals versus fish. Aquat. Toxicol. 82, 145–162.

Owen, S., Huggett, D., Hutchinson, T., Hetheridge, M., Kinter, L., Ericson, J., Sumpter, J., 2009. Uptake of propranolol, a cardiovascular pharmaceutical, from water into fish plasma and its effects on growth and organ biometry. Aquat. Toxicol. 93, 217–224.

Park, K.J., Cristinacce, A., 2006. Use of sewage treatment works as foraging sites by insectivorous bats. Anim. Conserv. 9, 259–268.

Park, K.J., Muller, C.T., Markman, S., Swinscow-Hall, O., Pascoe, D., Buchanan, K.L., 2009. Detection of endocrine disrupting chemicals in aerial invertebrates at sewage treatment works. Chemosphere 77, 1459–1464.

Pycke, B.F.G., Roll, I.B., Brownawell, B.J., Kinney, C.A., Furlong, E.T., Kolpin, D.W., Halden, R.U., 2014. Transformation products and human metabolites of triclocarban and triclosan in sewage sludge across the United States. Environ. Sci. Technol. 48, 7881–7890.

Ramirez, A.J., Brain, R.A., Usenko, S., Mottaleb, M.A., O'Donnell, J.G., Stahl, L.L., Wathen, J.B., Snyder, B.D., Pitt, J.L., Perez-Hurtado, P., Dobbins, L.L., Brooks, B.W., Chambliss, C.K., 2009. Occurrence of pharmaceuticals and personal care products in fish: results of a national pilot study in the United States. Environ. Toxicol. Chem. 28, 2587–2597.

Rand-Weaver, M., Margiotta-Casaluci, L., Patel, A., Panter, G., Owen, S., Sumpter, J., 2013. The read-across hypothesis and environmental risk assessment of pharmaceuticals. Environ. Sci. Technol. 47, 11384–11395.

Rattner, B.A., 2009. History of wildlife toxicology. Ecotoxicology 18, 773–783.

Rattner, B., Whitehead, M., Gasper, G., Meteyer, C., Link, W., Taggart, M., Meharg, A., Pattee, O., Pain, D., 2008. Apparent tolerance of turkey vultures (*Cathartes aura*) to the non-steroidal anti-inflammatory drug diclofenac. Environ. Toxicol. Chem. 27, 2341–2345.

Rattner, B., Lazarus, R., Elliott, J., Shore, R., van den Brink, N., 2014. Adverse outcome pathway and risks of anticoagulant rodenticides to predatory wildlife. Environ. Sci. Technol. 48, 8433–8445.

Richards, N., Cook, G., Simpson, V., Hall, S., Harrison, N., Scott, K., 2011. Qualitative detection of the NSAIDs diclofenac and ibuprofen in the hair of Eurasian otters (*Lutra lutra*) occupying UK waterways with GC-MS. Eur. J. Wildl. Res. 57 (5), 1107–1114.

Rivera, S., Nevarez, J.G., Maxwell, L.K., Barker, S.A., 2012. Pharmacokinetics of tetracycline after single-dose oral administration in the American alligator (*Alligator mississippiensis*). J. Zoo Wildl. Med. 43, 858–863.

Russell, R., Franson, J., 2014. Causes of mortality in eagles submitted to the National Wildlife Health Center 1975–2013. Wildl. Soc. Bull. 38, 697–704.

Sabourin, L., Beck, A., Duenk, P.W., Kleywegt, S., Lapen, D.R., Li, H.X., Metcalfe, C.D., Payne, M., Topp, E., 2009. Runoff of pharmaceuticals and personal care products following application of dewatered municipal biosolids to an agricultural field. Sci. Total Environ. 407, 4596–4604.

Safholm, M., Norder, A., Fick, J., Berg, C., 2012. Disrupted oogenesis in the frog *Xenopus tropicalis* after exposure to environmental progestin concentrations. Biol. Reprod. 86, 126.

Safholm, M., Ribbenstedt, A., Fick, J., Berg, C., 2014. Risks of hormonally active pharmaceuticals to amphibians: a growing concern regarding progestogens. Philos. Trans. R. Soc. B – Biol. Sci. 369. https://doi.org/10.1098/rstb.2013.0577.

Sato, T., Qadir, M., Yamamoto, S., Endo, T., Zahoor, A., 2013. Global, regional, and country level need for data on wastewater generation, treatment, and use. Agric. Water Manag. 130, 1–13.

Savarie, P.J., Shivik, J.A., White, G.C., Hurley, J.C., Clark, L., 2001. Use of acetaminophen for large-scale control of brown treesnakes. J. Wildl. Manag. 65, 356–365.

Schafer, E.W., Bowles, W.A., Hurlbut, J., 1983. The acute oral toxicity, repellency, and hazard potential of 998 chemicals to one or more species of wild and domestic birds. Arch. Environ. Contam. Toxicol. 12, 355–382.

Shore, R.F., Taggart, M.A., Smits, J., Mateo, R., Richards, N.L., Fryday, S., 2014. Detection and drivers of exposure and effects of pharmaceuticals in higher vertebrates. Philos. Trans. R. Soc. B. https://doi.org/10.1098/rstb.2013.0570.

Stankevičiūtė, J., Barkauskaitė, S., Brazaitis, G., 2015. Detection of xenobiotic substances in mute swans' (*Cygnus olor*) blood. In: Proceedings of the 7th International Scientific Conference Rural Development, Aleksandras Stuginskis University, Lithuania, p. 5.

Swan, G.E., Cuthbert, R., Quevedo, M., Green, R.E., Pain, D.J., Bartels, P., Cunningham, A.A., Duncan, N., Meharg, A.A., Oaks, J.L., Parry-Jones, J., Shultz, S., Taggart, M.A., Verdoorn, G., Walter, K., 2006a. Toxicity of diclofenac to *Gyps* vultures. Biol. Lett. 2, 279–282.

Swan, G.E., Naidoo, V., Cuthbert, R., Green, R.E., Pain, D.J., Swarup, D., Prakeash, V., Taggart, M., Bekker, L., Diekmann, D.D.J., Diekmann, M., Killian, E., 2006b. Removing the threat of diclofenac to critically endangered asian vultures. PLoS Biol. 4, e66.

Taggart, M.A., Richards, N., Kinney, C.A., 2015. Impacts of pharmaceuticals on terrestrial wildlife. In: Hester, R.E., Harrison, R.M. (Eds.), Pharmaceuticals in the Environment. Royal Society of Chemistry, pp. 216–254.

Tanoue, R., Nomiyama, K., Nakamura, H., Hayashi, T., Kim, J.W., Isobe, T., Shinohara, R., Tanabe, S., 2014. Simultaneous determination of polar pharmaceuticals and personal care products in biological organs and tissues. J. Chromatogr. A 1355, 193–205.

Thomas, N.J., 1999. Barbiturates. In: Friend, M., Franson, C.J. (Eds.), Field manual of wildlife diseases. US Department of the Interior, Washington, DC, pp. 349–350.

Tong, A.Y.C., Peake, B.M., Braund, R., 2011. Disposal practices for unused medications around the world. Environ. Int. 37, 292–298.

Toutain, P.-L., Ferran, A., Bousquet-Melou, A., 2010. Species differences in pharmacokinetics and pharmacodynamics. In: Cunningham, F., Elliot, J., Lees, P. (Eds.), Comparative and Veterinary Pharmacology, Handbook of Experimental Pharmacology. Springer-Verlag, Berlin, pp. 19–48.

USEPA, 2016a. Identification of Non-Hazardous Secondary Materials That Are Solid Waste. https://www.epa.gov/rcra/identification-non-hazardous-secondary-materials-are-solid-waste (accessed 9 August 2016).

USEPA, 2016b. Managing Pharmaceutical Waste: A 10 Step Blueprint for Healthcare Facilities in the United States. https://www.epa.gov/hwgenerators/management-pharmaceutical-hazardous-waste (accessed 9 August 2016).

USEPA, 2016c. Pollution Prevention: Hospital Pollution Prevention. https://www3.epa.gov/region9/waste/p2/hospart.html (accessed 9 August 2016).

USEPA, 2016d. Rapid Chemical Exposure and Dose Research. https://www.epa.gov/chemical-research/rapid-chemical-exposure-and-dose-research (accessed 24 July 2016).

USEPA, 2016e. Method 1694: Pharmaceuticals and Personal Care Products in Water, Soil, Sediment, and Biosolids by HPLC/MS/MS. http://nepis.epa.gov (accessed 24 July 2016).

Walters, E., McClellan, K., Halden, R.U., 2010. Occurrence and loss over three years of 72 pharmaceuticals and personal care products from biosolids-soil mixtures in outdoor mesocosms. Water Res. 44, 6011–6020.

Washburn, B.E., Begier, M.J., 2011. Wildlife responses to long-term application of biosolids to Grasslands in North Carolina. Rangel. Ecol. Manag. 64, 131–138.

Xue, J.C., Sasaki, N., Elangovan, M., Diamond, G., Kannan, K., 2015. Elevated accumulation of parabens and their metabolites in marine mammals from the United States coastal waters. Environ. Sci. Technol. 49, 12071–12079.

Zorrilla, I., Martinez, R., Taggart, M.A., Richards, N., 2015. Suspected flunixin poisoning of a wild Eurasian Griffon vulture from Spain. Conserv. Biol. 29, 587–592.

Chapter 7

Pharmaceuticals in the Environment and Human Health

Alistair B.A. Boxall
University of York, Heslington, United Kingdom

INTRODUCTION

While most work on pharmaceuticals in the environment has focused on impacts on aquatic organisms (see Chapter 4 by Brooks), it is recognized that humans can also be exposed to residues of pharmaceuticals in the environment. Consequently, concerns have been raised over the potential toxic effects of pharmaceutical residues in the environment on humans. This is a topic that has received significant media attention in North America and Europe. Indirect effects on human health and well-being may also be possible due to impacts of the compounds on ecosystem service delivery and, for antimicrobial substances, the potential effects on the levels and maintenance of antimicrobial resistance in bacterial populations in environmental media (the latter is discussed in more detail in the next chapter by Pruden). In this chapter, the routes of exposure of humans to residues of pharmaceuticals in the environment are described and the evidence for direct and indirect effects on human health discussed.

ROUTES OF HUMAN EXPOSURE TO PHARMACEUTICALS IN THE ENVIRONMENT

The main environmental pathways of exposure to pharmaceuticals will be through consumption of contaminated drinking water and food materials such as fruit and vegetable crops and fish and shellfish. Contact to pharmaceuticals associated with soils and dust and dermal exposure to contaminated river and coastal waters while swimming is also possible.

In developed countries, drinking water is typically sourced from rivers and aquifers and treated using different combination of processes including

Environmental Contaminants, Vol. 11. https://doi.org/10.1016/B978-0-444-63857-1.00007-3

coagulation and flocculation and sand filtration to remover colloidal materials, ozonation to break down dissolved organic compounds and granular activated carbon filtration to remove dissolved organic compounds and metals. Finally, the treated water is disinfected before being distributed to the consumer. Pharmaceuticals will be removed to different degrees during the different steps in the process. For example, studies into the efficacy of coagulation and flocculation for removing a range of pharmaceuticals indicate that very low removal occurs for the majority of substances in this stage of the process (e.g., Vieno et al., 2006). Studies into the removal of 45 organic compounds that were detected in source waters, including many pharmaceuticals, through a whole drinking water process involving clarification, carbon filtration, and disinfection showed removals of between 25% and 100% (Stackelberg et al., 2007). Carbon adsorption was found to contribute most to the observed removal of the pharmaceuticals followed by disinfection and clarification. The effectiveness of the removals of the different steps varied according to the chemical class with the more hydrophobic compounds being removed in the carbon filtration step and some of the more hydrophilic molecules reacting with free chlorine in the disinfection process.

Given that pharmaceuticals occur in source waters (see Chapter 2 by Oliviera) and that not all compounds are completely removed by drinking water treatment processes, it is not surprising that a wide range of active pharmaceutical ingredients have been detected in drinking waters around the world. Although there are fewer studies looking at occurrence of pharmaceuticals in drinking waters than in surface waters, a wide range of pharmaceutical active ingredients have been detected in drinking waters around the world including substances used as antibiotics, antidepressants, antianxiety agents, nonsteroidal anti-inflammatory drugs (NSAID), estrogens, β-blockers, stimulants, lipid regulators, anticonvulsants, and in x-ray contrast media (e.g., Benotti et al., 2009; Leung et al., 2013; Stackelberg et al., 2007; Table 1). Many pharmaceuticals that are used are not detected in drinking water but where they are; the concentrations detected are typically in the tens to hundreds of nanogram-per-liter range. High concentrations have been reported for the lipid regulator, clofibric acid, and the antiepileptic medicine, carbamazepine.

Pharmaceuticals in rivers and coastal systems may also be accumulated into fish and mollusks. Most pharmaceuticals are ionizable molecules, so the degree of accumulation is driven by factors such as the hydrophobicity and degree of dissociation of the pharmaceutical and the characteristics of the environment, most notably the pH of water (e.g., Meredith-Williams et al., 2012; Karlsson et al., 2017). While uptake into fish and mollusks has not been studied as much as drinking water, monitoring of pharmaceuticals in tissues of organisms in the aquatic environment has been done in a range of geographic regions, including North America and Asia. Anticonvulsants, antibiotics, antidepressants, NSAIDs, diuretics, lipid regulators, anti-itch agents,

TABLE 1 Maximum Reported Concentrations of Pharmaceuticals That Have Been Detected in Drinking Water, Plant Material, and Fish or Mollusks

	Therapeutic Class	ADI (μg/kg/d)	Drinking Water (ng/L)	Plants (ng/g dwt)	Fish and Mollusks (ng/g dwt)
Atenolol	β-Blocker	0.4	18	5.6	ND
Bezafibrate	Lipid regulator	2.9	27	27	–
Bleomycin	Chemotherapy drug	0.013	13	–	–
Clofibric acid	Lipid regulator	14.3	270	18	
Carbamazepine	Antiepileptic	2.9	258	800	6.3
Chloramphenicol	Antibiotic	11	–	22	–
Ciprofloxacin	Antibiotic	7.1	–	2.2	–
Citalopram	Antidepressant	0.29	–	–	2.3
Clopidogrel	Antiplatelet agent	1.1	–	–	0.2
Diatrizoate	x-Ray contrast agent	43	85	–	–
Diazepam	Antianxiety agent	0.03	10	–	1.8
Diclofenac	Anti-inflammatory	1.4	6	19	2.2
Dilantin	Antiepileptic	1.4	–	1.7	–
Diphenhydramine	Antihistamine	0.4	–	21	–
Erythromycin	Antibiotic	4.4	–	2.2	–
Fenofibric acid	Lipid regulator	0.7	42	–	–

Continued

TABLE 1 Maximum Reported Concentrations of Pharmaceuticals That Have Been Detected in Drinking Water, Plant Material, and Fish or Mollusks—Cont'd

	Therapeutic Class	ADI (μg/kg/d)	Drinking Water (ng/L)	Plants (ng/g dwt)	Fish and Mollusks (ng/g dwt)
Fluoxetine	Antidepressant	0.14	0.8	–	–
Gemfibrozil	Lipid regulator	7.8	70	8	–
Hydrochlorothiazide	Diuretic	0.36	–	–	10.5
Ibuprofen	Anti-inflammatory	11.4	3	50	–
Iopamidol	X-ray contrast agent	179	79	–	–
Ketoprofen	Anti-inflammatory	0.6	–	15	–
Lamotrigine	Antiepileptic	0.3	–	22	–
Meprobamate	Antianxiety agent	2.9	43	1.5	
Metoprolol	β-Blocker	0.3	–	29	0.7
Iopromide	X-ray contrast agent	8.6	86	–	–
Nadolol	β-Blocker	0.57	–	–	0.6
Naproxen	Anti-inflammatory	7.1	ND	113	–
Norfloxacin	Antibiotic	11.4	–	24	–
Norfluoxetine	Antidepressant	–	0.8	–	–
Phenazone	Anti-inflammatory	–	400	–	–
Phenytoin	Antiepileptic	4.3	19	–	–

Primidone	Antiseizure agent	0.7	–	2.4	–
Progesterone	Hormone	0.3	–	6.7	–
Propanolol	β-Blocker	1.1	–	–	0.5
Propyphenazone	Analgesic	7.1	120	–	–
Risperidone	Antipsychotic	0.03	2.9	–	–
Salbutamol	Asthma treatment	0.2	–	47.1	0.6
Sildenafil	Erectile dysfunction	0.3	–	30	–
Sulfamethazine	Antibiotic	10	–	1.9	–
Sulfamethoxazole	Antibiotic	5.7	3	0.25	–
Sulfapyridine	Antibiotic	3.3	–	1.25	–
Testosterone	Hormone	0.04	–	5.52	–
Tetracycline	Antibiotic	5.7	–	10.1	–
Triamterene	Antihypertension	0.4	–	0.064	–
Trimethoprim	Antibiotic	0.21	ND	3.4	–
Venlafaxine	Antidepressant	0.5	–	–	3.1

Data collated from Jones, O.A., Lester, J.N., Voulvoulis, N., 2005. Pharmaceuticals: a threat to drinking water? Trends Biotechnol. 23(4), 163–167; Webb, S., Ternes, T., Gibert, M., Olejniczak, K., 2003. Indirect human exposure to pharmaceuticals via drinking water. Toxicol. Lett. 157–167; Bruce, G., Pleus, R.C., Snyder, S.A., 2010. Toxicological relevance of pharmaceuticals in drinking water. Environ. Sci. Technol. 44, 5619–5626; Prosser, R.S., Sibley, P.K., 2015. Human health risk assessment of pharmaceuticals and personal care products in plant tissues due to biosolids and manure amendments and wastewater irrigation. Environ. Int. 75, 223–233; Christou, A., Karaolia, P., Hapeshi, E., Michael, C., Fatta-Kassinos, D., 2017. Long-term wastewater irrigation of vegetables in real agricultural systems: concentration of pharmaceuticals in soil, uptake and bioaccumulation in tomato fruits and human health risk assessment. Water Res. 109, 24–34; Moreno-González, R., Rodríguez-Mozaz, S., Huerta, B., Barceló, D. León, V.M., 2016. Do pharmaceuticals bioaccumulate in marine molluscs and fish from a coastal lagoon? Environ. Res. 146, 282–298.

and compounds for treating hypertension have been detected in fish tissues (e.g., Ramirez et al., 2007; Tanoue et al., 2015; Moreno-González et al., 2016). Studies into gastropod and bivalve mollusks have detected antidepressant and diuretic compounds (Moreno-González et al., 2016). Concentrations are generally in the low to sub-ng/g range with highest concentrations being reported for the diuretic medicine hydrochlorothiazide (Table 1). Given this compound is a hydrophilic molecule (log $K_{ow} = -0.071$), this observation is surprising.

Pharmaceuticals can also enter terrestrial food chains. Contamination of crops with pharmaceuticals results from uptake from soils that have been fertilized with biosolids that are produced during the wastewater treatment process or that have been irrigated with wastewater effluent. This pathway of exposure has been covered in detail in the chapter by Carter and Kinney (Chapter 5). A wide range of active ingredients including compounds from the antimicrobial, antidepressants, NSAID, β-blocker, anticonvulsant, and androgen classes have been detected in a range of plant types including tomatoes, carrots, corn, lettuce, cucumber, radish, sweet potatoes, and garlic (e.g., Carter et al., 2014; Malchi et al., 2014; Christou et al., 2017). The degree of uptake into plants varies across plant species and is thought to be driven by the degree of ionization of the pharmaceutical in the soil and the hydrophobicity of the molecule (Carter et al., 2014) as well as the properties of the soil in which the plants are growing. Concentrations in plant material range from low ng/g concentrations to concentrations in the hundreds of ng/g. Highest concentrations in plant tissues have been seen for the antiepileptic compound carbamazepine and NSAID naproxen. The high uptake of these two molecules is likely explained by their hydrophobicities. Naproxen has a log K_{ow} of 3.00, and carbamazepine has a log K_{ow} of 2.84. It is believed that substances that have a log K_{ow} around 3.00 have the greatest potential to be accumulated into aboveground plant tissue (Carter et al., 2014).

Given that pharmaceuticals have been detected in drinking waters and a range of materials that may be used as foodstuffs, it is inevitable that humans will be exposed to residues of these substances in the environment. A number of pharmaceuticals have been detected in broader food surveillance studies with concentrations in the low ng/g concentration range (Fussell et al., 2014). The nature of the food items in this study was such that the observed occurrence most likely occurred via the environment as direct treatment with pharmaceuticals was unlikely. Few studies however have actually demonstrated the transfer of pharmaceuticals from the environment to the human population. A study in Israel explored the concentrations of carbamazepine, an anticonvulsant that is accumulated by plants and has been detected in fish, in the urine of humans that had never received the drug (Paltiel et al., 2016). They explored two dietary scenarios: one where individuals consumed a normal diet and the other a diet sourced from water areas where wastewater reuse systems were present—that is, effluent from wastewater treatment plants was used to irrigate agricultural land. The study showed the presence of the carbamazepine in the urine of individuals exposed to both diets, but concentrations were higher in the urine of individuals on the water reuse diet. Data like these demonstrate that exposure is indeed occurring and that the source of a diet can have a big effect on the level of exposure. With the predicted increase in the water reuse to address water shortages around the world, this exposure pathway will become more and more important in the future.

The degree of exposure will also likely be driven by a range of other factors including the level of pharmaceutical use and the degree of wastewater and drinking water treatment in the region of interest, wastewater irrigation and biosolid application practices, river flows and characteristics, and the type of diet being consumed by a population. The recognition that pharmaceuticals occur in water and food and that they are being consumed by the population has resulted in concerns over the potential toxicological effects on humans. These effects and the risks of the effects being realized are discussed in the next section.

Toxicological Risks of Pharmaceutical Residues in the Environment

An ideal pharmaceutical would have no adverse effect on human health, and pharmaceutical companies and regulatory agencies work hard to ensure that these compounds are as safe as possible to humans. However, some classes of pharmaceuticals, such as cytotoxic and cytostatic drugs used in cancer treatments, need to be highly toxic to human cells in order to be effective. Other pharmaceuticals are known to cause toxicological side effects such as allergic reactions, blood disorders, kidney and liver toxicity, or effects on fetal development (Julien, 2001). Even though some pharmaceuticals are known to have adverse impacts, they are still used as it is believed that the therapeutic benefits outweigh these negative impacts. This does mean, however, that the presence of pharmaceuticals in drinking water supplies and food may have the potential to adversely affect human health.

A number of studies have attempted to establish the level of risk of human pharmaceuticals in the environment to humans (e.g., Schwab et al., 2005; Bercu et al., 2008; Bruce et al., 2010; Kummerer and Al-Ahmad, 2010; Leung et al., 2013; Carter et al., 2014). Most of these studies have focused on exposure to drinking water only with few considering the food route of exposure or the combined risks from food and drinking water consumption together. Typically, these studies have established the significance of exposure of a pharmaceutical in food and/or drinking water in terms of human health by comparing the daily intake that an individual will receive via consumption food or water with a safe level such as an acceptable daily intake (ADI) value. ADIs are expressed on a body mass basis and correspond to the amount of a substance that it is thought can be safely ingested by a standard human over their lifetime (Webb et al., 2003). The basis of an ADI varies: some are derived from toxicological data, such as no-observed-adverse-effect levels (NOAELs) for rodents (which companies will test to assess the safety of a new pharmaceutical), using a safety factor to account for uncertainties in extrapolating from a toxicological model system to humans; others might be derived based on information on the sensitivity of microbial organisms to the pharmaceutical; others are based on pharmacological information; and others are based on information on tumor incidences. Generally, in the studies comparing exposure to residues in the environment with ADIs, the predicted intakes of pharmaceuticals from consumption of food or water are at least an order of magnitude (and in most cases many orders of magnitude) lower than ADIs indicating that toxicological effects on humans are highly unlikely.

Using the data collated in Table 1, it is possible to compare and contrast the risks to human health arising from indirect exposure to pharmaceuticals via different routes. A comparison of predicted oral exposure of a 70kg human to the different active ingredients that have been detected in drinking water, plant material, and fish and mollusks is shown in Fig. 1. Assuming the individual eats a typical human diet comprising 56g of protein and 400g of vegetables or fruit and drinks 2L of water, the predicted daily exposures from each exposure route fell below the ADIs for the individual pharmaceuticals (Fig. 1). Based on maximum reported concentrations, bleomycin was the compound contributing the highest fraction of the ADI (2.9%) via drinking water consumption, carbamazepine was the compound contributing the highest fraction of the ADI (32%) via vegetable/fruit consumption, and diazepam was the compound contributing the highest fraction of the ADI (1.2%) via fish and mollusk consumption. Overall, these data support the conclusions from the previous studies and indicate that direct toxicity from indirect exposure to pharmaceuticals in the environment is unlikely to adversely affect human health. The data also indicate that exposure via fruit and vegetables from land receiving biosolids is probably a more significant exposure pathway than via consumption of drinking water and fish and mollusks, even though most of the risk assessment work to date has focused on the drinking water exposure route.

It is however important to recognize that the human health risk assessments of pharmaceuticals have typically considered risks to standard individuals and have not considered potential effects on sensitive subpopulations such as children, pregnant women and the fetus, and those with allergies who may be more prone to exposure (Touraud et al., 2011). Most of the monitoring studies for drinking water and food have explored risks in North America and Europe where the level of wastewater and drinking water treatment is good, and emissions of biosolids to land are more tightly regulated. Limited information is available on the risks to populations in low- or middle-income countries (LMICs). In these regions, sewerage connectivity

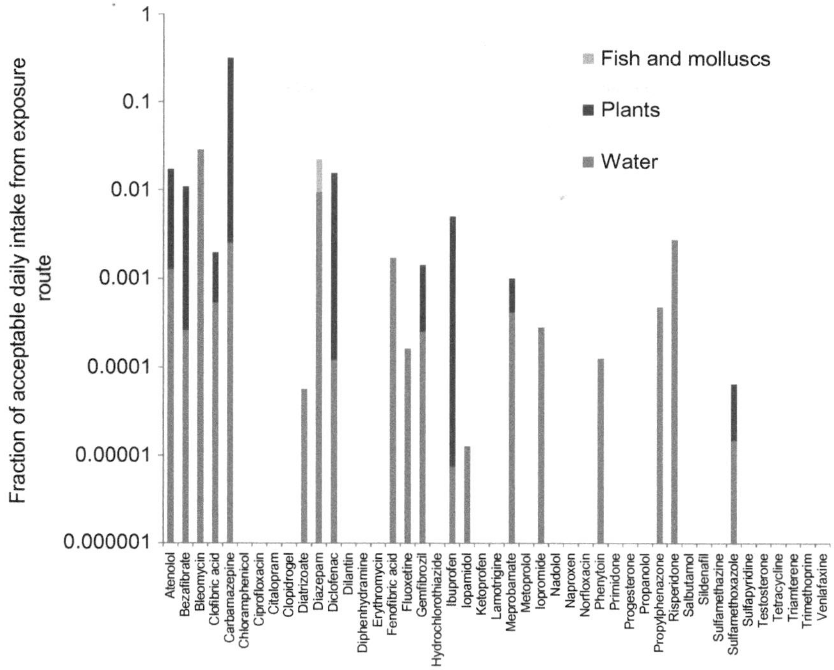

FIG. 1 Fraction of the acceptable daily intake of a 70 kg adult that could arise from indirect exposure to residues of pharmaceuticals in the environment via consumption of contaminated drinking water, plant material, and fish and mollusks. Fractions have been calculated based on maximum reported values in the literature and assuming that an individual drinks 2 L of water and eats 400 g (wwt) and 56 g (wwt) of plant material and protein each day. To convert from wet weight to dry weight, it was assumed that plants comprise 80% water and fish comprise 75% water.

is generally low, and in some areas, waste is collected predominantly in septic systems, and the degree of drinking water treatment may be low or nonexistent (Kookana et al., 2014). In rural areas of some LMICs, household septic systems and soakaways are in close proximity to aquifers that are used by the population as a water source. Concentrations in surface waters in these regions are often higher than in Europe and North America (Hughes et al., 2013). An assessment of the risks of environmental residues of pharmaceuticals to sensitive subpopulations and in regions with less developed regulations and treatment infrastructures is therefore a priority for future work.

POTENTIAL INDIRECT EFFECTS OF PHARMACEUTICAL RESIDUES ON HUMAN HEALTH AND WELL-BEING

While the risks from direct exposure to pharmaceuticals in the environment, in developed regions at least, are probably low, it is possible that pharmaceuticals in the environment can indirectly affect human health and well-being.

It is now recognized that ecosystem alterations, of the types described in other chapters in this book (see chapters by Brooks, Carter and Kinney, and Bean and Rattner), can adversely impact human health (e.g., Myers et al., 2013). Impacts of pharmaceuticals on things like crop productivity and quality, water quality, and biodiversity could adversely affect health by negatively impacting the provision of nutrition and safe water and the regulation of infectious diseases. This is however an area that has received limited attention from those researching the environmental impacts of pharmaceuticals.

A good illustration of the potential indirect effects of a human pharmaceutical on human health is provided by the case of diclofenac and vultures that has been mentioned previously in Chapter 6. In this particular case, diclofenac, which is an NSAID, was being used for veterinary purposes for the treatment of cattle in areas of India and Pakistan. In Europe and North America, the main use of this pharmaceutical is for humans. When vultures in the region consumed meat from the carcasses of the treated animals, they received a dose of diclofenac. Due to metabolic differences, the vultures were much more sensitive to diclofenac than humans receiving a similar dose and over time, exposure to the diclofenac resulted in a massive decline in the bird population and led to the near extinction of three *Gyps* species (Oaks et al., 2004). The decline in the bird population is believed to have resulted in impacts on human health. Vultures in the region provide a critical ecosystem service as they clear up the bodies of dead animals. It is thought that the disappearance of the birds removed this ecosystem service and therefore resulted in an increase in the feral dog population. It is thought that this increase in dog populations resulted in an increase in rabies incidence in humans due to an increase in the number of dog bites (Fig. 2; Markandya et al., 2008). Overall, the use of the diclofenac to treat the vultures is thought to have resulted in an extra 45,000 extra deaths in the area. While this is an extreme example, it does highlight the importance of considering the indirect effects of chemical contaminants in the environment on human health.

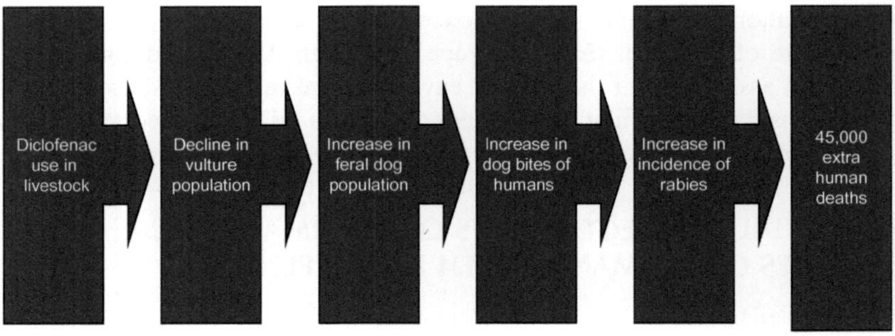

FIG. 2 The cascade of events that are thought to have occurred resulting in human deaths arising from the veterinary use of diclofenac in areas of India and Pakistan (Markandya et al., 2008).

Pharmaceutical contamination may also be contributing to the antimicrobial resistance problem. The O'Neill report recently predicted hundreds of thousands would have died from resistant infections in 2016 and millions might die annually by 2050 unless action is taken to curb AMR evolution and dispersal (DoH, 2016). There is a growing body of evidence that environmental residues of antibiotics resulting from the use in humans may be selecting for antibiotic resistance in bacteria (e.g., Martinez, 2009; see next chapter by Pruden). While the development of resistance in microbial communities may provide a benefit in that it could enable microbial communities to continue to deliver key functions in the presence of antibiotics, if this resistance is transferred into water supplies and the food chain and to human and animal pathogens, it could have implications for efficacy of antibiotic treatments and consequently affect the health of humans and livestock (Wellington et al., 2013). The importance of environmental exposure to antibiotics for resistance in clinically relevant bacteria is still however unclear, and this is a topic of significant ongoing research.

Research Needs

While the available data on human exposure to residues of pharmaceuticals in the environment are reassuring and situations like the vulture crisis are uncommon, there are still major gaps in our knowledge meaning that impacts on human health cannot be completely ruled out. In order to fully understand risks, work is needed in the following areas:

Monitoring of a wider range of pharmaceuticals in a wider range of locations and materials: As can been seen in Table 1, data are available on the concentrations of a handful of substances in drinking water and food items, yet over 1500 active pharmaceutical ingredients are in use by society. Future work should focus on understanding the occurrence (and the drivers of occurrence) of a wider range of pharmaceuticals in a wider range of matrices that the human population will come into contact with. The available data have also been generated for only a few geographic regions, most of which are well regulated and which have good infrastructure for water and wastewater treatment. It is likely exposure concentrations will be elevated in regions of the world that are less regulated and that are poor on nonexistent treatment infrastructure. There is a need to broaden out monitoring campaigns to establish the concentrations of pharmaceuticals in drinking water and food items in these other regions of the world and to assess the implications of these concentrations for human health.

Effects of long-term, low-level exposure to pharmaceuticals and on sensitive subpopulations: Exposure of humans to pharmaceuticals in drinking water and food will occur across the lifetime of an individual. Approaches used to derive ADIs for use in risk assessment of active ingredients may not be protective of such exposures. Work is therefore needed to better assess the implications of long-term exposure to pharmaceuticals on human health end points.

Indirect impacts of pharmaceutical exposure: More work is needed to understand the potential for antibiotics in the environment to select for resistance in bacteria and for this resistance then to be transferred to clinically relevant microorganisms so that the importance of environmental exposure can be weighed up against other exposure routes such as in the clinic. We also need to better understand the impacts of residues of different pharmaceuticals in the environment on ecosystem service delivery and to better characterize the relationships between ecosystem services and human health and well-being.

By exploring the above research topics, we will be much better placed to establish that, if any, pharmaceuticals are impacting human health as a result of environmental contamination and that regions and exposure scenarios are of most importance. This knowledge will allow some of the interventions that are described in other chapters of this book to be much better targeted.

REFERENCES

Benotti, M.J., Trenholm, R., Vanderford, B.J., Holady, J.C., Stanford, B.D., Snyder, S.A., 2009. Pharmaceuticals and endocrine disrupting compounds in U.S. drinking water. Environ. Sci. Technol. 43, 597–603.

Bercu, J.P., Parke, N.J., Fiori, J.M., Meyerhoff, R.D., 2008. Human health risks for three neuropharmaceutical compounds in surface waters. Regul. Toxicol. Pharmacol. 50, 420–427.

Bruce, G., Pleus, R.C., Snyder, S.A., 2010. Toxicological relevance of pharmaceuticals in drinking water. Environ. Sci. Technol. 44, 5619–5626.

Carter, L.J., Harris, E., Williams, M., Ryan, J.J., Kookana, R.S., Boxall, A.B.A., 2014. Fate and uptake of pharmaceuticals in soil-plant systems. J. Agric. Food Chem. 62 (4), 816–825.

Christou, A., Karaolia, P., Hapeshi, E., Michael, C., Fatta-Kassinos, D., 2017. Long-term wastewater irrigation of vegetables in real agricultural systems: concentration of pharmaceuticals in soil, uptake and bioaccumulation in tomato fruits and human health risk assessment. Water Res. 109, 24–34.

DoH, 2016. Tackling Drug-Resistant Infections Globally: Final Report and Recommendations. DoH, London, UK.

Fussell, R.J., Garcia Lopez, M., Mortimer, D.N., Wright, S., Sehnalova, M., Sinclair, C.J., Fernandes, A., Sharman, M., 2014. Investigation into the occurrence in food of veterinary medicines, pharmaceuticals, and chemicals used in personal care products. J. Agric. Food Chem. 62 (17), 3651–3659.

Hughes, S.R., Kay, P., Brown, L.E., 2013. Global synthesis and critical evaluation of pharmaceutical data sets collected from river systems. Environ. Sci. Technol. 47, 661–677.

Julien, R.M., 2001. A Primer of Drug Action. Owl Books, NY, USA.

Karlsson, M.V., Carter, L.J., Agatz, A., Boxall, A.B.A., 2017. Novel approach for characterizing pH-dependent uptake of ionizable chemicals in aquatic organisms. Environ. Sci. Technol. 51, 6965–6971.

Kookana, R.S., Williams, M., Boxall, A.B.A., Larsson, D.G.J., Gaw, S., Choi, K., Yamamoto, H., Thatikonda, S., Zhu, Y.-G., Carriquiriborde, P., 2014. Potential ecological footprints of active pharmaceutical ingredients: an examination of risk factors in low-, middle-and high-income countries. Philos. Trans. R. Soc. B: Biol. Sci. 369, 20130586.

Kummerer, K., Al-Ahmad, A., 2010. Estimation of the cancer risk to humans resulting from the presence of cyclophosphamide and ifosfamide in surface water. Environ. Sci. Pollut. Res. 17, 486–496.

Leung, H.W., Jin, L., Wei, S., Tsui, M.M.P., Zhou, B., Jiao, L., Cheung, P.C., Chun, Y.K., Murphy, M.B., Lam, P.K.S., 2013. Pharmaceuticals in tap water: human health risk assessment and proposed monitoring framework in China. Environ. Health Perspect. 121 (7), 839–846.

Malchi, T., Maor, Y., Tadmor, G., Shenker, M., Chefetz, B., 2014. Irrigation of root vegetables with treated wastewater: evaluating uptake of pharmaceuticals and the associated human health risks. Environ. Sci. Technol. 48, 9325–9333.

Markandya, A., Taylor, T., Long, A., Murty, M.N., Murty, S., Dhavala, K., 2008. Counting the cost of vulture decline—an appraisal of the human health and other benefits of vultures in India. Ecol. Econ. 67, 194–204.

Martinez, J.L., 2009. Environmental pollution by antibiotics and by antibiotic resistance determinants. Environ. Pollut. 157, 2893–2902.

Meredith-Williams, M., Carter, L.J., Fussell, R., Raffaelli, D., Boxall, A.B.A., 2012. Uptake of pharmaceuticals into aquatic invertebrates. Environ. Pollut. 165, 250–258.

Moreno-González, R., Rodríguez-Mozaz, S., Huerta, B., Barceló, D., León, V.M., 2016. Do pharmaceuticals bioaccumulate in marine molluscs and fish from a coastal lagoon? Environ. Res. 146, 282–298.

Myers, S.S., Gaffikin, L., Golden, C.D., Ostfeld, R.S., Redford, K.H., Ricketts, T.H., Turner, W.R., Osofsky, S.A., 2013. Human health impacts of ecosystem alteration. PNAS 110 (47), 18753–18760.

Oaks, J.L., Gilbert, M., Virani, M.Z., Watson, R.T., Meteyer, C.U., Rideout, B.A., Shivaprasad, H.L., Ahmed, S., Chaudhry, M.J., Arshad, M., Mahmood, S., Ali, A., Khan, A.A., 2004. Diclofenac residues as the cause of vulture population decline in Pakistan. Nature 427 (6975), 630–633.

Paltiel, O., Fedorova, G., Tadmor, G., Kleinstern, G., Maor, Y., Chefetz, B., 2016. Human exposure to wastewater-derived pharmaceuticals in fresh produce: a randomized controlled trial focusing on carbamazepine. Environ. Sci. Technol. 50, 4476–4482.

Ramirez, A.J., Mottaleb, M.A., Brooks, B.W., Chambliss, K., 2007. Analysis of pharmaceuticals in fish using liquid chromatography-tandem mass spectrometry. Anal. Chem. 79, 3155–3163.

Schwab, B.W., Hayes, E.P., Fiori, J.M., Mastrocco, F.J., Roden, N.M., Cragin, D., Meyerhoff, R.D., D'Aco, V., Anderson, P.D., 2005. Human pharmaceuticals in US surface waters: a human health risk assessment. Regul. Toxicol. Pharmacol. 42, 296–312.

Stackelberg, P.E., Gibs, J., Furlong, E.T., Meyer, M.T., Zaugg, S.D., Lippincott, R.L., 2007. Efficiency of conventional drinking-water-treatment processes in removal of pharmaceuticals and other organic compounds. Sci. Total Environ. 377, 255–272.

Tanoue, R., Nomiyama, K., Nakamura, H., Kim, J.-W., Isobe, T., Shinohara, R., Kunisue, T., Tanabe, S., 2015. Uptake and tissue distribution of pharmaceuticals and personal care products in wild fish from treated-wastewater-impacted streams. Environ. Sci. Technol. 49, 11649–11658.

Touraud, E., Roig, B., Sumpter, J.P., Coetsier, C., 2011. Drug residues and endocrine disruptors in drinking water: risk for humans? Int. J. Hyg. Environ. Health 214, 437–441.

Vieno, N., Tuhkanen, T., Kronberg, L., 2006. Removal of pharmaceuticals in drinking water treatment: effect of chemical coagulation. Environ. Technol. 27, 183–192.

Webb, S., Ternes, T., Gibert, M., Olejniczak, K., 2003. Indirect human exposure to pharmaceuticals via drinking water. Toxicol. Lett. 157–167.

Wellington, E.M.H., Boxall, A.B.A., Cross, P., Feil, E.J., Gaze, W.H., Hawkey, P.M., Johnson-Rollings, A.S., Jones, D.L., Lee, N.M., Otten, W., Thomas, C.M., Williams, A.P., 2013. The role of the natural environment in the emergence of antibiotic resistance in Gram-negative bacteria. Lancet Infect. Dis. 13 (2), 155–165.

FURTHER READING

Jones, O.A., Lester, J.N., Voulvoulis, N., 2005. Pharmaceuticals: a threat to drinking water? Trends Biotechnol. 23 (4), 163–167.

Kumar, A., Xagoraraki, I., 2010. Human health risk assessment of pharmaceuticals in water: an uncertainty analysis for meprobamate, carbamazepine and phenytoin. Regul. Toxicol. Pharmacol. 57, 146–156.

Prosser, R.S., Sibley, P.K., 2015. Human health risk assessment of pharmaceuticals and personal care products in plant tissues due to biosolids and manure amendments and wastewater irrigation. Environ. Int. 75, 223–233.

Shen, R., Andrews, S.A., 2011. Demonstration of 20 pharmaceuticals and personal care products (PPCPs) as nitrosamine precursors during chloramine disinfection. Water Res. 45, 944–952.

Wu, X., Dodgen, L.K., Conkle, J.L., Gan, J., 2015. Plant uptake of pharmaceutical and personal care products from recycled water and biosolids: a review. Sci. Total Environ. 536, 655–666.

Xie, Z., Lu, G., Yan, Z., Liu, J., Wang, P., Wang, P., 2017. Bioaccumulation and trophic transfer of pharmaceuticals in food webs from a large freshwater lake. Environ. Pollut. 222, 356–366.

Chapter 8

Antibiotics and Antibiotic Resistance: Closing the Loop Between Hospitals and the Environment

Amy Pruden
Civil and Environmental Engineering, Virginia Tech, Blacksburg, VA, United States

WHAT ARE ANTIBIOTICS?

Antibiotics have been described to be a "precious resource." Penicillin was the first antibiotic to be discovered, an accidental discovery by Sir Alexander Fleming in 1928. Shortly thereafter, sulfonamide antibiotics were synthesized in the laboratory and were the first to be commercially produced for medical purposes. Now, nearly 100 years later, we as a society have become wholly dependent upon antibiotics for treating and preventing infections. Antibiotics are largely hailed for providing a cure for deadly and debilitating infections, such as tuberculosis, pneumonia, and syphilis, and thus extending life expectancy and improving quality of life throughout the world. They are also critical for preventing serious infections following major surgery, such as knee and hip replacement, and are even used in dentistry to prevent bacterial contamination of the blood stream. Antibiotics have found wide use in agriculture, where they can prevent and cure illness in livestock and improve feed efficiency by promoting weight gain. They are also sometimes used in the production of produce to prevent rotting and extend shelf life. However, antibiotic resistance is a grave and growing threat to our ability to continue to rely on the precious resource of antibiotics on into the future. It is thus critical to understand what antibiotic resistance is, including its sources and causes, so that we can identify means of preventing its spread.

Appropriate antibiotic use is an essential aspect of preventing the spread of antibiotic use and one that is directly in the hands of health-care professionals. What is important to understand is that antibiotics are solely effective for killing bacterial infections and thus their application for treatment of viral

Environmental Contaminants, Vol. 11. https://doi.org/10.1016/B978-0-444-63857-1.00008-5

infection is often blamed as one critical form of misuse. Antibiotics belong to various classes, depending on their chemical structure and bacterial targets. Examples include beta-lactams, sulfonamides, macrolides, aminoglycosides, quinolones, and glycopeptides. Some are narrow spectrum, in that they are effective for killing a narrow range of bacteria, while others are broad spectrum and affect a wide range of bacteria. Most tend to affect either gram-positive or gram-negative bacteria, but some do affect both classes. One strategy for limiting the spread antibiotic resistance is to limit the use of broad-spectrum antibiotics and, where possible, to employ antibiotics that directly target the pathogen of concern (SRS Pharmaceuticals, 2017). This is challenging because it typically requires time to diagnose the precise cause of an infection. Regardless, the point is that there is a two-edged sword to antibiotic use in which, while the intention is to kill the pathogen and source of infection, strains that are resistant to the antibiotic can survive the treatment, proliferate with the lack of competition from sensitive strains, and spread. Thus, prudent antibiotic use, including prescribing only when medically necessary and completing the course of treatment once prescribed, is generally considered best practice.

WHAT IS ANTIBIOTIC RESISTANCE?

The ingenuity of the discovery of antibiotics for medical purposes lies in the fact that, in principle, they primarily target cell metabolism and machinery unique to bacteria and thus leave the patient unaffected. For example, a common target of antibiotic drugs is bacterial cell wall, given that human cells do not possess walls. However, bacteria mutate and evolve at an extremely accelerated pace relative to humans, often with generation times of a few minutes. Because of this, a mutation in the DNA that allows the bacterium to survive antibiotic treatment will provide an evolutionary advantage when subject to antibiotic treatment, and resistant cells will proliferate and become selectively enriched. In this context, antibiotics exert what is called a *selective pressure* on the bacterial populations, favoring survival and proliferation of resistant strains.

DNA segments that encode the ability of bacteria to resist antibiotics are called antibiotic resistance genes (ARGs). ARGs can protect bacteria from antibiotics by various means, including modifying the target, modifying the antibiotic itself so it is no longer functional, degrading the antibiotic, or pumping it out of the cell. Most antibiotics are actually natural compounds, such as the penicillin that is produced by the mold *Penicillium*. Such compounds are produced by microbes as a natural defense against attack from bacteria. In fact, it is thought that antibiotics have existed on Earth for as long as microbes themselves, and thus, it is not surprising that recent works by scientists have found ARGs even in remote environments, including Arctic soils that were frozen long before antibiotics were ever used by humans. What has dramatically changed is the scale at which antibiotics are now being produced, used, and excreted by people and livestock, thus placing an unprecedented level of selection pressure in favor of bacteria carrying ARGs.

Selection pressure in terms of enhanced survival of resistant cells is not the only concern. To complicate matters, bacteria possess the unique ability to share genes among themselves, through a process called *horizontal gene transfer*. Horizontal gene transfer includes three main processes: *conjugation, transformation, and transduction*. Conjugation is commonly referred to as bacterial mating and involves physical connection and exchange of DNA, typically on small circular pieces of DNA called *plasmids*. Transformation takes place when the DNA from a dead cell is absorbed and assimilated by a living cell. Transduction occurs when viruses that attack bacteria, called bacteriophages, inadvertently package and transfer bacterial DNA instead of viral DNA and transfer it to their next host bacterium. Mobile genetic elements, such as *plasmids*, *transposons*, and *integrons*, facilitate horizontal gene transfer and typically carry ARGs, sometimes multiple variants that confer resistance to multiple antibiotics. Thus, horizontal gene transfer is an important means by which bacterial strains can acquire multiantibiotic resistance. A grave example is the New Delhi metallo-beta-lactamase-1 ($bla_{\mathrm{NDM\text{-}1}}$) gene, which confers resistance to several antibiotics and has now spread to several different bacterial hosts, including *Klebsiella pneumoniae*, *Escherichia coli*, *Citrobacter freundii*, *Enterobacter cloacae*, and *Morganella morganii*, with corresponding infections considered to be virtually impossible to treat.

The problem of antibiotic resistance has become increasingly grave. In 2013, the US Centers for Disease Control estimated that 2 million people are infected and 23,000 people die per year as a result of resistant infections, but this is a very conservative estimate (U.S. Centers for Disease Control, 2013). In the EU, it is estimated that 25,000 people die per year because of antibiotic-resistant infections (EMA, 2017), while in India, 58,000 newborn babies alone die because of sepsis infections caused by resistant bacteria (Center for Disease Dynamics Economics and Policy, 2015). Likewise, diseases that were thought to be essentially eradicated, such as tuberculosis, are now making a comeback due to antibiotic resistance. Worse, general trends indicate that globally, resistance rates among pathogens are steadily increasing.

HOW TO COMBAT ANTIBIOTIC RESISTANCE

Given the threat of antibiotic resistance, numerous bodies and organizations have developed guidelines and strategies for combatting the spread of antibiotic resistance, for example, the EU (European Union Directorate General for Health and Consumers, 2017), the United States, and WHO. One broad recommendation is to limit the use of existing antibiotics for agricultural purposes and, to the extent possible, to situations where medically necessary. Nordic countries were the first to put such a ban in practice, with the EU following suit shortly thereafter. Other countries, such as the United States, are just beginning to implement policy to curb agricultural antibiotic use, while much of Asia still using antibiotics in an unrestricted fashion. As mentioned above, choosing antibiotics that are as narrow as possible in spectrum is also thought to limit the potential of selecting for resistant nontarget populations of bacteria and promoting the possibility of horizontal gene transfer to new hosts. Of course, hygiene and other good practices are essential for preventing the spread of antibiotic-resistant bacteria in hospitals, just as this is true for infection control in general. Strains of resistant bacteria that have become a scourge to hospitals as a source of hospital-acquired infection include methicillin-resistant *Staphylococcus aureus* (MRSA), vancomycin-resistant enterococci (VRE), multiantibiotic-resistant *K. pneumoniae*, and multiantibiotic-resistant *Pseudomonas aeruginosa*. In the context of this chapter, focused on the interplay between hospitals and the environment, *P. aeruginosa* is of particular interest because it tends to colonize hospital water systems and taps, serving as a relentless and difficult to control the source of infection (Hota et al., 2009). One area of current research deals with identifying the most effective means of disinfection, given that some disinfectants could also impart selective pressures for antibiotic resistance (Xi et al., 2009).

While hospital control will continue to be a front line in the war against antibiotic resistance, the remaining sections make the case for a much broader strategy for control of the spread of resistance. In particular environmental sources, sinks, and pathways should also be taken into consideration in order to limit the potential for antibiotic resistance to spread in the first place.

ANTIBIOTIC RESISTANCE GENES AS ENVIRONMENTAL "CONTAMINANTS"

In order to develop broad strategies for containing the spread of antibiotic resistance, it is important to consider where it comes from and the full array of pathways and selective pressures that can contribute to its spread. Because of horizontal gene transfer, here, we conservatively focus on limiting the spread of ARGs as the "contaminants" of concern, rather than individual pathogens as is traditionally the case (Pruden et al., 2006). Antibiotics themselves also typically co-occur with the source of antibiotic resistance and can be transported along the same pathways, exerting various levels of selective pressure and potentially stimulating horizontal gene transfer along the way (see Chapters 2 and 3 for a discussion of the occurrence and fate of antibiotics in the environment). Importantly, other selective pressures, such as metals and disinfectants, can also play a role. This is because of *cross-resistance* and *coresistance*. Cross-resistance is the case where an ARG has multiple functions, for example, pumping both antibiotics and metals out of the cell, and thus, both could select for bacteria carrying this type of ARG. Coresistance, on the other hand, happens when multiple ARGs or detoxification genes occur on the same genetic element and thus is selected together as part of a package even when the corresponding antibiotic is not being used. Thus, ARGs, along with bacterial hosts and selective pressures, including antibiotics and other substances, should be considered.

In considering sources, antibiotics themselves are given in hospitals, taken in the home, administered to livestock, and used for other agricultural purposes. While leftover drugs are sometimes flushed down the toilet or disposed to the trash, the vast majority of consumed antibiotics are excreted. In terms of human consumption, in developed countries, this translates to flushing into the sewage system and ending up in a wastewater treatment plant and ultimately, if not degraded or removed along the way, in impacted water bodies. In terms of livestock consumption, this translates into passage into manure, which may be composted or land-applied, where any antibiotics that escape can also end up in impacted water bodies following storms and watershed runoff. Of course, freshwater sources, including rivers, lakes, reservoirs, and groundwater are a critical resource for human survival and are used as drinking water and irrigation sources as well as for swimming and other recreational purposes.

ARGs are thought to follow similar pathways as antibiotics, and the above-noted routes could serve as pathways for human exposure and loop back to the hospital in terms of tap water and colonized patients carrying community-acquired strains. Bacteria carrying ARGs can be selectively enriched in the gut and on the skin and subsequently washed down the drain or into the watershed and ultimately impact water bodies. In fact, several studies have now noted a strong positive correlation between occurrence of ARGs in rivers and wastewater treatment plants and upstream livestock operations (Storteboom et al., 2010; Amos et al., 2015; Graham et al., 2010). Such studies demonstrate that human activities can serve to elevate these natural background levels of ARGs by10-, 100-, or even 1000-fold. Some have also made the case that genetic elements responsible for horizontal gene transfer could be just as much or even more of a concern as markers for the spread of resistance. Interestingly, Gillings et al. (2014) noted that the class 1 integron gene, *int*I1, which is widely known to facilitate horizontal gene transfer of ARGs, displays robust correlation with other indicators of human pollution.

Fig. 1 provides a conceptual diagram of sources and pathways of antibiotics and antibiotic resistance in the environment and potential routes of exposure.

At present, several scientific questions remain regarding exactly the extent to which antibiotics, ARGs, and bacterial hosts carrying ARGs are transported together in the environment. Although doses of antibiotics required to kill pathogens have been well defined for the purpose of human medicine, such as the *minimum inhibitory concentration*, this knowledge is not as well

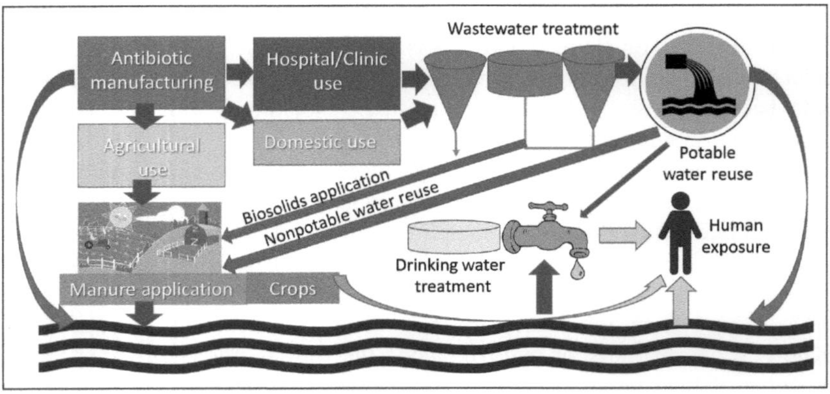

FIG. 1 Conceptual diagram of sources and pathways of antibiotics and antibiotic resistance in the environment and potential routes of exposure.

established when considering bacteria inhabiting feces, manure, soil, and water. Thus, it is difficult to discern what level of antibiotic contamination in the environment is a concern. Likewise, there is evidence that heavy metal pollution in the environment, such as lead, cadmium, copper, and zinc, could be even worse than antibiotics in terms of their potential to select for resistant strains (Song et al. 2017). This is especially important when considering alternatives to antibiotics on livestock operations, for example, antimicrobial copper baths for livestock.

OPPORTUNITIES TO MITIGATE THE SPREAD OF ANTIBIOTIC RESISTANCE VIA THE WATER CYCLE

A current challenge to the scientific community is translating the occurrence of ARGs to risk. *Risk assessment* is a means of translating an observation, such as a specific ARG and its concentration, into a quantifiable risk, such as the likelihood that someone will be exposed and subsequently become ill with an antibiotic-resistant infection. *Quantitative microbial risk assessment (QMRA)* models have been developed to a limited extent for some pathogens. For example, Schoen and Ashbolt (2011) developed a QMRA model to estimate how many *Legionella pneumophila* cells inhabiting a shower head biofilm are required to cause a human inhaling aerosols during showering to develop the severe pneumonia known as Legionnaires' disease. It is much more challenging to estimate risk for antibiotic resistance because ARGs can be transferred from nonpathogens to pathogens and because the resistant bacteria may be carried for months or years before the opportunity for infection actually occurs.

Until reliable risk assessment models can be developed, which may require several years, a prudent approach is to develop best management

TABLE 1 Critical Control Points Where Mitigation Strategies May Be Targeted in Order to Limit the Possibility of Spread of Antibiotic Resistance

Antibiotic manufacturing	Treat waste and remove antibiotics and resistant bacteria
Hospitals	Prudent antibiotic use, minimize drain disposal
	Maintaining and disinfecting water system
	On-site "hot-spot" treatment of sewage
Wastewater treatment plants	Biological processes that reduce gene transfer
	Ensure biodegradation and removal of antibiotics
	Filtering of effluent
	Disinfection of effluent to destroy ARGs
Recycled water treatment	Membrane treatment to remove antibiotics and ARGs
	Advanced oxidation processes to destroy DNA
Agriculture	Prudent antibiotic use
	Manure treatment and containment
	Crop selection and postharvest practice
	Limits to harvest time following manure or biosolid application
Drinking water treatment	Avoid disinfectants and processes that select for resistant bacteria

practices for limiting the spread of antibiotic resistance, continually informed and updated as new knowledge is developed (Berendonk et al., 2015; Pruden et al., 2013). Table 1 provides an overview of current opportunities for limiting the spread of antibiotic resistance via the water cycle.

"Hot-Spot" Treatment at Hospitals

Hospital sewage is a key waste stream of concern with respect to antibiotic resistance. Hospital sewage has been noted to be highly enriched in antibiotics (see Chapter 2), resistant pathogens, and ARGs relative to typical domestic waste streams (Czekalski et al., 2012; Novo and Manaia, 2010; Varela et al., 2013). Of particular concern are antibiotics of last resort, such as vancomycin, extended spectrum beta-lactams, and colistin, as well as the corresponding antibiotics. Thus, segregated treatment of hospital wastewater prior to discharge to the municipal systems is an important consideration (New Standard for Hospital Wastewater Treatment, 2015).

At the other end of the water cycle, hospital taps can be susceptible to colonization by multiantibiotic-resistant bacteria, such as *P. aeruginosa*. Multiantibiotic-resistant *P. aeruginosa* can be especially deadly to premature infants, burn patients, and immunocompromised individuals. Point-of-use disinfection, such as periodic pasteurization of taps and UV units, is used in some cases, though there is no universal means of preventing this type of contamination. The design of the hospital tap itself has been cited in some outbreaks. For example, it is thought that the mixing valve in hands-free faucets can provide an ideal warm water/high surface area environment for the growth of *P. aeruginosa* (Bédard et al., 2016).

Wastewater Treatment Plants and Recycled Water

Wastewater treatment plants are an essential aspect of sanitation and water infrastructure. Collection and treatment of sewage along with purifying and returning the water to the environment is a cornerstone to public health and safety. Wastewater treatment works by employing several physical, chemical, and biological treatment barriers. A typical wastewater treatment plant will first employ screens and grit chambers to remove larger and denser material followed by primary settling tanks to remove slower to settle material. The water that leaves these processes is largely particle-free and fed to a biological treatment stage where high-density populations of bacteria, called activated sludge, biodegrade the carbon and nutrients in the water. The purified water is sometimes disinfected by UV light or chlorination, depending on season and local practice, and returned to rivers, lakes, and streams as a means of water reclamation.

There are numerous opportunities to consider in adapting wastewater treatment to better control antibiotic resistance. There is some concern that the high concentrations of bacteria used in the activated sludge could increase the likelihood of pathogens flushed down the drain acquiring ARGs with the reservoir of bacteria present in the WWTP. As a precautionary practice, filtration or disinfection of the final effluent is being considered. However, such practices add cost to the treatment facility, and effective disinfection measures for destroying ARGs have yet to be identified. New wastewater treatment processes are also currently under development and may present opportunities, such as anaerobic processes that save cost on the requirement to aerate activated sludge. Ideally, new treatment processes should be evaluated for their ability to remove antibiotic-resistant bacteria and ARGs in the treated water.

With increasing population pressure and the need for sustainable water supplies, there is growing interest in the application of recycled water (see discussion in the previous chapter around the use of recycled water for crop irrigation). Recycled water is produced by employing additional treatment steps following conventional sewage treatment to further purify it to a level appropriate for the intended reuse. For example, additional filtration and disinfection steps may be added prior to piping the water for nonpotable applications such as irrigation of crops and recreational fields. If the water is intended for potable application, then more aggressive treatments may be employed. For example, in southern California, the treated wastewater effluent is subject to reverse osmosis, the most selective membrane available in that it is designed essentially to only allow water molecules to pass, prior to injecting the water into an underground aquifer where the drinking water supply is stored. Another promising water treatment technology is the use of *advanced oxidation processes (AOPs)*, which are more aggressive than traditional UV and chlorination disinfection and have the potential to break apart ARGs and gene transfer elements and thus limit the possibility that they will remain functional and spread downstream. Treatments such as reverse osmosis and AOPs are expensive, however, thus calling for research to help guide selection of the right process that balances cost and risk mitigation.

Agricultural Practices

In the United States and many parts of the world, the majority of antibiotic use is for agricultural purposes. Such practices evolved because they helped keep livestock healthy, improved weight gain, shortened time to market, and improved shelf life of certain produce. However, such practices have come under increased scrutiny globally, with the prudent approach considered to be reducing the use to the extent practical, while also finding alternatives to antibiotics to ensure productivity is maintained. In Nordic countries and the EU, antibiotic used for prophylactic or subtherapeutic purposes has been banned since 2006 or earlier. In the United States, a ban was only recently put in place, in 2017, to curb the use of antibiotics that are "critically important" to human health in livestock while also beginning to limit subtherapeutic use.

Livestock manure is typically highly enriched in antibiotic-resistant bacteria and ARGs. In some cases, even livestock that have not been treated with antibiotics have been noted to carry loads of ARGs on par with what has been observed for livestock that are receiving antibiotics. Manure management and other best practices thus make sense as means to limit flow of these ARGs to the environment. Research is currently underway to examine whether practices such as composting and methods of land application can lower the levels of ARGs in soils. Buffer zones for application of manure and manure products can also limit the potential for runoff to neighboring streams, a practice that is commonly implemented to limit fecal coliform and pathogen runoff.

An area of emerging interest is whether the kinds of practices that are implemented to minimize the likelihood that crops will be contaminated with manure-borne pathogens could also serve to reduce the likelihood of spread of antibiotic resistance, or if more rigorous approaches would be necessary. For example, the Food Safety and Modernization Act in the United States requires that vegetables cannot be harvested until at least 120 days after manure has been applied to soil. Also of interest would be whether the type of vegetable matters; for example, leafy vegetables have greater surface area for bacterial colonization, while root vegetables are more directly exposed to the soil. Likewise, certain soil types may be more effective for attenuating antibiotic resistance.

Aquaculture

Antibiotics, including tetracyclines, sulfonamides, and quinolones, are commonly used in aquaculture. These are used to fight fish diseases such as furunculosis. However, given that antibiotics are added directly to the water and at high concentrations, aquaculture practices in particular have been cited to be potentially of even greater concern than conventional agricultural use. Given that antibiotics are directly dosed to the water, reducing or eliminating use in aquaculture appears to be the only realistic option for minimizing the potential to spread antibiotic resistance, although improved treatment and containment of the water may also help (Responsible Use of Antibiotics in Aquaculture, 2004).

Antibiotic Manufacturing

In thinking about the full life cycle of antibiotics and antibiotic resistance, it is important to consider the manufacturing of antibiotics. Arguably, the greatest potential for high concentrations of antibiotics to enter the environment and exert selection pressure lies in the dealing with antibiotic manufacturing waste. This was brought to light in 2008 when Swedish researchers found levels of antibiotics in an Indian river neighboring an antibiotic production facility to be well above what is typically dosed to a patient (Fick et al., 2009). This is a critical environmental and public health justice issue to consider, given that pharmaceutical plants may be located in countries with poorer environmental regulations. This is an issue that would need to be taken up by geopolitical action along with consumer awareness regarding the origins of their pharmaceuticals (Pruden et al., 2013).

CLOSING THOUGHTS

Antibiotic resistance is likely to be the number one public health threat in the 21st century. The development of antibiotic resistance in bacteria closely follows antibiotic use. Thus, curbing the use of antibiotics along with developing appropriate means managing and treating antibiotic-laden waste streams, including forms excreted by humans and livestock, is key to implementing a comprehensive strategy to control the spread of antibiotic resistance. To proactively minimize the risk of spreading antibiotic resistance, environmental routes should be considered and means of limiting the spread of ARGs should be implemented, beyond traditional measures aimed at pathogen control.

REFERENCES

Amos, G.C.A., Gozzard, E., Carter, C.E., Mead, A., Bowes, M.J., Hawkey, P.M., Zhang, L., Singer, A.C., Gaze, W.H., Wellington, E.M.H., 2015. Validated predictive modelling of the environmental resistome. ISME J. 9, 1467. https://doi.org/10.1038/ismej.2014.237.

Antibiotic Resistance Threats in the United States. 2013. U.S. Centers for Disease Control. https://www.cdc.gov/drugresistance/threat-report-2013/.

Bédard, E., Prévost, M., Déziel, E., 2016. *Pseudomonas aeruginosa* in premise plumbing of large buildings. Microbiologyopen 5 (6), 937–956. https://doi.org/10.1002/mbo3.391.

Berendonk, T.J., Manaia, C., Merlin, C., Fatta-Kassinos, D., Cytryn, E., Walsh,F., Bürgmann, H., Sørum, H., Norström, M., Pons, M.-N., Kreuzinger, N., Huovinen, P., Stefani, S., Schwartz, T., Kisand, V., Baquero, F., and Martinez, J.L. (2015). Tackling antibiotic resistance: the environmental framework. Nat. Rev. Microbiol. 13, 310–317.

Communication from the Commission to the European Parliament and the Council Action Plan Against the Rising Threats From Antimicrobial Resistance. 2017. European Union Directorate General for Health and Consumers. https://mail.google.com/mail/u/0/#label/ARG+News/159ef95bb79777a2.

Czekalski, N., Berthold, T., Caucci, S., Egli, A., Bürgmann, H., 2012. Increased levels of multiresistant bacteria and resistance genes after wastewater treatment and their dissemination into Lake Geneva, Switzerland. Front. Microbiol. 3, 106. https://doi.org/10.3389/fmicb.2012.00106.

Fick, J., Soderstrom, H., Lindberg, R.H., Phan, C., Tysklind, M., Larsson, D.G.J., 2009. Contamination of Surface, Ground and Drinking Water from Pharmaceutical Production. Environmental Toxicology and Chemistry 28 (12), 2522–2527.

Gillings, M., Gaze, W., Pruden, A., Smalla, K., Tiedje, J., Zhu, Y.-G., 2014. Using the class 1 integron-integrase gene as a proxy for anthropogenic pollution. ISME J. 9 (6), 1269–1279.

Graham, D.W., Olivares-Rieumont, S., Knapp, C.W., Lima, L., Werner, D., Bowen, E., 2010. Antibiotic resistance gene abundances associated with waste discharges to the Almendares River near Havana, Cuba. Environ. Sci. Technol. 45 (2), 418–424. https://doi.org/10.1021/es102473z.

Hota, S., Hirji, Z., Stockton, K., Lemieux, C., 2009. Outbreak of multidrug-resistant *Pseudomonas aeruginosa* colonization and infection secondary to imperfect intensive care unit room design. Infect. Control Hosp. Epidemiol. https://doi.org/10.1086/592700.

NARROW SPECTRUM ANTIBIOTICS: Uses, Advantages & Disadvantages. 2017. SRS Pharmaceuticals, Pvt., Lmt. http://www.srspharma.com/narrow-spectrum-antibiotics.htm.

New Standard for Hospital Wastewater Treatment, 2015. Filtration + Separation. vol. 52(3), pp. 44–45.

Novo, A., Manaia, C., 2010. Factors influencing antibiotic resistance burden in municipal wastewater treatment plants. Appl. Microbiol. Biotechnol. 87, 1157–1166.

Pruden, A., Larsson, D.G.J., Amézquita, A., Collignon, P., Brandt, K.K., Graham, D.W., Lazorchak, J.R., Suzuki, S., Silley, P., Snape, J.R., Topp, E., Zhang, T., Zhu, Y.-G., 2013.

Management options for reducing the release of antibiotics and antibiotic resistance genes to the environment. Environ. Health Perspect. 121 (8), 1–9.

Pruden, A., Pei, R., Storteboom, H.N., Carlson, K.H., 2006. Antibiotic resistance genes (ARG) as emerging contaminants: studies in Northeastern Colorado. Environ. Sci. Technol. 40 (23), 7445–7450.

Responsible Use of Antibiotics in Aquaculture, 2004. FAO Fisheries Technical Paper 469.

Schoen, M., Ashbolt, N.J., 2011. An in-premise model for Legionella exposure during showering events. Water Res. 45 (18), 5826–5836.

Song, J., Rensing, C., Holm, P.E., Virta, M.P.J., Brandt, K.K., 2017. Comparison of metals and tetracycline as selective agents for development of tetracycline resistant bacterial communities in agricultural soil. Environ. Sci. Technol. 51 (5), 3040–3047. https://doi.org/10.1021/acs.est.6b05342.

Storteboom, H.N., Arabi, M., Davis, J.G., Crimi, B., Pruden, A., 2010. Tracking antibiotic resistance genes in the South Platte River Basin using molecular signatures of urban, agricultural, and pristine sources. Environ. Sci. Technol. 44 (19), 7397–7404.

The Bacterial Challenge: Time to React European Medicines Agency (EMA) and the European Center for Disease Prevention and Control. 2017. http://www.ema.europa.eu/ema/index.jsp?curl=pages/special_topics/general/general_content_000439.jsp.

The State of the World's Antibiotics. 2015. Center for Disease Dynamics Economics and Policy. http://cddep.org/publications/state_worlds_antibiotics_2015.

Varela, A.R., Ferro, G., Vredenburg, J., Yanık, M., Vieira, L., Rizzo, L., Lameiras, C., 2013. Vancomycin resistant enterococci: from the hospital effluent to the urban wastewater treatment plant. Sci. Total Environ. 450, 155–161.

Xi, C., Zhang, Y., Marrs, C.F., Ye, W., Simon, C., Foxman, B., Nriagu, J., 2009. prevalence of antibiotic resistance in drinking water treatment and distribution systems. Appl. Environ. Microbiol. 75 (17), 5714.

FURTHER READING

Global Strategy to Contain Antimicrobial Resistance. 2017. World Health Organization. http://www.who.int/drugresistance/WHO_Global_Strategy.htm/en/.

National Action Plan to Combat Antimicrobial Resistant Bacteria. 2017. The United States White House Archives. https://obamawhitehouse.archives.gov/the-press-office/2015/03/27/fact-sheet-obama-administration-releases-national-action-plan-combat-ant.

Chapter 9

Antimicrobial Use and Ecotoxicological Risks from Pandemics and Epidemics

Andrew C. Singer
NERC Centre for Ecology & Hydrology, Wallingford, United Kingdom

INTRODUCTION

Influenza pandemics have been reported nearly every 40 years for the last 500 years.(Taubenberger and Morens, 2010) Modern influenza pandemics stand apart from those of the past because of the quantity and range of drugs available to treat primary and secondary infections, that is, antivirals and antibiotics. A new class of influenza antivirals, neuraminidase inhibitors (Alves Galvao et al., 2014), greatly improves on the largely ineffective antivirals developed in the mid-1960s (Davies et al., 1964). The large number of antibiotics in use today was not available during the last pandemic and was totally absent from the "toolbox" during the 1918 Spanish pandemic that caused the death of over 50 million people worldwide. This armory of drugs to fight infection has significant implications to not only human health but also the wider environment because of its inevitable release into the waste water through urine and feces and passage into sewage works and ultimately the rivers. The combination of huge volumes of drugs, the short span of time in which they are consumed, and the novelty of some of these drugs create a unique "downstream" risk to society, critical infrastructure, and the environment. This chapter aims to highlight rare but catastrophic events such as epidemics and pandemics and the threats they pose to the environment and human health from coordinated medical responses.

CASE STUDY: INFLUENZA PANDEMIC 2009

Many nation states were exceptionally well prepared for the 2009 influenza A (H1N1) pandemic, but paradoxically, the pandemic came as a total surprise. The surprise came from the fact that the world had its eyes focused on the

Environmental Contaminants, Vol. 11. https://doi.org/10.1016/B978-0-444-63857-1.00009-7

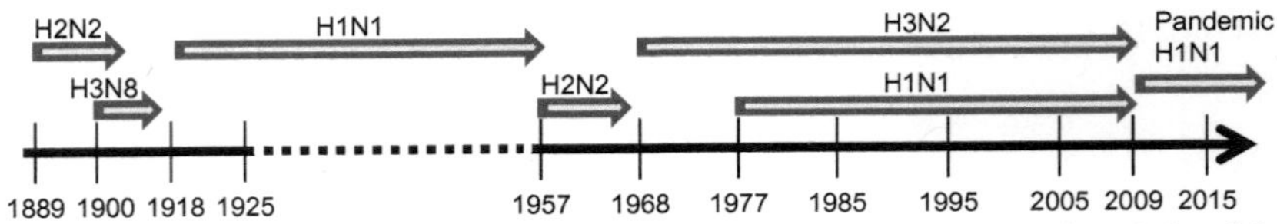

FIG. 1 Timeline of major influenza strains and the span of time they dominated the globe (Nicoll, 2010).

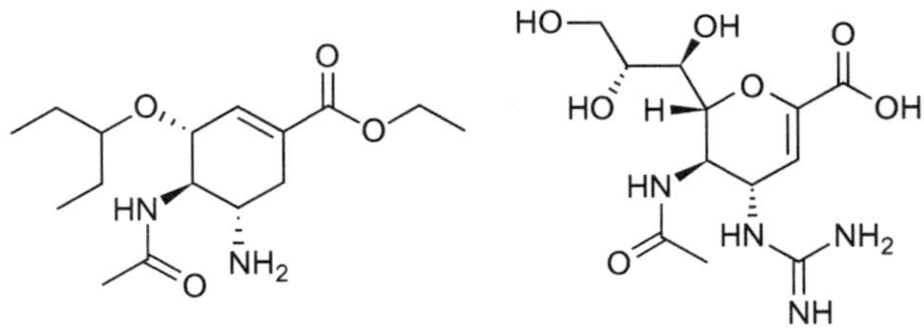

FIG. 2 Left, oseltamivir (Tamiflu), and right, zanamivir (Relenza), the two neuraminidase inhibitors available at the time of the 2009 influenza pandemic.

highly pathogenic influenza A (H5N1) virus circulating for years in Southeast Asia since the end of the 1990s, not the influenza A (H1N1) virus that rapidly emerged out of Mexico in April of 2009 (Team PIP, 2011a). The mortality rate of the H5N1 in Southeast Asia back in 2005 was over 70%. It was this frightening mortality statistic that sparked the development and refinement of national pandemic preparedness plans (Health UDo, 2005), which included the stockpiling of antivirals. Hence, the degree to which society was prepared for the 2009 pandemic was a direct consequence of the looming threat from H5N1.

The growing problem of influenza resistance to amantadine and rimantadine, the only influenza antivirals available prior to the new millennium (De Clercq, 2006), helped spur the development of two novel antivirals, Relenza and Tamiflu (Fig. 2). Relenza (zanamivir; developed by Biota Holdings and manufactured and distributed by GlaxoSmithKline) was the first antiviral in a new influenza antiviral class called neuraminidase inhibitors (NAI). It was approved for the treatment of influenza A in the United States in 1999. The second drug in this class was Tamiflu (oseltamivir carboxylate; developed by Gilead and distributed by Hoffmann-La Roche; Fig. 2), which was also approved for therapeutic use in the United States in 1999. For those nations choosing to develop an influenza antiviral stockpile, most chose Tamiflu (Fig. 3).

A major contributing reason for the preference for Tamiflu over Relenza, speaking directly for the United Kingdom's decision to stockpile Tamiflu in lieu of Relenza, was the familiarity people had with consuming capsules (pills), as opposed to the disc inhaler used to administer Relenza (Scientific

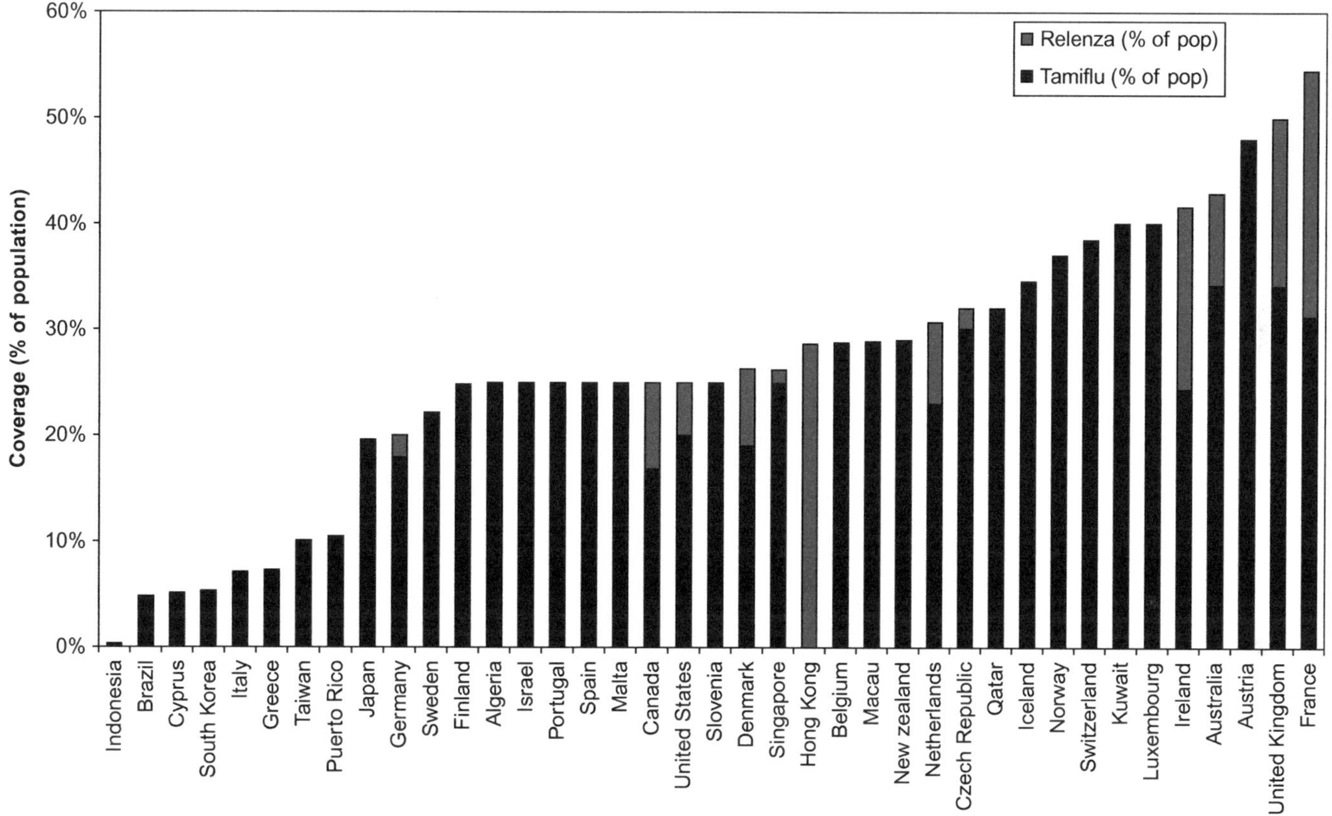

FIG. 3 Tamiflu *(black bar)* and Relenza *(grey bar)* stockpiles in national pandemic plans expressed in percent of population coverage (Singer and Schmitt, 2011; Singer et al., 2011).

Pandemic Influenza Advisory Committee, 2008). The risk of under- or misdosing using the disc inhaler was seen as a major barrier for Relenza, especially in those patients that find it hard to breathe and/or have asthma. However, Relenza was recognized as being more appropriate for children with renal insufficiency and pregnant women (Scientific Pandemic Influenza Advisory Committee, 2008). In addition, a stockpile of Relenza was seen as appropriate to prepare for the possible emergence of Tamiflu resistance. Relenza was also planned for use to prevent and treat influenza in frontline health workers at the earliest outset of Tamiflu resistance emergence (Scientific Pandemic Influenza Advisory Committee, 2008). On the balance of these factors, the United Kingdom acquired sufficient stockpiles of Tamiflu to treat 25% of the population, including prophylaxis and therapeutic prescription of the antiviral. This stockpile was further increased to cover up to 50% of the population at the outset of the pandemic (Fig. 3).

Antivirals were seen as a very important tool within pandemic preparedness plans for limiting morbidity and mortality. Preexposure prophylaxic dosing was implemented to prevent infections before individuals were exposed to an infected contact, and postexposure prophylaxis dosing was implemented to prevent infections after an individual has been exposed to an infected contact. The clinical benefits acquired from antiviral treatment were maximal when given within 48 h of symptom onset, most notably reducing the time to symptom alleviation by roughly half of 1 day, increasing to 1 day in cases where influenza has been confirmed by laboratory tests (Team PIP, 2013). Early treatment with NAIs compared with late treatment significantly reduced mortality in hospitalized patients by 62% (Team PIP, 2013). Antiviral prophylaxis was more than 80% effective in preventing laboratory-confirmed influenza when initiated within 48 h of initial contact (Team PIP, 2013). The evidence clearly supported the widespread use of antivirals for as much of the population as could be afforded.

The benefits to society could be further enhanced during a severe pandemic with the combined use of postexposure prophylaxis, school closures, and the use of antibiotics, thereby reducing the clinical attack rate by 50% (i.e., the percentage of the whole population that becomes infected at some stage during the pandemic) and case fatality ratio by 80% (i.e., the percentage of those who become ill and subsequently die as a direct result of the illness). In the event that the influenza virus acquired reduced susceptibility to Tamiflu, it was expected that this challenge would be met by increasing the dose (Scientific Pandemic Influenza Advisory Committee, 2008). Hence, dosing would increase from the recommended dose for treatment (75 mg, twice daily, for 5 days) or prophylaxis (1 dose of 75 mg, once daily, for 10 days) to a dose to be determined empirically during the pandemic. In the event that antiviral resistance was seen to emerge, both NAIs may have been coadministered (Scientific Pandemic Influenza Advisory Committee, 2008); however, this did not occur, fortunately.

Considerable speculation in the academic community was raised about the efficacy of NAIs and whether they reflected sufficient value for money (Jefferson and Doshi, 2014). However, in the "evidence review," the UK Department of Health writes "The research findings since the 2009-10 pandemic offer retrospective endorsement of the UK's use of NAIs and increase the weight of evidence in favour of stockpiling for a future pandemic (Team PIP, 2013)." Hence, it should be expected that in the absence of any further innovation in influenza antivirals, that similar stockpiling will exist ahead of future influenza pandemics (assuming there is sufficient warning as we were fortunate enough to have this last pandemic).

SECONDARY BACTERIAL INFECTIONS

An emerging point from the literature in the 2000s was the extent to which secondary bacterial infections (i.e., bacterial pneumonia) contributed to the overall death rate in the 1918 pandemic and the role it was likely to play in future pandemics (Brundage and Shanks, 2008; Gupta et al., 2008). Rates of secondary bacterial pneumonia among patients admitted to hospital with influenza have varied from 0.6% to 46% (Team PIP, 2011b). Influenza infection predisposes to secondary bacterial infection due to a combination of respiratory mucosal damage and a reduced capacity to mount an immune response to bacterial invasion (Team PIP, 2011b). Bacterial pneumonia can usually be easily treated with a course of antibiotics—a luxury that didn't exist until the latter half of the 20th century.

The evidence prior to the 2009 pandemic strongly supported the use of NAI antivirals as it was apparent from the literature that the likelihood of requiring antibiotics was reduced by 60% if antivirals are given within 48 h on the appearance of symptoms (Team PIP, 2013). Hence, an increased use of antivirals would lead to a potential reduction in bacterial pneumonia, which would translate into a reduction in antibiotic use as compared with a pandemic without the use of antivirals. The UK pandemic preparedness plans did not recommend stockpiling of antibiotics as it was felt that these drugs, which are already in high use, were already in plentiful supply (Singer et al., 2011; Lim, 2007).

ENVIRONMENTAL RELEVANCE OF MEDICAL RESPONSE

Prior to the pandemic, influenza antivirals were not frequently prescribed; for most, influenza is a self-limiting disease. The decision to use these novel drugs during the pandemic presented a novel ecotoxicological challenge (Singer et al., 2007). Determining the quantity of antiviral to be used and developing scenarios for its potential effects within the waste water system and the wider environment became the focus of subsequent research in the late 2000s to the present day.

Unlike antivirals, antibiotics are in constant use throughout the world. One might conclude that given the widespread use of antibiotics, an increase in their prescription on the order of 1%–5% might pose a negligible increased ecotoxicological risk to sewage work operation and the ecosystem services of the wider river environment. On the other hand, given that the primary mode of action of antibiotics is the inhibition and death of bacteria, it would be reasonable to assume that there is a concentration of antibiotics, above which, one might witness the loss of microorganisms performing essential functions in sewage works and the wider environment (i.e., a tipping point). The loss of these key bacteria from increased antibiotic uses during a pandemic was (and remains) the main ecotoxicological concern from pandemic usage of antibiotics.

Most antibiotics are persistent, which is to say that most of the antibiotics that are consumed are excreted in their biologically active form (Table 1). The same is essentially true for antivirals (Table 1), with the only exception being that Tamiflu is a prodrug. Upon consumption, approximately 80% of Tamiflu is converted to the active antiviral (oseltamivir carboxylate), with the remaining 20% leaving in the feces unchanged. There is considerable uncertainty with regard to the removal rate of any one drug in any one sewage works at any one time. It is expected that both antibiotics and antivirals would reach sewage works and pass right through them with relatively minimal loss (Jain et al., 2013). Antivirals, in particular, are seen as highly resistant to biodegradation (Prasse et al., 2010) and photodegradation (Gonçalves et al., 2011; Fick et al., 2007). It is realistic to assume that antivirals could persist in the environment for sufficiently long enough to enter the drinking water system (Wang et al., 2016; Drinking Water Inspectorate, n.d.). For most modeling purposes and for predicting a realistic worst-case scenario, it is common practice to assume that sewage works and the wider environment do not result in significant loss in the drugs within the first 24 h.

Several questions arise when considering the use and fate of drugs used during a pandemic: Do the drugs persist in the sewage system? To what extent do they partition to the sludge or remain in the aqueous phase? Do these drugs have an impact on the operation of the sewage works? Do these drugs inhibit the growth and/or function of microorganisms within the receiving rivers that are important for its integrity and function? Are there acute or chronic toxicity issues relating to the exposure of aquatic organisms to environmentally relevant concentrations of these drugs? Do the drugs increase the prevalence of antiviral or antibiotic resistance in the environment with potential human health impacts? Do the drugs pose a threat to drinking water quality? Very few of these questions were addressed prior to the 2009 pandemic (Singer et al., 2008).

Evidence has emerged in the literature to suggest a mechanism by which microbial biofilms could indeed be disrupted by exposure to Tamiflu within sewage works (Singer et al., 2008; Parker et al., 2009; Gut et al., 2011;

TABLE 1 Percentage of Parent Compound Excreted in the Feces and Urine (Singer and Schmitt, 2011)

Common Name	Total Excretion (%)[a]
Tetracyclines	
Tetracycline	91
Doxycycline	80
Oxytetracycline	35
Chlortetracycline	20
β-Lactams	
Floxacillin (flucloxacillin)	80
Amoxicillin	75
Clavulanic acid	38
Cephalexin	100
Cefuroxime	95
Cefaclor	85
Cefotaxime	61
Ceftriaxone	100
Imipenem	70
Sulfonamide and trimethoprim	
Trimethoprim	100
Sulfamethoxazole	100
Imidazoles	
Metronidazole	100
Macrolides	
Erythromycin	100
Roxithromycin	60
Azithromycin	85
Clarithromycin	55
Clindamycin	14
Fluoroquinolones	
Ciprofloxacin	100
Levofloxacin	96

Continued

TABLE 1 Percentage of Parent Compound Excreted in the Feces and Urine (Singer and Schmitt, 2011)—Cont'd

Common Name	Total Excretion (%)
Norfloxacin	40
Ofloxacin	85
Moxifloxacin	100
Influenza antivirals	
Tamiflu-prodrug	20
Tamiflu-active[2]	80
Zanamivir	100
Amantadine	100
Rimantadine	100

[a]Total excretion includes the biologically active chemicals found in the urine and feces. In most cases, this is the same as the parent chemicals.
[2]80% of oseltamivir phosphate is metabolised to the active antiviral. The remaining 20% of the parent chemical and all of the active antiviral oseltamivir carboxylate are excreted.

Slater et al., 2011). Hence, the risk of sewage work failure from exposure to elevated levels of Tamiflu during an influenza pandemic remains theoretically possible (Slater et al., 2011). The extent to which sewage works "fail" remains the next question. Total failure to treat raw sewage would mean the release of untreated sewage into the receiving rivers, some of which are the source of drinking water for major cities, such as the River Thames is for London. Such a release would result in the loss of fish and other sensitive aquatic lives downstream of any failing sewage works. It remains unclear as to the likelihood of pandemic quantities of drug resulting in incomplete treatment of sewage before discharge into the river.

Antibiotics are designed to have a biostatic or lethal effect on a wide range of microorganisms. They are thus capable of inhibiting the growth and function of microorganisms in sewage works (Singer and Schmitt, 2011; Singer et al., 2011). The degree to which this inhibition could and would be observed remains unclear as there are numerous complicating factors that make predicting inhibition of microorganisms in sewage works based on antibiotic concentrations extremely difficult to determine. All sewage works receive relatively high concentrations of a very broad mix of antibiotics all the time (Singer and Schmitt, 2011; Singer et al., 2014). This constant influx of these drugs and antibiotic-resistant microorganisms into sewage works contributes to the resilience of sewage works as much as it can contribute to the risk of its failure.

The release of the active antiviral of Tamiflu (oseltamivir carboxylate) into the receiving rivers of the world had the potential to select for antiviral resistance in wildfowl that cohabits sewage-impacted rivers (Singer et al., 2007; Fick et al., 2007). Wildfowl is attracted to the nutrient-rich, warm sewage outflows, exposing them to relatively high environmental concentrations of Tamiflu. The colocation of wildfowl and Tamiflu means that there is a risk of the avian influenza virus, naturally found in many wildfowl, to develop Tamiflu resistance. The concentration of Tamiflu predicted to be present in highly sewage-impacted rivers, such as in the United Kingdom, was expected to be sufficiently high to select for the Tamiflu-resistance gene in avian influenza, which would ultimately increase the risk of the next pandemic influenza strain containing the Tamiflu-resistance gene (Gillman et al., 2015; Orozovic et al., 2014). Application of ozonation to the sewage effluent would potentially reduce the load of antiviral released into the environment (Azuma et al., 2015); however, relatively few sewage treatment plants utilize such a treatment option, particularly in the United Kingdom, owing to the high cost and energy requirements. Few options for mitigation of the identified risks were available at the time, and little has changed since the pandemic.

In summary, the 2009 influenza A (H1N1)pdm09a virus, as it is officially called, generated a relatively small number of fatalities as compared with severe pandemics like the 1918 "Spanish flu," which meant that the medical response was proportionately lower than would have been expected in a moderate or severe influenza pandemic (Singer et al., 2011). Hence, few negative effects to sewage work function and ecosystem services were apparent. The lack of evidence of an ecotoxicological effect might have as much to do with the fact that the research community were not looking/monitoring for such effects, as it has to do with the low pathogenicity of the pandemic influenza strain.

PREPARING FOR THE NEXT PANDEMIC

The influenza pandemic of 2009–10 remains the only case study of a pandemic for which large quantities of pharmaceuticals were deployed worldwide within a very short time frame. One would be forgiven for thinking that pandemics are rare; however, at any one time, there is at least one ongoing. Depending on how one characterizes a pandemic, there are several right now! There are many well-known disease-causing pathogens that cause epidemics or pandemics, such as cholera, typhoid, tuberculosis, malaria, HIV, MERS, SARS, Ebola, chikungunya, and Zika virus.

Some pathogens are not prone to go pandemic for reasons that will be explained; others, despite being perfectly suited for a pandemic, have yet to go beyond an epidemic level. The following discussion is intended to raise awareness of the epidemic and pandemic risks of various high-profile human pathogens—speculate on how a pandemic might develop—and highlight some of the current treatment options.

Cholera is caused by a bacterium, *Vibrio cholerae*, and transmitted through unsafe water, most often contaminated with feces from others suffering or carrying the bacterium. There have been seven separate pandemics since 1816 and several notable outbreaks of cholera since 1991. Treatment for cholera can include antibiotics (doxycycline or azithromycin), but it is often not necessary if rehydration strategies are employed. It is unlikely that cholera will be anything more than an epidemic looking forward, as the availability of antibiotics is no longer limiting and suboptimal hygiene and water treatment standards can often be improved given adequate infrastructure. The biggest risk from cholera typically follows major losses in critical infrastructure, such as loss of water and sanitation following natural disasters, such as earthquakes, severe storms, or tsunami.

Typhoid fever is caused by a bacterium *Salmonella typhi* and transmitted by consuming contaminated food or water and through person-to-person contact. Treatment includes doxycycline, ciprofloxacin, or a number of other antibiotics. Typhoid has a similar etiology as cholera and thus remains endemic and limited to areas of poor sanitation and water treatment (Karkey et al., 2016).

Tuberculosis (Tb) is currently found in every country in the world making it a global pandemic. It is estimated that one-third of the world's population is infected with *Mycobacterium tuberculosis*, which represents the leading infectious cause of death worldwide (approximately 1.5 million per year). Tb is spread from person to person through tiny droplets released in coughs and sneezes, a mechanism similar to influenza. With tuberculosis, one must take antibiotics for at least 6–9 months. A sensitive strain may be treated with only one antibiotic, while a resistant strain will require several antibiotics all at once. These include isoniazid, rifampin, ethambutol, pyrazinamide, and fluoroquinolones. The emergence of multidrug-resistant tuberculosis is one of the highest priorities in international public health. With the continued loss of antibiotic efficacy as resistance spreads, there is a real threat of multidrug-resistant Tb becoming an even more severe public health threat.

Malaria is caused by *Plasmodium*, a parasite transmitted by the bite of an infected *Anopheles* mosquito. The World Health Organization estimates that there were approximately 198 million cases of malaria and 584,000 deaths, mostly among children under the age of five, just in 2013. Depending on the parasite, the antimalarial drugs include chloroquine, quinine sulfate, hydroxychloroquine, mefloquine, and a combination of atovaquone and proguanil (Malarone). Resistance to chloroquine has rendered the drug ineffective. The recommended treatment for malaria is currently a combination of antimalarial medications that include an artemisinin and either mefloquine, lumefantrine, or sulfadoxine/pyrimethamine. Quinine along with doxycycline may be used if an artemisinin is not available. Malaria is limited to areas where the Anopheles mosquito can live and reproduce, making the wider spread of this disease somewhat limited; however, the range of the mosquito vector can expand in a changing climate, increasing the risk to regions that might have previously been free of malaria.

Human immunodeficiency virus (HIV) causes acquired immunodeficiency syndrome (AIDS) and is currently in a pandemic state. As of 2012, approximately 35.3 million people are living with HIV globally, with 1.8 million deaths from HIV/AIDS in 2010. HIV is transmitted from person to person through sexual contact, through blood, and during pregnancy to the child. There are several anti-HIV drugs: nonnucleoside reverse transcriptase inhibitors, which disable the protein needed by HIV to replicate (efavirenz, etravirine, and nevirapine); nucleoside or nucleotide reverse transcriptase inhibitor, which forces the HIV to make faulty copies of itself (abacavir and combinations: emtricitabine-tenofovir (Truvada) and lamivudine-zidovudine (Combivir)); protease inhibitors, which disable required protein production (atazanavir, darunavir, fosamprenavir, and indinavir); entry or fusion inhibitors that block the HIV's entry into the target CD4 cells (enfuvirtide and maraviroc); and integrase inhibitors, which prevent HIV from integrating its genome into CD4 cells (raltegravir, elvitegravir, and dolutegravir). The use of these drugs is sufficiently high that they are now among the most frequently recovered drugs in waste water of countries heavily impacted by HIV/AIDS and reach concentrations over 100 μg/L (K'Oreje et al., 2012, 2016). The implications of very high concentrations of antiviral in the environment and potentially drinking water have yet to be studied.

Middle East respiratory syndrome coronavirus (MERS-CoV) was first discovered in Saudi Arabia in 2012; MERS-CoV severely disrupts respiratory functions, leading to pneumonia, shortness of breath, coughing, and fever. As of July 2015, MERS-CoV cases have been reported in over 21 countries. Based on recent World Health Organization data, about 40% of reported cases have resulted in death. There is no curative treatment for MERS-CoV. Symptomatic treatment includes convalescent plasma-containing MERS-CoV antibodies (i.e., blood from patients that have raised antibodies to MERS-CoV) (Publich Health England, 2015) and lopinavir (ABT-378), an antiretroviral of the protease inhibitor class that is nearly 100% bound to plasma proteins. Coadministering lopinavir with ritonavir is known as highly active antiretroviral therapy (HAART), which is also used to treat HIV/AIDS. The fact that MERS-CoV is easily transmissible and has limited treatment options makes it a serious pandemic risk.

Severe acute respiratory syndrome (SARS) is a viral respiratory disease caused by the SARS coronavirus that has symptoms nearly identical to influenza. SARS is a zoonosis and likely originated from infected bats that are sold as food in China, but it has since been found in several other mammals. SARS infection comes with a high risk of secondary infections such as pneumonia. SARS only emerged in 2002 and in 9 months infected 8273 people causing 775 deaths in 15 countries. Antibiotics are ineffective, as SARS is a viral disease, but much like in the treatment of influenza, antibiotics would be employed to treat bacterial pneumonia. Much like MERS-CoV, SARS maintains a high pandemic risk.

Ebola is a filovirus that causes severe hemorrhagic fever with a fatality rate often near 90%. Transmission occurs through contact with infected body fluids such as blood. A massive epidemic of Ebola spread through three West African countries, Guinea, Liberia, and Sierra Leone, in early 2014 resulting in 28,602 cases and 11,301 deaths, as of early 2016. Recent research has indicated the potential for a broad-spectrum antiviral for such RNA viruses. An RNA polymerase-inhibiting molecule called BCX4430 (Immucillin-A), an adenosine analogue, is incorporated into growing viral RNA strands, preventing further RNA synthesis and therefore viral replication. The biocompatibility of this drug makes it potentially at risk of impacting environmental organisms if the drug was to be excreted intact. The ease by which Ebola is transmitted and the difficulty in treating infections make Ebola a serious pandemic threat.

Chikungunya is an RNA virus of the *Alphavirus* genus that causes an estimated 3 million infections each year. The disease causes sudden onset of fever with accompanying joint pains that can last weeks to years. Fewer than 1 in 1000 of infected individuals die from the disease, mostly the elderly and those with underlying chronic medical problems. The virus is most frequently transmitted from human to human within the blood carried by infected *Aedes albopictus* or *Aedes aegypti* mosquito. There are a number of animal reservoirs of the virus, including monkeys, birds, cattle, and rodents. This is in contrast to dengue, for which primates are the only hosts. There is no treatment for chikungunya other than for the fever and joint pain. There remains no vaccine for the disease. As the transmission of chikungunya requires a mosquito vector, it will remain constrained to the geographic range of the mosquito vector; however, it can still reach pandemic levels as the range of the vector is quite substantial (Kraemer et al., 2015).

Dengue is an RNA virus from the genus *Flavivirus*. It is transmitted via *A. aegypti* and *A. albopictus,* identical to chikungunya from human to human in the blood meal. Dengue can be found in other primates and was the original source of the dengue virus. The symptoms of dengue include high fever, headache, vomiting, muscle and joint pains, and a characteristic skin rash. The incidence of dengue has increased fourfold in the last 20 years. There are no licensed therapeutics available for treatment of dengue. Several antivirals have been investigated for their efficacy in treating dengue, including balapiravir, celgosivir, and chloroquine; however, none have shown beneficial effects on viremia or clinical outcome (Yacoub et al., 2016). The inability to control the day-biting anthropophilic mosquitoes is the main reason for the global spread of dengue (Yacoub et al., 2016). Promising avenues for breaking the transmission cycle of dengue have been to infect the mosquito with an intracellular bacterium, which reduces viral replication in the mosquitoes (Yacoub et al., 2016). The virus' reliance on a geographically restricted mosquito vector limits infection to those who live and visit tropical environments (Kraemer et al., 2015). Despite that limitation on its range, dengue is common in more than 110 countries and infects 50–528 million people worldwide a year, leading to half a million hospitalizations.

Zika is an RNA virus of the genus *Flavivirus*, which first emerged from the Zika forest of Uganda in 1948. It more recently emerged in 2014 in French Polynesia to take center stage among the most serious emerging public health issues in the Americas after it was transmitted to South America. The virus is transmitted by the same mosquito as the chikungunya and dengue viruses, typically causing relatively minor ailments such as rash, fever, and headaches within a week of being bitten by the mosquito. The virus has spread to over 17 countries in the Americas by early 2016. The possible link to microcephaly in unborn fetuses has put this relatively minor viral infection on center stage. There are no vaccines for the virus. No antivirals have been reported to be effective for Zika, but one must assume that the experimental treatments used to treat dengue and chikungunya will be tested for their efficacy in this related virus. Zika virus is arguably at the cusp of being considered a global pandemic risk, with increasing cases of its transmission via sexual contact, thereby forgoing the otherwise geographically limited mosquito vector.

THE SPECIAL CASES OF MOSQUITO TRANSMISSION

The fact that Zika, chikungunya, dengue, and malaria are spread by mosquitoes introduces another strategy for combating the disease, the use of insecticides. Unlike pharmaceuticals, which must be prescribed in controlled clinical settings, insecticides can be used in truly staggering quantities in very large areas. It remains to be seen what the implications to human health and the environment will be from the recent campaigns against the spread of Zika virus, particularly in the run up to the 2016 Summer Olympic Games. Lessons from history, that is, DDT, indicate that there will indeed be consequences to its widespread use, to both the virus, human health, and the environment.

Those living and traveling in the mosquito vector's range are urged to use the strongest mosquito repellant both day and night. Hence, it is to be expected that the amount of DEET and picaridin used in these areas will increase substantially. It remains largely unstudied as to the consequences of widespread use of repellant. Will the repellants select for insecticide-tolerant mosquitoes? What is known is that increased use of insect repellants, such as DEET, will greatly increase in concentration in the environment—as it is already a routinely measured pollutant in rivers around the globe (Costanzo et al., 2007). Further consideration is the routine exposure of air travelers in mosquito-infested areas to insecticides as part of aircraft disinfection (i.e., permethrin, phenothrin, and etofenprox) (World Health Organization, 2013).

THE KNOWN AND UNKNOWN UNKNOWNS

Currently, 1415 pathogens are reported to cause disease in humans, including 538 bacteria and rickettsia, 307 fungi, 66 protozoa, and 287 helminths (Taylor et al., 2001). The most abundant source of human emerging pathogens comes from the organisms classified as viruses (44%) followed by bacteria (30%) (Jones et al., 2008). The majority of pathogen species causing disease in humans are zoonotic (61% of the total) (Taylor et al., 2001). By contrast, 616 pathogens have been identified in livestock and 374 in domestic carnivores (Cleaveland et al., 2001). Although morbidity and mortality statistics are available for some human pathogens, it is more often the case that pathogens remain completely unquantified with regard to their impact on human health (Taylor et al., 2001). This reality stems from the fact that (1) most viral diseases are treated symptomatically owing to the challenge and cost of demonstrating the causative organism, (2) many illnesses are self-limiting and do not undergo analysis for the causative agent, and (3) the health-care profession will typically look for the "known knowns" and not seek to screen beyond what is expected. Without thorough unbiased surveillance, emerging pathogens will continue to surprise public health services.

PANDEMICS OF CHRONIC DISEASES

It is also important to realize that pandemics do not have to be infectious or contagious diseases; they may take the form of chronic diseases, such as heart disease, diabetes, cancer, and obesity. The World Health Organization is monitoring the global pandemic of heart disease, which is being combated with the use of statins, among a wide range of other drugs. Statins are used for prevention of all the main cardiovascular events, reducing the risk of acute myocardial infarction, cardiovascular revascularization, stroke, cardiovascular mortality, and all-cause mortality. Statins are now the most prescribed drug in the world (Hobbs et al., 2016). Unlike episodic pathogen pandemics, noncommunicable disease pandemics pose a constant (and increasing) pollution threat to the environment and human health.

CONCLUSION

It is clear that society has much to gain from a well-coordinated medical response to a pandemic. It is likely that most will agree that a thorough and coordinated medical response to a pandemic is preferred over the application of a precautionary approach because of the potentially real effects of these chemicals on critical infrastructure and the wider environment. However, even if we are (nearly) all agreed that humans should be prioritized over the wider environment, this does not preclude society from having an understanding of what is *actually* at risk and determining whether anything can be done to mitigate these perceived threats.

REFERENCES

Alves Galvao, M.G., Rocha Crispino Santos, M.A., Alves da Cunha, A.J., 2014. Amantadine and rimantadine for influenza a in children and the elderly. Cochrane Database Syst. Rev. 11, CD002745.

Azuma, T., Nakada, N., Yamashita, N., Tanaka, H., 2015. Prediction, risk and control of anti-influenza drugs in the Yodo River Basin, Japan during seasonal and pandemic influenza using the transmission model for infectious disease. Sci. Total Environ. 521–522, 68–74.

Brundage, J.F., Shanks, G.D., 2008. Deaths from bacterial pneumonia during 1918–19 influenza pandemic. Emerg. Infect. Dis. 14 (8), 1193–1198.

Cleaveland, S., Laurenson, M.K., Taylor, L.H., 2001. Diseases of humans and their domestic mammals: pathogen characteristics, host range and the risk of emergence. Philos. Trans. R. Soc. Lond. B Biol. Sci. 356 (1411), 991–999.

Costanzo, S.D., Watkinson, A.J., Murby, E.J., Kolpin, D.W., Sandstrom, M.W., 2007. Is there a risk associated with the insect repellent DEET (*N*,*N*-diethyl-*m*-toluamide) commonly found in aquatic environments? Sci. Total Environ. 384 (1–3), 214–220.

Davies, W.L., Grunert, R.R., Haff, R.F., McGahen, J.W., Neumayer, E.M., Paulshock, M., et al., 1964. Antiviral activity of 1-adamantanamine (amantadine). Science 144 (3620), 862–863.

De Clercq, E., 2006. Antiviral agents active against influenza A viruses. Nat. Rev. Drug Discov. 5 (12), 1015–1025.

Drinking Water Inspectorate, n.d. Desk Based Review of Current Knowledge on Pharmaceuticals in Drinking Water and Estimation of Potential Levels (Defra Project Code: CSA 7184/WT02046/DWI70/2/213). http://googl/nkBVlP2007.

Fick, J., Lindberg, R.H., Tysklind, M., Haemig, P.D., Waldenstrom, J., Wallensten, A., et al., 2007. Antiviral oseltamivir is not removed or degraded in normal sewage water treatment: implications for development of resistance by influenza a virus. PLoS One 2(10), e986.

Gillman, A., Muradrasoli, S., Söderström, H., Holmberg, F., Latorre-Margalef, N., Tolf, C., et al., 2015. Oseltamivir-resistant influenza A (H1N1) virus strain with an H274Y mutation in neuraminidase persists without drug pressure in infected mallards. Appl. Environ. Microbiol. 81 (7), 2378–2383.

Gonçalves, C., Pérez, S., Osorio, V., Petrovic, M., Alpendurada, M.F., Barceló, D., 2011. Photofate of oseltamivir (Tamiflu) and oseltamivir carboxylate under natural and simulated solar irradiation: kinetics, identification of the transformation products, and environmental occurrence. Environ. Sci. Technol. 45 (10), 4307–4314.

Gupta, R.K., George, R., Nguyen-Van-Tam, J.S., 2008. Bacterial pneumonia and pandemic influenza planning. Emerg. Infect. Dis. 14 (8), 1187–1192.

Gut, H., Xu, G., Taylor, G.L., Walsh, M.A., 2011. Structural basis for *Streptococcus pneumoniae* NanA inhibition by influenza antivirals zanamivir and oseltamivir carboxylate. J. Mol. Biol. 409 (4), 496–503.

Health UDo, 2005. UK Health Departments' Influenza Pandemic Contingency Plan. http://wwwdhgovuk/assetRoot/04/10/44/37/04104437pdf.

Hobbs, F.R., Banach, M., Mikhailidis, D.P., Malhotra, A., Capewell, S., 2016. Is statin-modified reduction in lipids the most important preventive therapy for cardiovascular disease? A pro/con debate. BMC Med. 14 (1), 1–8.

Jain, S., Kumar, P., Vyas, R.K., Pandit, P., Dalai, A.K., 2013. Occurrence and removal of antiviral drugs in environment: a review. Water Air Soil Pollut. 224(2), 1410.

Jefferson, T., Doshi, P., 2014. Multisystem failure: the story of anti-influenza drugs. BMJ 348, g2263.

Jones, K.E., Patel, N.G., Levy, M.A., Storeygard, A., Balk, D., Gittleman, J.L., et al., 2008. Global trends in emerging infectious diseases. Nature 451 (7181), 990–993.

Karkey, A., Jombart, T., Walker, A.W., Thompson, C.N., Torres, A., Dongol, S., et al., 2016. The ecological dynamics of fecal contamination and *Salmonella typhi* and *Salmonella paratyphi A* in municipal Kathmandu drinking water. PLoS Negl. Trop. Dis. 10(1), e0004346.

K'Oreje, K.O., Demeestere, K., De Wispelaere, P., Vergeynst, L., Dewulf, J., Van Langenhove, H., 2012. From multi-residue screening to target analysis of pharmaceuticals in water: development of a new approach based on magnetic sector mass spectrometry and application in the Nairobi River basin, Kenya. Sci. Total Environ. 437, 153–164.

K'Oreje, K.O., Vergeynst, L., Ombaka, D., De Wispelaere, P., Okoth, M., Van Langenhove, H., et al., 2016. Occurrence patterns of pharmaceutical residues in wastewater, surface water and groundwater of Nairobi and Kisumu city, Kenya. Chemosphere 149, 238–244.

Kraemer, M.U.G., Sinka, M.E., Duda, K.A., Mylne, A.Q.N., Shearer, F.M., Barker, C.M., et al., 2015. The global distribution of the arbovirus vectors *Aedes aegypti* and *Ae. albopictus*. eLife 4, e08347.

Lim, W.S., 2007. Pandemic flu: clinical management of patients with an influenza-like illness during an influenza pandemic. Thorax 62 (Suppl. 1), 1–46.

Nicoll, A., 2010. A new decade, a new seasonal influenza: the Council of the European Union Recommendation on seasonal influenza vaccination. Euro Surveill. 15(1), 19458.

Orozovic, G., Orozovic, K., Järhult, J.D., Olsen, B., 2014. Study of oseltamivir and zanamivir resistance-related mutations in influenza viruses isolated from wild mallards in Sweden. PLoS One 9(2), e89306.

Parker, D., Soong, G., Planet, P., Brower, J., Ratner, A.J., Prince, A., 2009. The NanA neuraminidase of *Streptococcus pneumoniae* is involved in biofilm formation. Infect. Immun. 77 (9), 3722–3730.

Prasse, C., Schlüsener, M.P., Schulz, R., Ternes, T.A., 2010. Antiviral drugs in wastewater and surface waters: a new pharmaceutical class of environmental relevance? Environ. Sci. Technol. 44 (5), 1728–1735.

Publich Health England, 2015. Treatment of MERS-CoV: Information for Clinicians. Clinical Decision-Making Support for Treatment of MERCS-CoV Patients. https://wwwgovuk/government/uploads/system/uploads/attachment_data/file/459835/merscov_for_clinicians_sept2015pdf.

Scientific Pandemic Influenza Advisory Committee, 2008. SPI Statement 2008/01 on Neuraminidase Stockpiling. Available from: http://webarchivenationalarchivesgovuk/20130107105354/http://wwwdhgovuk/prod_consum_dh/groups/dh_digitalassets/@dh/@ab/documents/digitalasset/dh_095702pdf. Accessed 16 July 2008.

Singer, A.C., Colizza, V., Schmitt, H., Andrews, J., Balcan, D., Huang, W.E., et al., 2011. Assessing the ecotoxicologic hazards of a pandemic influenza medical response. Environ. Health Perspect. 119 (8), 1084–1090.

Singer, A.C., Howard, B.M., Johnson, A.C., Knowles, C.J., Jackman, S., Accinelli, C., et al., 2008. Meeting report: risk assessment of Tamiflu use under pandemic conditions. Environ. Health Perspect. 116 (11), 1563–1567.

Singer, A.C., Jarhult, J.D., Grabic, R., Khan, G.A., Lindberg, R.H., Fedorova, G., et al., 2014. Intra- and inter-pandemic variations of antiviral, antibiotics and decongestants in wastewater treatment plants and receiving rivers. PLoS One 9(9), e108621.

Singer, A.C., Nunn, M.A., Gould, E.A., Johnson, A.C., 2007. Potential risks associated with the proposed widespread use of Tamiflu. Environ. Health Perspect. 115 (1), 102–106.

Singer, A.C., Schmitt, H., 2011. Antibiotic use during an influenza pandemic: downstream ecological effects and antibiotic resistance. In: Antimicrobial Resistance in the Environment. John Wiley & Sons, Inc, Hoboken, NJ, pp. 503–537.

Slater, F.R., Singer, A.C., Turner, S., Barr, J.J., Bond, P.L., 2011. Pandemic pharmaceutical dosing effects on wastewater treatment: no adaptation of activated sludge bacteria to degrade the antiviral drug oseltamivir (Tamiflu®) and loss of nutrient removal performance. FEMS Microbiol. Lett. 315 (1), 17–22.

Taubenberger, J.K., Morens, D.M., 2010. Influenza: the once and future pandemic. Public Health Rep. 125 (Suppl. 3), 16–26.

Taylor, L.H., Latham, S.M., Woolhouse, M.E.J., 2001. Risk factors for human disease emergence. Philos. Trans. R. Soc. Lond. B Biol. Sci. 356 (1411), 983–989.

Team PIP, 2011a. Scientific Summary of Pandemic Influenza & Its Migigation: Scientific Evidence Base Review. https://wwwgovuk/government/uploads/system/uploads/attachment_data/file/215666/dh_125333pdf.

Team PIP, 2011b. Use of Antibiotics in an Influenza Pandemic: Scientific Evidence Base Review. https://wwwgovuk/government/uploads/system/uploads/attachment_data/file/215669/dh_125424pdf.

Team PIP, 2013. Use of Antiviral in an Influenza Pandemic: Scientific Evidence Base Review. https://wwwgovuk/government/uploads/system/uploads/attachment_data/file/316206/Use_of_antivirals_Evidence_Reviewpdf.

Wang, H., Wang, N., Wang, B., Zhao, Q., Fang, H., Fu, C., et al., 2016. Antibiotics in drinking water in Shanghai and its contribution to antibiotic exposure of school children. Environ. Sci. Technol. 50 (5), 2692–2699.

World Health Organization, 2013. International Programme on Chemical Safety. Aircraft Disinsection Insecticides. http://wwwwhoint/ipcs/publications/ehc/ehc243pdf?ua=1.

Yacoub, S., Mongkolsapaya, J., Screaton, G., 2016. Recent Advances in Understanding Dengue [version 1; referees: 3 approved].

Chapter 10

Management of Waste from the Health-Care Sector

Amy Wolff
Melbury Environmental Ltd

THE NEED FOR HEALTHCARE WASTE MANAGEMENT

Alongside massive advances in medical science and lifesaving treatments, a highly sophisticated waste sector must develop to manage the range of wastes created by these activities. In addition to the large swathes of household-type wastes produced by health-care establishments, wastes are generated that are highly offensive in appearance or smell or that pose a significant and hazardous threat to human health or the environment. Contaminants can include infectious microorganisms; biologically active pharmaceutical ingredients, some of which are cytotoxic or cytostatic and contain genotoxic properties (Chartier et al., 2014); infected or contaminated anatomical wastes; contaminated sharps such as hypodermic needles or surgical instruments; hazardous chemical waste; mercury; and even radioactive materials (Department of Health, 2013). During treatment, the waste industry can go onto produce harmful secondary emissions and new wastes that possess new hazards. At every level of the chain, from waste handling to transport, treatment, and disposal, health-care waste needs careful management to keep the environment and people safe.

DEFINING "HEALTHCARE" WASTE

The World Health Organization estimates 85% of waste outputs from health-care facilities are comparable with household-type wastes (Chartier et al., 2014) including things like batteries, waste electronic and electrical equipment (WEEE), dry recyclables, food waste, construction and demolition waste, and residual nonrecyclable, nonhazardous waste (Department of Health, 2013). This means that only a small percentage are wastes that have distinctly arisen in a health-care environment and it is these that this chapter

Environmental Contaminants, Vol. 11. https://doi.org/10.1016/B978-0-444-63857-1.00010-3

focuses on. Worldwide health-care wastes are produced not only by hospitals, dental practices, or national or private health services but also by private clinics, podiatrists/chiropodists, tattooists, nursing and care homes, alternative practices (such as acupuncturists or osteopaths), veterinary practices, research and laboratory facilities, pharmaceutical plants, and pharmacies (The Chartered Institute of Wastes Management (CIWM), 2014).

UK Definition of Healthcare Waste

As the definition of health-care waste is so complex (as almost any kind of waste can come from a health-care establishment), it varies hugely across the world, so this chapter has honed in on the United Kingdom by way of example. In the United Kingdom, the different types of health-care waste are broken down into wastes that are "hazardous" or "nonhazardous" and "clinical" or "nonclinical," both of which ultimately impact the regulatory controls that need to apply (Department of Health, 2013). The definition of "hazardous" across Europe and therefore the United Kingdom (until it is clear how the United Kingdom's exit from the EU will impact its own regulatory regime) is provided by the European Waste Framework Directive 2008/98/EC (European Parliament, 2008), which provides 15 groups of "hazards" that a waste could possess (not just health-care waste). These include categories such as mutagenic, explosive, infectious, or ecotoxic. Any waste that possesses one of these 15 properties at levels that could cause harm is defined as hazardous. For years, UK legislation has defined clinical waste as (Blenkharn, 2015).

- *...any waste which consists wholly or partly of human or animal tissue, blood or other bodily fluids, excretions, drugs or other pharmaceutical products, swabs or dressings, syringes, needles, or other sharp instruments, being waste which unless rendered safe may provide hazardous to any person coming into contact with it; and*
- *any waste arising from medical, nursing, dental, veterinary, pharmaceutical or similar practice, investigation, treatment, care, teaching or research, or the collection of blood for transfusion, being waste which may cause infection to any person coming into contact with it (Blenkharn, 2015, p. 2).*

In England and Wales, recently updated legislation in the form of the Controlled Waste (England and Wales) Regulations 2012 (The Stationary Office, 2012a) has amended this definition to say:

> "clinical waste" means waste from a health-care activity (including veterinary health care) that
> **(a)** contains viable microorganisms or their toxins that are known or reliably believed to cause disease in humans or other living organisms,
> **(b)** contains or is contaminated with a medicine that contains a biologically active pharmaceutical agent,

(c) is a sharp or a body fluid or other biological materials (including human and animal tissue) containing or contaminated with a dangerous substance and waste of a similar nature from a non-health-care activity (the Controlled Waste (England and Wales) Regulations 2012, Schedule 1, Regulation 4, p. 3) (The Stationary Office, 2012a).

For the most part in the United Kingdom, clinical means "hazardous,"[a] but the definition of clinical does not cover all the wastes arising from a distinctly health-care environment. Other "nonclinical" wastes that arise in a health-care environment might include dental amalgam and X-ray fixers and developers (which are also hazardous wastes), as well as wastes that are simply offensive in appearance or smell such as incontinence pad waste but are not contaminated with infectious microorganisms, medicines, or chemicals (Department of Health, 2013). "Classifying a waste using the European Waste Catalogue" section further explains how wastes are categorized in the EU and the United Kingdom.

HEALTHCARE WASTE MANAGEMENT IN ECONOMICALLY LESS DEVELOPED COUNTRIES

The problems relating to safe management of health-care waste are most prominent in economically less developed countries due to the lack of resources to both physically manage the waste and develop a legal system to ensure everyone abides by the rules of its safe management. Today, "over half the world's population are now at risk from occupational, environmental or public health threats from improperly treated medical waste" (Harey et al., 2009, pp. 1414–1417 cited in Stringer, 2012, p. 8). Images of rag pickers, segregating syringe bodies and intravenous bags for reuse on the black market, are not uncommon (Stringer, 2012). Poor practices in health-care waste management mean that injuries caused through needlesticks or inhalation of aerosols directly from the waste or through crude treatments such as open burning or unabated incineration are much more common and life-threatening (Stringer, 2012). The lack of an in-depth categorization system means the majority of residual "domestic" wastes emerging from health-care practices can become contaminated with infectious microorganisms and hypodermic needles, rendering the whole waste stream hazardous (Chartier et al., 2014).

---

[a]With the exception being for the following:

- Noncytotoxic or noncytostatic medicines from animal and human health-care facilities, manufacturing facilities, and/or outpatient returned medicines. These are a "clinical waste" but are also "nonhazardous."
- Clinical waste that is not associated with health-care activities (such as needles and swabs from cosmetic body art or piercing or sharps drug waste). These are only classed as nonhazardous, because the European Waste Catalogue Codes (EWC) do not have a hazardous code in the household waste category.

These issues are difficult to resolve as a huge amount of capital investment is required to start a decent collection, treatment, recovery, and disposal industry. There is also a need for the provision of training and access to knowledge, especially with regard to routes of exposure to infectious microorganisms, chemicals, pharmaceuticals, or radioactive materials (Stringer, 2012). More advanced technologies that do exist have high operational running costs and suffer if there is intermittent electricity, lack of spare parts, or lack of trained operators (Stringer, 2012). Safe management of health-care waste also requires policy and legislation within the health-care environment and at waste treatment sites to ensure that good practice is upheld across the waste management chain.

The World Health Organization (WHO) directs and coordinates international health within the United Nation's system (World Health Organization (WHO), 2016). In 1995, in collaboration with the WHO's European Center for Environment and Health (Chartier et al., 2014), an international working group was established to address problems associated with health-care waste management in developing and emerging economies. This resulted in a handbook targeted at these nations, now in its second edition, commonly known as "the blue book" (Chartier et al., 2014). This was written for all those who have an active interest in management of health-care waste, from health-care workers, to policy makers, to environmental and development organizations. Many of the guiding principles provided by the blue book originate from the experience of Europe and other economically developed nations in the safe management of health-care waste (Chartier et al., 2014). Taking the United Kingdom as an example of a country that has implemented European guidelines and law, the remainder of this chapter explores how one nation's health-care waste management system has developed to ensure the protection of the environment and human health.

THE BASICS OF HEALTHCARE WASTE MANAGEMENT

Before exploring how legislation shapes and guides the health-care waste sector, specifically in the United Kingdom, some of the basic concepts of good waste management practice are outlined here. The principle of all waste management is to prevent harm to human health or the environment, from creation to disposal. The ultimate aim is to be able to manage waste, so all outputs, including heat, energy, and materials, are safely reincorporated back into our cycles of production, with nothing wasted—a concept that is becoming known as "the circular economy" (The Waste and Resources Action Programme (WRAP), 2016). The latter requires a change not only within the waste sector but also in the way certain products are made, to eliminate the use of harmful ingredients such as persistent organic pollutants (POPs), which are unsuitable for recycling (GOV.UK, 2015b).

The basics of safe management of health-care wastes are as follows:

- **Decisions regarding the management of health-care waste should follow the "waste hierarchy."** This is the classic *reduce, reuse,* and *recycle*, often heard in relation to waste. If it is not possible to do any of these three, the next best option for health-care waste is to *recover* it (usually in the form of energy or heat given off by processes such as clinical waste incinerators), and the final and least favorable option is to *dispose* it (i.e., bury in a landfill or incinerate without energy recovery).
- **All those who handle waste have a "duty of care" to prevent it causing harm** by ensuring all those in the waste management chain, from collectors, brokers, treatment sites, or landfills, are capable and authorized to safely deal with that waste. Duty of care measures are designed to tackle waste crime, by defining the paperwork trail that should track movements of waste (in the United Kingdom, this is in the guise of waste transfer notes and consignment notes) and making everyone responsible for ensuring illegal operators that are avoided (GOV.UK, 2015a).
- **Segregation and categorization of wastes are critical to minimize risk during handling and identify safe transport and treatment requirements**. In the health-care environment, segregation is also a vital part of infection control (Department of Health, 2013). If wastes are not categorized well, they can pose severe risk to handlers; for example, "sharps" need containment in rigid, leakproof, puncture-resistant boxes to prevent injury. In Europe, the European Commission has provided a waste classification system to provide common terminology across the region (European Commission, 2016b) known as the European Waste Catalogue (EWC) (transposed into the List of Wastes (LoW) in the United Kingdom), which fully covers both hazardous and nonhazardous wastes.
- **Proper transport and packaging prevent waste producers, handlers, the environment, and public coming to harm**. Legislation should support a consistent approach to waste transport to ensure waste ends up where it should, and its journey is traceable, and waste is packaged and transported in ways that prevent its release into the environment. Hazardous wastes are considered to be dangerous goods, and the United Nations has developed mechanisms to harmonize the classification and transport of dangerous goods (not just wastes) by road, rail, and inland waterways (The United Nations Economic Commission for Europe (UNECE), 2016). In Europe, this is implemented via the European Agreement concerning the International Carriage of Dangerous Goods by Road (or the "ADR agreement") (The United Nations Economic Commission for Europe (UNECE), 2015b), meaning transport of hazardous wastes needs to be identified and packaged as per these requirements (The United Nations Economic Commission for Europe (UNECE), 2015a).
- **Additional containment measures are required for use of biological agents or genetically modified organisms** as these could cause severe

human disease or put the environment at severe risk. Specific legislation governing containment requirements for laboratories or research facilities using biological agents or GMOs is very stringent in Europe, requiring wastes to be inactivated or denatured in the laboratory suite or even in the laboratory itself (depending on the level of risk) before being moved off-site for further treatment/disposal.

- **Treatment/recovery and disposal routes should be appropriate for the waste type**. An infectiously contaminated waste may simply need denaturing or sterilizing before the residues or "flocks" are sent to landfill or used as a solid recovered fuel to generate energy. Comparatively, hazardous chemical waste or pharmaceutically contaminated waste may be unsuitable for treatment via lower temperature treatment techniques (such as autoclaves that generally operate below 130°C), as this would not be sufficient to destroy the contaminants and therefore require destruction in high-temperature processes such as an incinerator (Environment Agency, 2011). Liquid wastes (e.g., intravenous saline bags) (Chartier et al., 2014) on the other hand may not be suitable for landfill as they risk increasing pollutants to surface and ground waters—in Europe, it is not possible to landfill liquid wastes under the EU Landfill Directive 1991/31 EC7 (Council Directive, 1999). Other wastes may have a defined treatment route for ethical reasons—such as anatomical wastes, which must be destroyed in the United Kingdom via clinical waste incineration rather than sterilized in an autoclave or similar (Environment Agency, 2011).
- **All waste treatment, recovery, and disposal sites should be able to handle waste without posing risk of harm to human health and the environment**. In industrialized nations, even landfill sites are no longer just holes in ground but are highly engineered to prevent escape of contaminants into the environment—plus there are increasing restrictions on what can be disposed of into landfill. Under European law, in the United Kingdom, all waste sites require either an environmental permit (England and Wales) (The Environmental Industries Commission (EIC) and B P Collins LLP, 2017) or a waste management license or pollution prevention control permit (Scotland and Northern Ireland) (The Stationary Office, 2003, 2011a, 2012b, 2013b) to operate to ensure sites fully comply with environmental legislation. These permits or licenses help implement the multiple layers of legislation relevant to different treatment technologies to ensure that emissions to air, sewer, soil, and waters are minimized.

OPTIONS FOR HEALTHCARE WASTE TREATMENT AND DISPOSAL

There are a range of options for treatment, recovery, or disposal of health-care wastes, which are determined by the properties of the waste.

Landfill

Landfill is considered to be the least desirable option for all waste, as this is simply a disposal activity under the waste hierarchy with no energy or materials being recovered (apart from landfill gas at some sites). In Europe, the landfilling of wastes is governed by the Landfill Directive 1999/31/EC (Council Directive, 1999), which classifies sites into hazardous, nonhazardous, or inert. Most health-care wastes cannot be landfilled under this directive including (Environment Agency, 2010) the following:

- Infectiously contaminated waste (unless it has been pretreated to render it nonhazardous)
- Liquid wastes
- Chemical substances from research and development or teaching activities (i.e., lab wastes) that are new and the impact on human health or the environment that is unknown
- Waste pharmaceuticals
- Sharps
- Blood or anatomical wastes

This means only offensive hygiene wastes (such as incontinence waste) or the residues of health-care wastes produced by other treatment processes generally end up in landfill in the United Kingdom (Health and Safety Executive (HSE), 2014b). Health-care wastes that end up in the household waste stream (such as unused medications) could by default also be sent to landfill.

Destruction Processes

Processes that fully destroy waste, leaving behind ash or char, are ideal for health-care wastes that contain pharmaceutical or chemical residues as these require full breakdown of constituents to prevent their release into the environment. Most of these processes also generate energy in the form of electricity and heat either through energy released by burning the waste directly or from the synthetic gases produced by treating the waste via high temperatures. The majority of process that destroy waste in industrialized nations involves high temperatures and includes the following:

- **Incineration** destroys the waste in the presence of oxygen in a primary combustion chamber, followed by a secondary combustion engine to further destroy breakdown products. Primary and secondary abatement measures that are then used to ensure the air and sewer emissions are clean enough to be emitted. In Europe, under the European Industrial Emissions Directive 2010/75/EU (European Parliament, 2010), technically and economically viable best available techniques must be used to further reduce emissions, even if environmental quality standards are already being met (European Commission, 2016a). Most health-care wastes can be incinerated at normal temperatures (800–1100°C) (Department of Health, 2013), but some wastes, including cytotoxic and cytostatic wastes, low-level radioactive waste, and hazardous chemically contaminated wastes, must be incinerated via "high-temperature incineration"(Environment Agency, 2011) to ensure complete destruction of these materials. This means the secondary combustion chamber must operate at above 1100°C for a minimum of 2 seconds (Environment Agency, 2011; Department of Health, 2013; European Parliament, 2010).

- **Pyrolysis** involves the thermal decomposition of waste at 540–1000°C (Environment Agency, 2011) in the absence of oxygen, producing gas, solid, and liquid residues, followed by a secondary chamber that normally combusts the resulting synthetic gases or oils often generating electricity or usable heat in the process. Because combustion is used in the secondary chamber, the process must comply with emission limits and high-temperature threshold requirements under the Industrial Emissions Directive 2010/75/EU (European Parliament, 2010) for the treatment of defined hazardous wastes (Environment Agency, 2011). In many cases, gasification follows the pyrolysis phase (prior to secondary combustion) to maximize the amount of energy released from the waste.
- **Gasification** is similar to pyrolysis, again operating at 900–1100°C, but uses a tiny amount of oxygen (not enough for combustion) to degrade the waste, normally followed by combustion of the resulting gases in a secondary chamber (Department of Health, 2013). Again, due to the combustion required in the secondary chamber, the process must comply with the Industrial Emissions Directive 2010/75/EU (European Parliament, 2010) that is implemented in the United Kingdom through the Environmental Permitting (England and Wales) Regulations 2016 (The Environmental Industries Commission (EIC) and B P Collins LLP, 2017) and the Pollution Prevention and Control (Scotland) Regulations 2012 (The Stationary Office, 2012b) and the Pollution Prevention and Control (Industrial Emissions) Regulations (Northern Ireland) 2013 (The Stationary Office, 2013b).
- **Plasma technology** involves putting an electric current through an inert gas (such as argon) to produce a plasma at up to 6000°C (Department of Health, 2013). The clinical waste is treated at between 1300 and 1700°C that converts the waste into glassy rock and synthesis gas, the latter of which is combusted in a secondary chamber or engine generator for power. Again, because the secondary chamber involves combustion in the presence of oxygen, the process must comply with the Industrial Emissions Directive 2010/75/EU (European Parliament, 2010).

Another technology that fully destroys clinical waste and is approved as a treatment process for animal and human tissue waste in the EU, although is not currently deployed in the United Kingdom, is as follows:

- **Alkaline hydrolysis** that exposes waste to hot alkali at around 150°C for several hours, converting animal, human, and microbial tissue into a sterile, neutral, and aqueous solution suitable for disposal to sewer. It sterilizes and digests in one operation, reducing waste volume and weight by up to 97% and destroying known pathogens. It also eliminates radioactive, cytotoxic, and cytostatic tissues and contaminated tissue through neutralizing toxic fixing agents (Biosafe Engineering, 2016).

Low Temperature Sterilization Processes

For infectiously contaminated wastes, inactivation or sterilization of the pathogens can take place in what is commonly known in the United Kingdom and United States as alternative (*to incineration*) treatment technologies (ATTs) (Environment Agency, 2011). These technologies do not destroy the waste but rather use chemicals, heat, pressure, or microwaves to "reduce the number of organisms present in the waste to a level such that no additional precautions are needed to protect workers or the public against infection by the waste" (Environment Agency, 2010, p. 37; Environment Agency, 2011). A number of examples include (NHS Estates, 1998) the following:

- **Autoclaves** that use saturated steam at pressure, with temperatures reaching between 110 and 137°C to sterilize infectious microorganisms (Tuttnauer, 2013). They are used at a very small scale in the health-care environment to sterilize instruments or glass or lab wastes and on a larger scale for waste treatment. Large-scale processes often use maceration before treatment to increase the exposure of the waste to the steam (Environment Agency, 2011). Smaller laboratory-sized autoclaves do not tend to do this, although they often have added features especially if treating high-risk materials, to ensure full containment of biohazardous wastes.
- **Steam augers** treat waste using low-pressure steam, essentially enabling continued heating of the waste until all pathogens are sterilized (Environment Agency, 2011). They can be batch or continuous systems. Continuous systems pre shred the waste while exposing it to steam, before moving it up an enclosed screw-feed conveyor belt, surrounded by a superheated jacket, which exposes waste to boiling point temperatures for at least 60 min. Material is then typically dried, and the flocks, which contain a lot of plastics, are either used as solid recovered fuel, landfilled, or in some parts of the United States converted to biodiesel (STI Biosafe: A Division of Biosafe Engineering, 2016).
- **Chemical disinfection** uses chemicals such as "sodium hypochlorite, chlorine dioxide, peracetic acid, glutaraldehyde and quaternary ammonium compounds" (Department of Health, 2013, p. 163; Environment Agency, 2011). Waste is first shredded, to enable maximum exposure to chemicals, before being disinfected. As there are groups of microorganisms resistant to chemical disinfection (Giroult et al., 1999), it is vital to ensure the chemicals can kill all the pathogens and ensure the waste has not been rendered hazardous by the end of the process.

- **Microwave irradiation** causes the water within the waste to be rapidly heated and the infectious organisms destroyed through heat conduction (Giroult et al., 1999).

As the performance or efficacy of these processes is difficult to predict, a testing method developed by the US State and Territorial Association on Alternative Treatment Technologies (STAATT), in collaboration with United Kingdom and other international experts (The State and Territorial Association on Alternative Treatment Technologies (STAATT), 1998), was developed to demonstrate efficacy of the different ATTs. These tests utilize biological indicator strips, which are inserted into the equipment, after which they are tested to ensure that the required log reduction in spores has been achieved.

STAATT level III inactivation requires the demonstration of a $4\log_{10}$ reduction in numbers of *Bacillus atrophaeus (subtilis)* or *Geobacillus (bacillus) stearothermophilus* spores (Environment Agency, 2011; The State and Territorial Association on Alternative Treatment Technologies (STAATT), 1998), while level IV requires a $6\ \log_{10}$ reduction in the number of *G. stearothermophilus* spores (Environment Agency, 2011; The State and Territorial Association on Alternative Treatment Technologies (STAATT), 1998). Level IV inactivation is only required for plants treating defined biohazardous wastes that require additional measures to ensure the safety of the waste. In the United Kingdom, use of sterilization equipment within laboratories or research establishments treating biohazardous wastes is not generally required to demonstrate efficacy through the STAATT testing methods, as this equipment is heavily tested through other methods.

THE LEGISLATIVE FRAMEWORK OF THE HEALTHCARE WASTE SECTOR IN THE UNITED KINGDOM

The Global Dimension

Sign-up to international agreements and conventions, whose aim is to protect human health and the environment on a global scale, has quite an impact on health care waste management within the United Kingdom. By way of example, the EU implements international agreements and conventions through the following legislation:

- **The Basel Convention** controls the transboundary movements of hazardous (including health care) waste, including more recently prohibiting the movement of hazardous waste from industrialized to economically less developed countries (Basel Convention, 2011). In Europe, this is implemented and described in Regulation (EC) No. 1013/2006 on shipments of waste (European Parliament, 2006)—a piece of regulation that restricts and guides the movement of waste between countries and is directly applicable in all EU member states (Chartier et al., 2014).

- **The European Agreement concerning the International Carriage of Dangerous Goods by Road (ADR)** was completed in Geneva in September 1957 under the auspices of the United Nations (UN). This has since been amended several times and only enables dangerous goods to be transported on roads if specified packaging and labeling requirements are followed as well as requirements for the vehicles themselves (The United Nations Economic Commission for Europe (UNECE), 2015a). In the United Kingdom today, this agreement is directly implemented through the Carriage of Dangerous Goods and Use of Transportable Pressure Equipment Regulations 2009 (The Stationary Office, 2009; Health and Safety Executive (HSE), 2016a). Although the regulations are not specifically related to waste materials, the movement of all dangerous goods by road must be identified by a four-digit UN number and description (proper shipping name) and be assigned to a "class" of dangerous goods (Environment Agency, 2011), which is relevant for transport of all hazardous wastes.
- **The Stockholm Convention** on persistent organic pollutants (POPs) is designed to reduce or eliminate unintentional releases from POPs, which are released, for example, through health-care waste incineration (Chartier et al., 2014; European Parliament, 2013). There is a whole range of legislative mechanisms in the EU to help meet this convention—in the case of waste incinerators, the EU Industrial Emissions Directive (IED) 2010/75/EU (European Parliament, 2010) ensures plants are fitted with primary and secondary abatement measures to limit the release of POPs from air and wastewater emissions, forcing operators to use "best available techniques" (BAT) to do this. The convention also means priority consideration is given to ATTs that do not release or form these chemicals when treating waste (European Parliament, 2013). Regulation (EC) 850/2004 on POPs (European Parliament, 2004) is directly applicable in UK law and implements the most important obligations of both the Stockholm Convention and the UNECE POPs Protocol, prohibiting the production, use, and marketing of the POPs listed in the annexes of both instruments (European Parliament, 2013).

Post Brexit, some legal effort by the United Kingdom may be required to ensure the United Kingdom remains signatories to these treaties (The Environmental Industries Commission (EIC) and B P Collins LLP, 2017).

How European Legislation Works in the United Kingdom and Implications for Brexit

The United Kingdom joined the European Union (EU) in 1973, and until this relationship ends, it is subject to EU legislation. Following a referendum in June 2016, the United Kingdom has now indicated that it will leave the EU, a process formally known as "Brexit," (UK Parliament, 2016; GOV.UK, 2013) with the negotiation deadline expected in March 2019 (The Environmental Industries Commission (EIC) and B P Collins LLP, 2017). Until Brexit completes, it can be said that virtually all environmental and waste management legislation in the United Kingdom derives directly from the EU. The Institute for European Environmental Policy estimates that there are over 500 directives, regulations, and decisions that have been introduced to cover waste management, pollution, and the environment.

EU legislation comes in the form of *regulations* that are directly applicable in EU countries, such as Regulation (EC) No. 1013/2006 on shipments of waste (which implements the Basel Convention) or *directives* such as the Waste Framework Directive 2008/98/EC that are addressed to EU countries and must be transposed by them into their own national laws. Transposition is a process by which the European Union's member states give force to a directive by passing appropriate implementation measures, through primary legislation (i.e., an act or order—such as the UK's Environmental Protection Act 1990) and/or secondary legislation (i.e., regulations—such as the Waste (England and Wales) (Amendment) Regulations 2014). Through transposition, we can see much variance across the EU with regard to how these directives are interpreted and implemented.

At the time this book was published, Brexit negotiations were still in the infancy stages, so it is difficult to fully speculate the true impact of the UK's exit from the EU on the health-care waste sector. As waste and material resource management has been so transformed by EU legislation, equally, Brexit could have a significant impact on the future of the waste and resources sector itself. The United Kingdom's exit will mean that unless it has made specific commitments to apply EU laws as part of negotiating a new arrangement with the EU, laws can be repealed or amended at the will of UK parliament (and in some instances the devolved administrations) (The Environmental Industries Commission (EIC) and B P Collins LLP, 2017).

The UK Regulatory Authorities

There are four administrations in the United Kingdom, and the following regulators oversee responsibility for the regulation of waste:

- The Environment Agency in England (EA)
- Natural Resources Wales (NRW)
- The Scottish Environment Protection Agency (SEPA)
- The Northern Ireland Environment Agency (NIEA)

Each is responsible for regulating the waste sector (among other things) and ensuring all environmental regulations have been fully implemented at sites. They have the power for formal enforcement and sanctions on companies who are not compliant.

In addition, the Health and Safety Executive and the Health and Safety Executive for Northern Ireland are responsible for "the prevention of death, injury and ill health in Great Britain's workplaces" (Health and Safety Executive (HSE), 2016b, p. 1). They deal exclusively with the health and safety concerns within the waste management industry. Inspectors check that waste collections are being implemented correctly and sorting, processing, and treatment techniques are safe, including for sites such as landfills, incinerators, and ATTs. They also oversee the inspection of "contained laboratories," which form part of the health-care sector undertaking work with harmful biohazardous microorganisms and genetically modified organisms (Health and Safety Executive (HSE), 2016b).

Applying the Waste Hierarchy in the Healthcare Sector

In Europe, the waste hierarchy is embedded into the legal framework via the Waste Framework Directive 2008/98/EC (European Parliament, 2008), which is implemented in part in the United Kingdom through the Waste (England and Wales) (Amendment) Regulations 2014 (The Stationary Office, 2014c), the Waste Regulations (Northern Ireland) (The Stationary Office, 2011b) and the Waste (Scotland) Regulations 2012 (The Stationary Office, 2012c). All producers of waste must sign a declaration when they transfer waste to someone else that they have first considered the waste hierarchy (this is both on consignment notes and waste transfer notes) (360 Environmental, 2011). Sites that treat waste, under an environmental permit (England and Wales) or a waste management license or pollution prevention control permit (Northern Ireland and Scotland), are increasingly forced to apply the hierarchy through these legislative mechanisms (Fig. 1).

In terms of how the waste hierarchy is applied in the health-care sector, there is always, *always*, opportunity to *reduce* waste. For example, a recent report by NHS England (Hazell and Robson, 2015) quoted that £300 million of NHS prescribed medicines are wasted each year, concluding "that

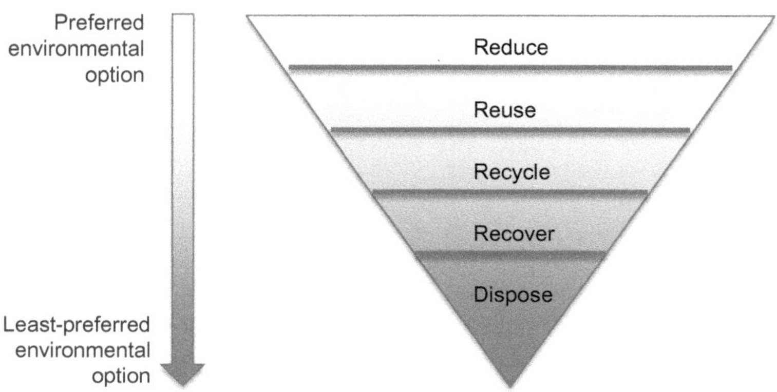

FIG. 1 The waste hierarchy.

[pharmaceutical] waste is a by-product of the 'business as usual' activities" (Hazell and Robson, 2015, p. 6). The report provides a starting point in identifying how a change in policy could help the NHS reduce these wastes and thus massively reduce their environmental and economic impact.

The health-care sector can also apply the hierarchy by *reducing* the use of dangerous chemicals within its products. Increasingly, this is occurring through product legislation, which in turn impacts the waste stream. For example, EU regulation (EC) 847/2012 on registration, evaluation, authorization, and restriction of chemicals (REACH) (Chorlton, 1982) that is a piece of EU regulation directly applicable in the United Kingdom specifically restricts the use of mercury in devices such as fever thermometers, manometers, sphygmomanometers, and barometers. This legislation helps prevent mercury being released from landfills and incinerators (Environment Agency, 2015) and helps implement our commitment to the Stockholm Convention on POPs (European Parliament, 2013).

In terms of *reuse*, there are obvious reasons why this is a hotly debated topic in the health-care sector (Chartier et al., 2014), specifically with regard to reuse of hypodermic needles and syringes, which has made single-use instruments very popular and in many cases rightly so. There are other cases where instruments are safely reused, for example, in operating theaters, where instruments are effectively sterilized in various types of device, using heat, chemicals, or microwaves to inactivate or kill microorganisms (NHS Estates, 2003).

An example of *reuse* practice, which is becoming more commonplace in Europe and other industrialized nations, is the reuse of sharps containers, which are normally mass-produced, used once, and then incinerated with their contents (releasing harmful emissions that then need to be controlled). Innovations such as Sharpsmart™ are gradually being introduced to hospitals worldwide (Sharpsmart, 2010). This concept allows sharp containers (which in turn comply with EU/UN regulations on transport of dangerous goods) to be filled on-site, transported to a centralized facility, emptied robotically, sterilized, and returned to the health-care facility to reuse. A reusable design also means that the container can have more complex safety features, making it safer for the user—benefitting both the environment and the user or handler (Sharpsmart, 2010) (Fig. 2).

Recovery happens during health-care waste treatment. Incinerators, for example, can use the heat released when waste is treated for useful purposes—such as for heat or energy—but the amount of heat recovered varies greatly depending on the efficiency of the process and calorific value of the waste. Under the Waste Framework Directive 2008/98/EC (European Parliament, 2008), there is a formula to calculate whether a waste incinerator is a "recovery" or "disposal" operation (European Commission, 2011). This can only be applied to the treatment of household waste as it is generally accepted that the additional needs of an incinerator to treat hazardous waste (which must burn much hotter in the secondary chamber to destroy hazardous pollutants) would not be able to meet this definition of recovery. Instead, the

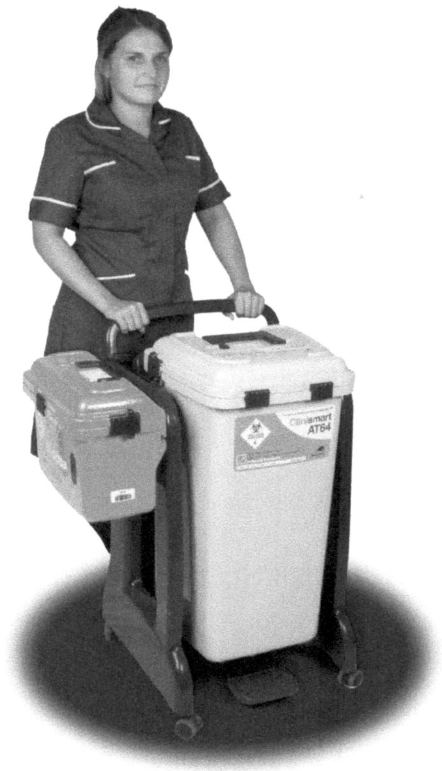

FIG. 2 Sharpsmart™ system for reuse of sharps boxes. Source: *Courtesy of Sharpsmart™*.

Waste Framework Directive 2008/98/EC (European Parliament, 2008) has established principles to differentiate between recovery and disposal operations, which means criteria such as the combustion of waste must generate more energy than the process itself consumes and must be applied to determine where the process sits on the waste hierarchy (The Chartered Institution of Wastes Management (CIWM), 2014; European Commission, 2011). In addition, ashes and metals are also *recovered* post incineration, which reduces landfill costs and ensures these materials become resources.

Other large-scale ATTs are increasingly enabling sterilized residues to be *recovered* as a solid recovered fuel. One of the United Kingdom's largest health-care waste companies, SRCL, claims that 99% of their residual, sterilized, waste "flock" from these technologies is now used as an alternative to fuel in cement production (SRCL, 2015)—although this is not the case across the whole sector.

The last and least desirable part of the waste hierarchy is waste *disposal*. This is where nonrecoverable residues from ATTs and incinerators (including hazardous ashes) end up. Offensive wastes (e.g., incontinence pad waste) tend to end up buried deep inside landfills ("deep" due to its offensive nature and odors), as there is a lack of recovery technologies available. In the United Kingdom, the only plant dedicated to recycling absorbent hygiene products, run by Knowaste, was closed in 2014 (Reece, 2013) although the company is promising to relocate and reopen. In the United Kingdom, it is assumed that this waste stream is uncontaminated with pharmaceuticals or hazardous residues, hence why it can be landfilled.

The Circular Economy Approach

The concept of the "circular economy" is gradually being introduced across Europe. The circular economy moves away from a linear economy (make, use, and dispose) in theory enabling products and materials to last longer and virtually eliminating waste (The Waste and Resources Action Programme (WRAP), 2016). Europe has recently adopted a circular economy "closing the loop" package, which will eventually revise legislation across Europe, to bring these principles into regulatory mechanisms (European Commission, 2015a).

The "Duty of Care" Principle

The "duty of care" principle is about the ethical and legal duty of "any person who imports, produces, carries, keeps, treats or disposes of controlled waste, or is a broker" (The Environmental Protection Act, 1990, Section 34) (The Stationary Office, 1990) to prevent it from causing harm during production, storage, transport, treatment, and final disposal. Specific statutory requirements for the duty of care regarding *nonhazardous waste* in the United Kingdom are covered by the following:

- *Primary legislation* in the form of
  - Section 34 of Environmental Protection Act 1990 (The Stationary Office, 1990),
  - Section 5 of the Waste and Contaminated Land (Northern Ireland) Order 1997 (The Stationary Office, 1997; Northern Ireland Environment Agency (NIEA), 2014).
- *Secondary legislation* in the form of
  - the Waste (England and Wales) (Amendment) Regulations 2014 (The Stationary Office, 2014c),
  - the Controlled Waste and Duty of Care Regulations (Northern Ireland) 2013 (The Stationary Office, 2013a),
  - the Environmental Protection (Duty of Care) (Scotland) Regulations 2014 (The Stationary Office, 2014a).

For *hazardous wastes*, duty of care is met through the same primary and secondary legislation as nonhazardous wastes and

- the Hazardous Waste (England and Wales) Regulations 2005 (The Stationary Office, 2005a),

- the Hazardous Waste Regulations (Northern Ireland) 2005 (The Stationary Office, 2005b),
- the Special Waste Amendment (Scotland) Regulations 2004 (The Stationary Office, 2004).

These legislative requirements mean all waste must be accompanied by a legal document in the form of a waste transfer note for nonhazardous waste or a consignment note for hazardous waste. Hazardous wastes also need to be identified by a unique UN identification number for movements on roads. Records need to be kept for a minimum of 2 years and must be shared with the relevant regulatory authorities if requested. More recently, it is now a legal requirement for producers of controlled waste to declare on waste transfer or consignment documents that the waste hierarchy has also been adhered too (GOV.UK, 2015a; Northern Ireland Environment Agency (NIEA), 2014; Scottish Environment Protection Agency (SEPA), 2009).

Classifying a Waste Using the European Waste Catalogue

The European Commission has provided a common classification system to provide common terminology across Europe for the classification of waste (European Commission, 2016b). This is known as the European Waste Catalogue (EWC). This has been transposed into the List of Wastes (LoW) regulations in the United Kingdom and fully covers both hazardous and nonhazardous wastes. In Europe, the list is currently under review, and revisions will need to be transposed into UK law at a later point, provided that this still moves forward post Brexit (The Environmental Industries Commission (EIC) and B P Collins LLP, 2017; European Commission, 2016b). As classification and assessment of a waste are not straightforward, the Scottish Environment Protection Agency (SEPA), the Environment Agency, the Department of Environment (in Northern Ireland), and the Natural Resources Wales have produced joint technical guidance, known as "Guidance on the Classification and Assessment of Waste Technical Guidance WM3" (Jones and Tansey, 2015). This helps guide producers through classification to ensure they remain complaint with the variety of legislative mechanisms to which classification applies.

The basics steps in this guidance are as follows (Jones and Tansey, 2015):

- **Identify if the waste needs to be classified**—almost all health-care waste would need to be classified unless the material either is not waste or is identified in Article 2 of the Waste Framework Directive 2008/98/EC as being excluded as it is covered by other legislation (e.g., radioactive waste or gaseous effluents emitted into the atmosphere).
- **Identify the EWC or LoW code that may apply to the waste**—wastes are categorized via 20 chapters, which are defined by the source of the waste or waste type. "Within each chapter, each type of waste is described using a six-digit numerical code:

- The first two digits relate to the EWC of LoW chapter;
- The second two relate to subgrouping; and
- The final two digits are unique to the waste"(Department of Health, 2013, p. 13).

  To see an example of a typical health-care waste code, see Table 1.
- **Determine the chemical composition of the waste** to determine if it has a hazardous property, which can be found through product manufacturer's safety data sheets, knowledge of the waste if it originates from a well-understood industrial process, or sampling or analysis of the waste (a procedure that must follow strict guidelines).
- **Identify if the substances are "hazardous substances" or "POPs"**—a full list of POPs for waste classification purposes in the United Kingdom is provided in WM3.
- **Assess the hazardous properties of the waste** to see if it contains any of the following hazardous descriptions at relevant concentrations (Jones and Tansey, 2015):
  - H1: Explosive
  - H2: Oxidizing
  - H3A: Highly flammable
  - H3B: Flammable
  - H4: Irritant
  - H5: Harmful
  - H6: Toxic
  - H7: Carcinogenic
  - H8: Corrosive
  - H9: Infectious
  - H10: Toxic for reproduction
  - H11: Mutagenic
  - H12: Substances that release toxic gases
  - H13: Sensitizing
  - H14: Ecotoxic
  - H15 Waste capable by any means, after disposal, of yielding another substance, for example, a leachate, which possesses any of the characteristics from H1 to H14, including H14 for the first time.
- **Conduct an assessment to select the classification**:
  - "Absolute hazardous entries" are wastes that are always hazardous (notified by superscript "b"), for example, cytotoxic or cytostatic medicines.[b]
  - "Absolute nonhazardous entries" are wastes that are always nonhazardous (notified by superscript "a"), such as human or animal medicines

[b]A cytotoxic or cytostatic medicine is defined as this, if it possesses one or more of the following hazardous properties: H6, toxic; H7, carcinogenic; H10, toxic for reproduction; and/or H11, mutagenic.

TABLE 1 Extract From EWC or LoW Chapter Most Relevant to Health-Care Wastes (Jones and Tansey, 2015)

18	Wastes From Human or Animal Health-Care and/or Related Research (Except Kitchen and Restaurant Wastes Not Arising From Immediate Health Care)	Entry Type
18 01	*Wastes from natal care, diagnosis, treatment, or prevention of disease in humans*	
18 01 01[a]	Sharps (except 18 01 03)	AN
18 01 02[a]	Body parts and organs including blood bags and blood preserves (except 18 01 03)	AN
18 01 03[b]	Wastes whose collection and disposal are subject to special requirements in order to prevent infection	AH
18 01 04[a]	Wastes whose collection and disposal are not subject to special requirements in order to prevent infection (e.g., dressings, plaster casts, linen, disposable clothing, and diapers)	AN
18 01 06[c]	Chemicals consisting of or containing hazardous substances	MH
18 01 07[d]	Chemicals other than those mentioned in 18 01 06	MN
18 01 08[b]	Cytotoxic and cytostatic medicines	AH
18 01 09[a]	Medicines other than those mentioned in 18 01 08	AN
18 01 10[b]	Amalgam waste from dental care	AH
18 02	*Wastes from research, diagnosis, treatment, or prevention of disease involving animals*	
18 02 01[a]	Sharps (except 18 02 02)	AN
18 02 02[b]	Waste whose collection and disposal are subject to special requirements in order to prevent infection	AH
18 02 03[a]	Wastes whose collection and disposal are not subject to special requirements in order to prevent infection	AN
18 02 05[c]	Chemicals consisting of or containing hazardous substances	MH
18 02 06[d]	Chemicals other than those mentioned in 18 02 05	MN
18 02 07[b]	Cytotoxic and cytostatic medicines	AH
18 02 08[a]	Medicines other than those mentioned in 18 02 07	AN

[a]*AN means "absolute nonhazardous."*
[b]*AH means "absolute hazardous."*
[c]*MH means "mirror hazardous."*
[d]*MN means "mirror nonhazardous."*

that do not contain cytotoxic or cytostatic ingredients (if the producer doesn't know, the worst case should be assumed, and it should be assigned hazardous).
  - "Mirror hazardous" (notified by superscript "c") are wastes that contain
    - hazardous substance(s) *at or above* levels that cause it to display a hazardous property,
    - persistent organic pollutants *at or above* the concentration limits that cause it to be hazardous.
  - "Mirror nonhazardous" (notified by superscript "d") are wastes that contain
    - hazardous substance(s) *below* levels that cause it to display a hazardous property,
    - persistent organic pollutants *below* the concentration limit that cause it to be hazardous.
- **Assign classification code and describe the hazardous properties (if it is hazardous**). The final classification helps determine if the waste should be transported with a waste transfer note or consignment note.

Health-care practices can produce a range of wastes across the EWC or LoW catalog, although the majority of "health-care wastes" fall into Chapter 18, as shown in Table 1. In addition, Chapter 20 of the LoW catalog focuses on municipal wastes (household waste) and includes descriptions for separately collected fractions of medicinal and other wastes (including cytotoxic or cytostatic medicines).

Color Coding of Waste Containers

To assist in waste classification and ensure that the waste goes to the right treatment/disposal facility, a color-coding segregation system is used in the United Kingdom. Although these are not regulatory requirements, it is a very widely applied, standardized system used by the health-care sector and waste industry and very clearly laid out in the Department of Health's health technical memorandum on the safe management of health-care waste (Department of Health, 2013). Following this, color classification enables procurement departments to purchase waste containers that are already labeled with the necessary codes to meet transport regulations.

Color codes for box lids, bins, or bags are as follows (Department of Health, 2013):

- **Yellow is waste that requires disposal by incineration**. This is usually infectious waste but has additional properties such as medicinal contamination (not cytotoxic or cytostatic).
- **Orange is waste that may be treated via ATTs or incineration.** This includes infectiously contaminated sharps (which must be boxed separately), bedding or swabs, and dressings but must not contain anatomical waste or medicinally contaminated waste.

- **Purple is cytotoxic and cytostatic waste that requires disposal by incineration**. This requires high temperatures for processing of secondary emissions to ensure destruction of hazardous properties.
- **Yellow/black is offensive hygiene waste** that can be recovered or disposed of via incineration or landfill.
- **Red is anatomical waste for incineration**. This must be destroyed in this way for ethical reasons (Environment Agency, 2011).
- **Blue is medicinal waste for incineration**. Cytotoxic or cytostatic contaminants cannot be mixed into this waste stream.
- **White is dental amalgam** waste for recovery.
- **Black is domestic waste**. This is disposed of in the same way as municipal waste, which means recyclables should be removed with the remainder going to energy from waste plants or landfill.

In addition, radioactive wastes are contained and marked with the radioactive symbol. Radioactive wastes from health-care establishments are considered to be low-level radioactive waste and are generated from radionuclides used in therapeutic and diagnostic medicine (Department of the Environment et al., 2012). In the United Kingdom, each regulator, including the EA, NRW, SEPA, and NIEA, is responsible for ensuring that the storage and use of radioactive materials adhere to the Environmental Permitting (England and Wales) Regulations 2016 (The Environmental Industries Commission (EIC) and B P Collins LLP, 2017) and the Radioactive Substance Act in Scotland and Northern Ireland (The Stationary Office, 1993a,b). If the waste contains radioactive substances below threshold, it may be exempt from these regulations. Obviously, transport of radioactive substances by road must comply with the ADR regulations. In addition, the Ionizing Radiation (Medical Exposure) Regulations 2000 (The Stationary Office, 2000) and the Ionizing Radiation (Medical Exposure) (Amendment) Regulations 2006 (The Stationary Office, 2006) specify that a radiation protection advisor is appointed to oversee the use and exposure to radioactive materials. They are also responsible for liaising on waste issues to ensure their safety.

Research and Laboratory Wastes

As there are additional risks associated with wastes arising from laboratories or research establishments, especially with regard to those that may contain pathogens or genetically modified microorganisms, the following legislation applies:

- The European Council Directive 2009/41/EC on the contained use of genetically modified microorganisms (GMMs) that is transposed by the United Kingdom through the Genetically Modified Organisms (Contained Use) Regulations 2014 (The Stationary Office, 2014b)

- The European Council Directive 2000/54/EC on the protection of workers from risks related to biological agents at work that is transposed in the United Kingdom through the Control of Substances Hazardous to Health (COSHH) Regulations 2002 (The Stationary Office, 2002)

Both sets of regulatory regimes are designed to contain the risks associated with these activities, to ensure that there are adequate barriers between these facilities and the outside world. This impacts not only the design of the laboratories and the rules workers must adhere to but also the ways in which wastes, effluents, and air emissions must be denatured or sterilized before leaving the site or lab. The regimes outline four containment levels (CL), with CL4 facilities requiring the highest containment requirements and CL1 the least. The CL is linked to the level of risk posed by the biological agents[c] or GMMs/GMOs (Health and Safety Executive (HSE), 2014a, 2005).

Under the COSHH regulations, biological agents are divided into four hazard groups (HGs), with HG1 presenting the lowest risk (i.e., unlikely to cause human disease) and HG4 presenting the highest risk (i.e., causes severe human disease and is a serious hazard to employees; it is likely to spread to the community, and there is usually no effective prophylaxis or treatment available) (Health and Safety Executive (HSE), 2005). Similarly for GMMs, there are four risk classes: Class 1 presents no or negligible risk to human health or the environment, and class 4 presents the highest risk to human health or environment (Health and Safety Executive (HSE), 2014a, 2005).

The HSE and Defra in England and Wales, the Scottish ministers, and the Department of Environment in Northern Ireland are the competent authorities that ensure compliance to these regulations. The HSE takes the lead on behalf of all four devolved administrations, taking charge of enforcement, inspections, notifications, and enquiries. The two main nongovernment bodies that provide expert scientific advice and guidance to the HSE, particularly with regard to hazard identification and risk assessment of the different biological agents and GMMs/GMOs, are the Scientific Advisory Committee on Genetic Modification (Contained Use) (SACGM(CU)) and the Advisory Committee on Dangerous Pathogens (ACDP) (Health and Safety Executive (HSE), 2014a, 2005).

With regard to waste management, the CL of the facility dictates the level of deactivation required. For example, in a CL4 facility, an autoclave must be located in the laboratory and use a double door system, so waste can be

[c]Under the COSHH regulations, a biological agent is defined as "a micro-organism, cell culture, or human endoparasite, whether or not genetically modified, which may cause infection, allergy, toxicity or otherwise create a hazard to human health" (Health and Safety Executive, 2005, p. 6; Harey et al., 2009a,b). However, not all GMOs would be classed as biological agents—only those that are hazardous to human health. The Genetically Modified Organisms (Contained Use) Regulations 2014 also considers environmental risk, whereas COSHH does not (Harey et al., 2009a,b).

inserted on the "dirty" side of a containment barrier, sterilized, and then taken off-site from the "clean" side of the containment barrier. Comparatively, a CL1 facility dictates that the autoclave must simply be on-site. Wastes leaving the site are no longer considered to be infectious or hazardous, rather offensive, although due to public sensitivity material is often sent for further treatment rather than to landfill (Department of Health, 2013). In very exceptional circumstances, such as if an autoclave has broken down and cannot be repaired, waste can be transported off-site for incineration or treatment via an ATT (if the ATT meets the required STAATT testing level). In these rare cases, wastes must be transported as the highest risk category A substance as per the ADR regulations (Department of Health, 2013; Environment Agency, 2011).

Environmental Permitting and Waste Management Licensing

As waste treatment or disposal sites could potentially harm human health or the environment if not properly controlled, EU directives dictate that operators must obtain a permit or authorization to operate. In the United Kingdom, these fall under the following regulations:

- The Environmental Permitting (England and Wales) Regulations 2016 (The Environmental Industries Commission (EIC) and B P Collins LLP, 2017)
- The Waste Management Licensing Regulations (Northern Ireland) 2003 (The Stationary Office, 2011b)
- The Pollution Prevention and Control (Industrial Emissions) Regulations (Northern Ireland) 2013 (The Stationary Office, 2013b)
- The Waste Management Licensing Regulations (Scotland) 2011 (The Stationary Office, 2011a)
- The Pollution Prevention and Control (Scotland) Regulations 2012 (The Stationary Office, 2012b)

Essentially, this means that all health-care waste sites, including landfill sites, recycling plants, incinerators (or other high-temperature treatment processes), and ATTs, will either hold an environmental permit (England and Wales) or a waste management license or pollution prevention control permit (Northern Ireland and Scotland). Some low-risk activities are exempt from the need of a permit or license, for example, storage of various wastes.

This regime is designed to implement multiple EU directives, international agreements, and government policy to prevent these sites causing pollution. In England and Wales, each directive or policy area within the Environmental Permitting (England and Wales) Regulations 2016 has a specific "schedule," which identifies particular requirements (usually article by article, in the case of EU directives) that should be delivered through the permitting regime. Certain schedules cover the Industrial Emissions Directive, while others by way of example cover the Waste Framework Directive, the Landfill Directive, and water discharge and groundwater activities (Department for Environment, Food and Rural Affairs (Defra) and The Welsh Government, Department of Energy and Climate Change, 2013).

In the case of Scotland and Northern Ireland, the Waste Management Licensing Regulations applies to the majority of waste facilities, while the pollution prevention control (PPC) regime deals with more high-risk activities that falls outside of waste management licensing (e.g., landfilling, incineration, and disposal of hazardous waste). The PPC regime therefore specifically implements the Industrial Emissions Directive and the Landfill Directive and Waste Framework Directive (the latter of which both regimes implement). In England and Wales, there used to be separate waste management licenses and pollution prevention and control (PPC) permits, but these were integrated when the Environmental Permitting (England and Wales) Regulations were first introduced in 2007 (Scottish Environment Protection Agency (SEPA), 2009, 2016). Notwithstanding, each regime seeks to authorize all waste sites and ensure that they comply with the ever-changing landscape of environmental legislation.

CONCLUSIONS—LEGISLATION SHAPING AND CREATING A HEALTHIER HEALTHCARE WASTE SECTOR

This chapter has helped explain the inextricable link between a legislation and an evolving and safer health-care waste sector, using the United Kingdom as an example. In the days predating environmental legislation, the decade of the 1960s witnessed 90% of unsegregated wastes in the United Kingdom being dumped in local tips (Willets, 2007). "Tips" were not sophisticated like today's controlled landfill sites and would have leached all kinds of pollutants into both surface and groundwater. Furthermore, poor segregation practices meant it was not uncommon for needlestick injuries to be encountered by both refuse collectors and hospital staff (Jones and Tansey, 2015).

The Control of Pollution Act of 1974 (European Parliament, 2009) was one of the first pieces of environmental legislation to move the United Kingdom away from local dumps to controlled landfill. However, in the early 1980s, the disposal of health-care waste hit the headlines when a water-skier crashed into bags of blood-filled clinical waste in the Thames Estuary, and more clinical waste marked "warning hepatitis" was found dumped in London's Grosvenor Dock (Chorlton, 1982). Up until this point, there was no clear system for segregation of wastes in hospitals, and in most cases, everything went into one bin (Jones and Tansey, 2015). This decade also saw the majority of clinical wastes in the United Kingdom being incinerated on-site at health-care or hospital premises. Unbelievably, none of these incinerators were required to have any pollution abatement, as they were protected by crown immunity (Jones and Tansey, 2015), shielding them from legislation designed to reduce the impact of waste management. Perversely, this would have meant the health-care sector contributed to poor human health in any populations coming into contact with their pollutants. In 1990, this crown immunity was finally lifted, which meant a more sophisticated, high-end waste management sector could take over.

The United Kingdom joined the European Union in 1973, and since this time, the Institute for European Environmental Policy estimates that there has been over 500 directives, regulations, and decisions introduced to cover waste management, pollution, and the environment. European legislation has massively driven the shape of the UK waste sector, with some European regulations being directly applicable and others being "directive" (i.e., addressed to EU countries with requirements to transpose them into their own national laws). Undoubtedly, this has driven a more resourceful and resource-efficient economy and pulled the United Kingdom away from its landfill habits. Following a referendum in June 2016, the United Kingdom has now indicated that it will leave the EU, a process formally known as "Brexit," with the negotiation deadline expected in March 2019 (The Environmental Industries Commission (EIC) and B P Collins LLP, 2017). This "exit" will undoubtedly change the shape of the UK health-care waste sector again, but at the time of print, it is difficult to predict in what ways.

The historical issues that have been have witnessed in the United Kingdom with regard to unsafe management of health-care wastes are sadly echoed across today's world. "Over half the world's population are now at risk from occupational, environmental or public health threats from improperly treated medical waste" (Harey et al., 2009, pp. 1414–1417 cited in Stringer, 2012, p. 8). Populations are at risk if they come directly into contact with the waste not only through needlestick injuries or inhalation of aerosols coming directly from the waste but also through crude treatments such as unabated incineration. Sadly, in order to prevent health-care waste from causing harm, a huge level of capital investment, sophistication and will is required by governments and policy makers to help develop a common legal language in regionalized waste sectors to drive better segregation, better care, and better treatments across the sector as a whole. In regions where this occurs, we see that waste management's impact on the environment and human health are greatly reduced.

REFERENCES

360 Environmental, 2011. Defra Publish Guidance on Applying the Waste Hierarchy. http://www.360environmental.co.uk/news/375/. Accessed 17 February 2016.

Basel Convention, 2011. Controlling Transboundary Movement of Hazardous Wastes and Their Disposal. http://www.basel.int. Accessed 17 February 2016.

Biosafe Engineering, 2016. Alkaline Hydrolysis. http://biosafeeng.com/alkaline-hydrolysis. Accessed 16 February 2016.

Blenkharn, J.I., 2015. Healthcare Wastes. Reference Module in Earth Systems and Environmental Sciences. (e-journal) Available from: ScienceDirect (Elsevier) website, http://www.sciencedirect.com/. Accessed 17 February 2016.

Chartier, Y., Emmanuel, J., Pieper, U., Prüss, A., Rushbrook, P., Stringer, R. et al., (Eds.), 2014. Safe Management of Wastes from Health-Care Activities, second ed. World Health Organization, Geneva, Switzerland. Available from: http://www.searo.who.int/srilanka/documents/safe_management_of_wastes_from_healthcare_activities.pdf. Accessed 17 February 2016.

Chorlton, P., March 25, 1982. Protest on medical waste disposal. The Guardian Cited in Jones, E.M., Tansey, E.M., (Eds.), 2015. The development of waste management in the UK c. 1960–c.2000. Wellcome Witnesses to Contemporary Medicine, vol. 56. Queen Mary University London, London. Available from: http://www.histmodbiomed.org/sites/default/files/W56LoRes.pdf. Accessed 4 April 2016.

Council Directive, 1999. Council Directive 1999/31/EC of 26 April 1999 on the landfill of wastes. Off. J. L 182.

Department for Environment, Food and Rural Affairs (Defra), The Welsh Government, Department of Energy and Climate Change, 2013. Environmental Permitting Guidance – Core Guidance – For the Environmental Permitting (England and Wales) Regulations 2010. Defra, London. Available from: https://www.gov.uk/government/uploads/system/uploads/attachment_data/file/211852/pb13897-ep-core-guidance-130220.pdf. Accessed 2 December 2016.

Department of Health, 2013. Health Technical Memorandum 07-01: Safe Management of Healthcare Waste. Department of Health, London. Available from: https://www.gov.uk/government/uploads/system/uploads/attachment_data/file/167976/HTM_07-01_Final.pdf. Accessed 17 February 2016.

Department of the Environment, Department of Energy and Climate Change, The Scottish Government, the Welsh Government, 2012. Strategy for the Management of Solid Low Level Radioactive Waste from the Non-Nuclear Industry in the United Kingdom – Part 1 – Anthropogenic Radionuclides. Department of Energy and Climate Change, London. Available from: https://www.gov.uk/government/uploads/system/uploads/attachment_data/file/48291/4616-strategy-low-level-radioactive-waste.pdf. Accessed 4 January 2016.

Environment Agency, 2010. Waste Acceptance at Landfills – Guidance on Waste Acceptance Procedures and Criteria. Environment Agency, Bristol. Available from: https://www.gov.uk/government/uploads/system/uploads/attachment_data/file/296422/geho1110btew-e-e.pdf. Accessed 17 February 2016.

Environment Agency, 2011. How to Comply With Your Environmental Permit – Additional Guidance for Clinical Waste (EPR 5.07 Version 1.1). Environment Agency, Bristol. Available from: https://www.gov.uk/government/uploads/system/uploads/attachment_data/file/296418/geho0710bsvi-e-e.pdf. Accessed 17 February 2016.

Environment Agency, 2015. Mercury in Measuring Devices Guidance Note. Environment Agency, Bristol. Available from: https://www.gov.uk/government/uploads/system/uploads/attachment_data/file/481145/LIT_7498.pdf. Accessed 17 February 2016.

European Commission, 2011. Guidelines on the Interpretation of the R1 Energy Efficiency Formula for Incineration Facilities Dedicated to the Processing of Municipal Solid Waste According to Annex II of Directive 2008/98/EC on Waste. Available from: http://ec.europa.eu/environment/waste/framework/pdf/guidance.pdf. Accessed 17 February 2016.

European Commission, 2015a. Circular Economy Strategy. http://ec.europa.eu/environment/circular-economy/index_en.htm. Accessed 22 February 2016.

European Commission, 2016a. The Industrial Emissions Directive. http://ec.europa.eu/environment/industry/stationary/ied/legislation.htm. Accessed 17 February 2016.

European Commission, 2016b. Waste Framework Directive – The European List of Wastes. http://ec.europa.eu/environment/waste/framework/list.htm. Accessed 17 February 2016.

European Parliament, 2004. Regulation (EC) 850/2004 of the European Parliament and of the Council of 29 April 2004 on persistent organic pollutants and amending Directive 79/117/EEC. Off. J. L 158, 7.

European Parliament, 2006. Regulation (EC) 1013/2006 of the European Parliament and of the Council on 14 June 2006 on shipments of waste. Off. J. L 190.

European Parliament, 2008. Directive 2008/98/EC of 19 November 2008 on waste and repealing certain Directives (text with EEA relevance). Off. J. L 312, 3.

European Parliament, 2009. Directive 2009/41/EC of the European Parliament and of the Council of 6 May 2009 on the contained use of genetically modified micro-organisms (re-cast) (text with EEA relevance). Off. J. L 125, 75.

European Parliament, 2010. Directive 2010/75/EU of the European Parliament and of the Council of 24 November 2010 on industrial emissions (integrated pollution prevention and control) (re-cast) (text with EEA relevance). Off. J. L 334, 17.

European Parliament, 2013. Decision No 1386/2013/EU of the European Parliament and of the Council of 20 November 2013 on a General Union Environmental Action Programme to 2020 'Living well, within the limits of our planet' (text with EEA relevance). Off. J. L 354, 171.

Giroult, E., Prüss, A., Rushbrook, P. (Eds.), 1999. Treatment and Disposal Technologies for Health-Care Waste. World Health Organisation, Geneva, pp. 77–112. http://apps.who.int/iris/bitstream/10665/42175/1/9241545259.pdf. Accessed 22 March 2016.

GOV.UK, 2013. Devolution of Powers to Scotland, Wales and Northern Ireland. https://www.gov.uk/guidance/devolution-of-powers-to-scotland-wales-and-northern-ireland. Accessed 30 March 2016.

GOV.UK, 2015a. Business and Commercial Waste – Duty of Care. https://www.gov.uk/managing-your-waste-an-overview/overview. Accessed 30 March 2016.

GOV.UK, 2015b. Using Persistent Organic Pollutants (POPs). https://www.gov.uk/guidance/using-persistent-organic-pollutants-pops. Accessed 30 March 2016.

Harey, et al., 2009. Trop. Med. Int. Health 14 (11), 1414–1417. Cited in Stringer, R., 2012. Safe and Sustainable Healthcare Waste Treatment – An Essential Element of Healthcare Without Harm. (Microsoft Powerpoint) Health Care Without Harm. Available from: http://www.unep.org/ietc/Portals/136/Events/HealthcareWasteWorkshop-July2012/8-Stringer.pptx. Accessed 17 February 2016.

Harhay, M.O., Halpern, S.D., Harhay, J.S., Olliaro, P.L., 2009. Health care waste management: a neglected and growing public health problem worldwide. Trop. Med. Int. Health 14 (11), 1414–1417. Available from: http://onlinelibrary.wiley.com/doi/10.1111/j.1365-3156.2009.02386.x/pdf. Accessed 17 February 2016.

Hazell, B., Robson, R., 2015. Pharmaceutical Waste Reduction in the NHS. NHS, England. Available from: https://www.england.nhs.uk/wp-content/uploads/2015/06/pharmaceutical-waste-reduction.pdf. Accessed 17 February 2016.

Health and Safety Executive (HSE), 2005. Biological Agents: Managing the Risks in Laboratories and Healthcare Premises – Advisory Committee on Dangerous Pathogens. Health and Safety Executive (HSE), Suffolk. Available from: http://www.hse.gov.uk/biosafety/biologagents.pdf. Accessed 16 February 2016.

Health and Safety Executive (HSE), 2014a. The Genetically Modified Organisms (Contained Use) Regulations 2014 – Guidance on Regulations. Health and Safety Executive (HSE), Suffolk. Available from: http://www.hse.gov.uk/pubns/priced/l29.pdfhttp://www.hse.gov.uk/pubns/priced/l29.pdf. Accessed 2 December 2016.

Health and Safety Executive (HSE), 2014b. Managing Offensive/Hygiene Waste Safely. Health and Safety Executive. Available from: http://www.hse.gov.uk/pubns/waste22.pdf. Accessed 17 February 2016.

Health and Safety Executive (HSE), 2016a. Carriage Regulations. http://www.hse.gov.uk/cdg/regs.htm. Accessed 16 February 2016.

Health and Safety Executive (HSE), 2016b. HSE and Waste. http://www.hse.gov.uk/waste/how.htm. Accessed 16 February 2016.

Jones, E.M., Tansey, E.M. (Eds.), 2015. The development of waste management in the UK c. 1960–c.2000. In: Wellcome Witnesses to Contemporary Medicine, In: 56. Queen Mary University London, London. Available from: http://www.histmodbiomed.org/sites/default/files/W56LoRes.pdf. Accessed 4 April 2016.

NHS Estates, 1998. Health Technical Memorandum 2075: Clinical Waste Disposal/Treatment Technologies Alternatives to Incineration. The Stationery Office, London.

NHS Estates, 2003/2013. A Guide to the Decontamination of Reusable Surgical Instruments. NHS Estates. Available from: http://www.wales.nhs.uk/sites3/documents/254/instrdecontamguide.pdf. Accessed 17 February 2016.

Northern Ireland Environment Agency (NIEA), 2014. Waste Management the Duty of Care – A Code of Practice for Northern Ireland. Department for the Environment. Available from: https://www.doeni.gov.uk/sites/default/files/publications/doe/waste-policy-duty-of-care-code-of-practice-2014.pdf. Accessed 16 March 2016.

Reece, A., 2013. Knowaste West Bromwich Plant to Close. Resource Media.http://resource.co/article/Latest/Knowaste_West_Bromwich_plant_close-3126. Accessed 17 February 2016.

Scottish Environment Protection Agency (SEPA), 2009. A Guide to Waste Management Licensing. SEPA, Edinburgh. Available from: http://www.sepa.org.uk/media/28977/guide-to-waste-management-licensing.pdf. Accessed 2 February 2016.

Scottish Environment Protection Agency (SEPA), 2016. Pollution Prevention and Control. https://www.sepa.org.uk/regulations/pollution-prevention-and-control/. Accessed 2 March 2016.

Sharpsmart, 2010. http://sharpsmart.co.uk/. Accessed 17 February 2016.

SRCL, 2015. Environment. http://www.srcl.com/about/environment/. Accessed 17 February 2016.

STI Biosafe: A Division of Biosafe Engineering, 2016. STI Biosafe. Available from: http://biosafeeng.com/lib/sitefiles/pdf/STI-Brochure.pdf. Accessed 15 February 2016.

Stringer, R., 2012. Safe and Sustainable Healthcare Waste Treatment – An Essential Element of Healthcare Without Harm. (Microsoft Powerpoint). Health Care Without Harm. Available from: http://www.unep.org/ietc/Portals/136/Events/HealthcareWasteWorkshop-July2012/8-Stringer.pptx. Accessed 17 February 2016.

The Chartered Institution of Wastes Management (CIWM), 2014. An Introductory Guide to Healthcare Waste Management in England & Wales. The Chartered Institution of Wastes Management, Northampton. Available from: http://www.ciwm-journal.co.uk/downloads/Healthcare-Waste-WEB.pdf. Accessed 17 February 2016.

The Environmental Industries Commission (EIC), B P Collins LLP, 2017. Brexit: Implications for Waste and Resources Legislation. EIC, London. Available from: http://www.eic-uk.co.uk/Documents/Files/Waste_Legislation_Eng_Wales_landscape%20(2).pdf (Accessed 17 July 2017).

The State and Territorial Association on Alternative Treatment Technologies (STAATT), 1998. Technical Assistance Manual: State Regulatory Oversight of Medical Waste Treatment Technologies. EPRI, Palo Alto, CA. (TR-112222).

The Stationary Office, 1990. Environmental Protection Act 1990. (c.43). The Stationary Office, London.

The Stationary Office, 1993a. The Radioactive Substances Act 1993 (Amendment) Regulations (Northern Ireland) 2011 SI 2011/290. The Stationary Office, London.

The Stationary Office, 1993b. The Radioactive Substances Act 1993 Amendment (Scotland) Regulations 2011 SI 2011/207. The Stationary Office, London.
The Stationary Office, 1997. The Waste and Contaminated Land (Northern Ireland) Order 1997 SI 1997/2778 (N.I. 19). The Stationary Office, London.
The Stationary Office, 2000. The Ionising Radiation (Medical Exposure) Regulations, 2000. SI 2000/1059. The Stationary Office, London.
The Stationary Office, 2002. The Control of Substances Hazardous to Health Regulations 2002 SI 2002/2677. The Stationary Office, London.
The Stationary Office, 2003. The Waste Management Licensing Regulations (Northern Ireland) 2003 SI2003/493. The Stationary Office, London.
The Stationary Office, 2004. The Special Waste Amendment (Scotland) Regulations 2004 SI 2004/112. The Stationary Office, London.
The Stationary Office, 2005a. The Hazardous Waste (England and Wales) Regulations 2005 SI 2005/894. The Stationary Office, London.
The Stationary Office, 2005b. The Hazardous Waste Regulations (Northern Ireland) 2005 SI 2005/300. The Stationary Office, London.
The Stationary Office, 2006. The Ionising Radiation (Medical Exposure) (Amendment) Regulations, 2006 SI 2006/2523. The Stationary Office, London.
The Stationary Office, 2009. The Carriage of Dangerous Goods and Use of Transportable Pressure Equipment Regulations 2009 SI 2009/1348. The Stationary Office, London.
The Stationary Office, 2011a. The Waste Management Licensing (Scotland) Regulations 2011 SI 2011/228. The Stationary Office, London.
The Stationary Office, 2011b. The Waste Regulations (Northern Ireland) 2011 SI 2011/127. The Stationary Office, London.
The Stationary Office, 2012a. The Controlled Waste (England and Wales) Regulations 2012 SI 2012/811. The Stationary Office, London.
The Stationary Office, 2012b. The Pollution Prevention and Control (Scotland) Regulations 2012 SI 2012/360. The Stationary Office, London.
The Stationary Office, 2012c. The Waste (Scotland) Regulations 2012 SI 2012/148. The Stationary Office, London.
The Stationary Office, 2013a. The Controlled Waste and Duty of Care Regulations (Northern Ireland) 2013 SI 2013/255. The Stationary Office, London.
The Stationary Office, 2013b. The Pollution Prevention and Control (Industrial Emissions) Regulations (Northern Ireland) 2013 SI 2013/160. The Stationary Office, London.
The Stationary Office, 2014a. The Environmental Protection (Duty of Care) (Scotland) Regulations 2014 SI 2014/4. The Stationary Office, London.
The Stationary Office, 2014b. The Genetically Modified Organisms (Contained Use) Regulations 2014 S2014/1663. The Stationary Office, London.
The Stationary Office, 2014c. The Waste (England and Wales) (Amendment) Regulations 2014 SI 2014/656. The Stationary Office, London.
The United Nations Economic Commission for Europe (UNECE), 2015a. About the ADR. http://www.unece.org/trans/danger/publi/adr/adr_e.html. Accessed 17 February 2016.
The United Nations Economic Commission for Europe (UNECE), 2015b. European Agreement Concerning the International Carriage of Dangerous Goods by Road (ADR 2015). http://www.unece.org/index.php?id=36678&L=0. Accessed 25 March 2016.
The United Nations Economic Commission for Europe (UNECE), 2016. Dangerous Goods. http://www.unece.org/trans/danger/danger.html. Accessed 30 March 2016.

The Waste and Resources Action Programme (WRAP), 2016. WRAP and the Circular Economy. http://www.wrap.org.uk/content/wrap-and-circular-economy. Accessed 30 March 2016.

Tuttnauer, 2013. Medical Waste Autoclaves. Tuttnauer. Available from: https://tuttnauer.com/sites/tuttnauer.com/files/brochures/Medical-waste-en-autoclave-tuttnauer-21-10-13_0.pdf. Accessed 16 February 2016.

UK Parliament, 2016. Europe. http://www.parliament.uk/about/how/role/relations-with-other-institutions/europe/. Accessed 17 February 2016.

Willets, S., 2007. In: Everything you wanted to know about waste management legislation, but were afraid to ask. Practical Waste Management Training Course, 6–8 February 2007, Northampton. Chartered Institute of Waste Management (CIWM) and Environmental Services Association (ESA).

World Health Organization (WHO), 2016. About WHO. http://www.who.int/about/en/. Accessed 18 February 2016.

FURTHER READING

Commission Regulation, 2012. Commission Regulation (EU) 847/2012 of 19 September 2012 amending Annex XVII to Regulation (EC) No 1907/2006 of the European Parliament and of the Council on the Registration, Evaluation, Authorisation and Restriction of Chemicals (REACH) as regards mercury (text with EEA relevance). Off. J. L 253, 1.

Department for Environment, Food and Rural Affairs (Defra), 2012. National Implementation Plan for the Stockholm Convention on Persistent Organic Pollutants (Draft). Defra, London. Available from: https://www.gov.uk/government/uploads/system/uploads/attachment_data/file/82680/consult-nip-pop-doc-20121121.pdf. Accessed 17 February 2016.

EUR-Lex, 2015. The Direct Effect of European Law. http://eur-lex.europa.eu/legal-content/EN/TXT/?uri=uriserv%3Al14547. Accessed 2 October 2016.

European Commission, 2015b. Environmental Action Programme to 2020. http://ec.europa.eu/environment/action-programme/. Accessed 4 April 2016.

European Parliament, 2000. Directive 2000/54/EC of the European Parliament and of the Council of 18 September 2000 on the protection of workers from risks related to exposure to biological agents at work (seventh individual directive within the meaning of Article 16(1) of Directive 89/391/EEC). Off. J. L 262.

Health and Safety Executive (HSE), 2016c. Inside HSE. http://www.hse.gov.uk/aboutus/insidehse.htm. Accessed 17 February 2016.

Institute for European Environmental Policy (IEEP), Earthscan, 2010. Press Release – On the Online Manual of European Environmental Policy, A Comprehensive Guide on the EU Legislation and the Environment. http://www.ieep.eu/assets/668/Press_Release_Manual_Launch.pdf. Accessed 2 December 2016.

Legislation, 2013. Faculty of Law and Bodleian Law Library, Legal Research and Mooting Skills Programme (LRMSP). https://www3.law.ox.ac.uk/lrsp/overview/legislation.php. Accessed 3 January 2016.

Natural Resources Wales, Scottish Environment Protection Agency (SEPA), Northern Ireland Environment Agency (NIEA), Environment Agency, 2015. Waste Classification: Guidance on the Classification and Assessment of Waste, first ed. Natural Resources Wales/Scottish Environment Protection Agency (SEPA)/Northern Ireland Environment Agency (NIEA)/Environment Agency, Cardiff/Stirling/Belfast/Bristol. Technical Guidance WM3. Available from: https://www.gov.uk/government/uploads/system/uploads/attachment_data/file/427077/LIT_10121.pdf. Accessed 17 February 2016.

Scottish Government, 2012. Duty of Care – A Code of Practice. Scottish Government, Edinburgh. Available from: http://www.gov.scot/resource/0040/00404095.pdf. Accessed 4 January 2016.

The Stationary Office, 1974. The Control of Pollution Act 1974 SI 1974/40. The Stationary Office, London.

The Stationary Office, 2016. The Environmental Permitting (England and Wales) Regulations 2016 SI 2016/1154. The Stationary Office, London.

UK Parliament, 2013. Living Heritage Parliament and Europe. http://www.parliament.uk/about/living-heritage/evolutionofparliament/legislativescrutiny/parliament-and-europe/overview/britain-and-eec-to-single-european-act/. Accessed 16 February 2016.

Chapter 11

Stewardship Approaches to Reducing Health-Care Contaminants

Karin P.M. Helwig
Glasgow Caledonian University, Glasgow, Scotland

INTRODUCTION

As discussed in Chapter 2, the health-care sector, like any other industrial or service sector, inevitably gives rise to the release of some contaminants into the environment. They include a wide range of chemicals, such as cleaning agents and disinfectants, pesticide residues from food, radioactive substances, and substances used in medical devices and pharmaceuticals, which can negatively affect organisms and ecosystems. As the sewer is an important pathway for their release, risk to the aquatic compartment is of particular concern. This chapter explores the extent to which the concept of stewardship can play a role in the reduction of health-care-specific contaminants in the environment.

ENVIRONMENTAL STEWARDSHIP

The term "stewardship" implies careful management toward a particular set of outcomes. It has connotations of responsibility and care. Environmental stewardship can include making informed choices, as many people do in their daily lives, that are good (or less bad) for the environment. Concretely, it can describe a move from pollution control to pollution prevention (EPA Innovation Action Council, 2005). It can be argued that stewardship solutions may offer the best approach because they frequently address root causes (EPA Innovation Action Council, 2005) rather than "fix a problem." In the case of health-care contaminants, environmental stewardship can, for example, include a shift in focus from end-of-pipe solutions to meet environmental regulation, to a reduction of inputs of contaminants that may also apply to contaminants that are not regulated. Finally, it can be a useful concept to identify opportunities to

Environmental Contaminants, Vol. 11. https://doi.org/10.1016/B978-0-444-63857-1.00011-5

go beyond legislation and to strive for continuous improvement, commonly aspired to in environmental management more generally. With regard to pharmaceuticals and personal care products, environmental stewardship has been described as "...pollution prevention via voluntary actions as an alternative to conventional pollution control via prescribed standards" (Daughton, 2003) and has been mostly discussed for the domain of prescribers and patients. In this chapter, the sphere of stewardship as related to pharmaceuticals and other health-care contaminants is defined wider: actions of the legislator, which define the limits of voluntary action, are also considered from a stewardship perspective, and the entire life cycle of pharmaceuticals, from design to release into the environment, is considered.

THE MEDICINAL PRODUCT CHAIN

The extent to which health-care contaminants reach the environment is determined by a complex network of factors: human decisions and behaviors at multiple levels of governance, chemical properties of the compounds in question, the existing built infrastructure, and factors specific to the receiving environment. The pathway of pharmaceuticals into the environment, of particular relevance because they are designed to be bioactive, can be described as a "medicinal product chain" (Grinten et al., 2016) from drug design to discharge into the environment. The stakeholder field thus includes the pharmaceutical industry (both the "innovators" and the generics industry and including both design and manufacturing aspects); the licensing and regulatory authorities at international (e.g., EU) and national levels; wholesalers, suppliers, and importers; health-care organizations and hospital management; medical staff in hospitals, general practitioners, and other prescribers (e.g., dentists), as well as their professional bodies; pharmacies and their professional bodies; patient organizations and other NGOs; patients and the general public; environmental regulators; waste managers (e.g., waste collectors and landfill managers) and wastewater treatment companies; and farmers and drinking water companies (Fig. 1).

Understanding the drivers behind actions and decisions of these stakeholders can help identify opportunities for its reduction. As we consider to what extent stewardship can help reduce health-care contaminants, it is important to reflect on the fact that the direction and priorities of stewardship are not necessarily the same for all stakeholders. Financial stewardship, health outcomes, and environmental stewardship are weighed differently by different actors, and priorities in stewardship can be aligned or conflicting. Outcomes of decision-making depend on this strategic direction of stewardship; for example, financial stewardship and environmental stewardship both involve responsibility and careful management but may result in very different decisions. Where directions of stewardship are aligned, we may find "low-hanging fruit": areas where changes can be made that are noncontroversial or beneficial both internally to each stakeholder and externally to all stakeholders.

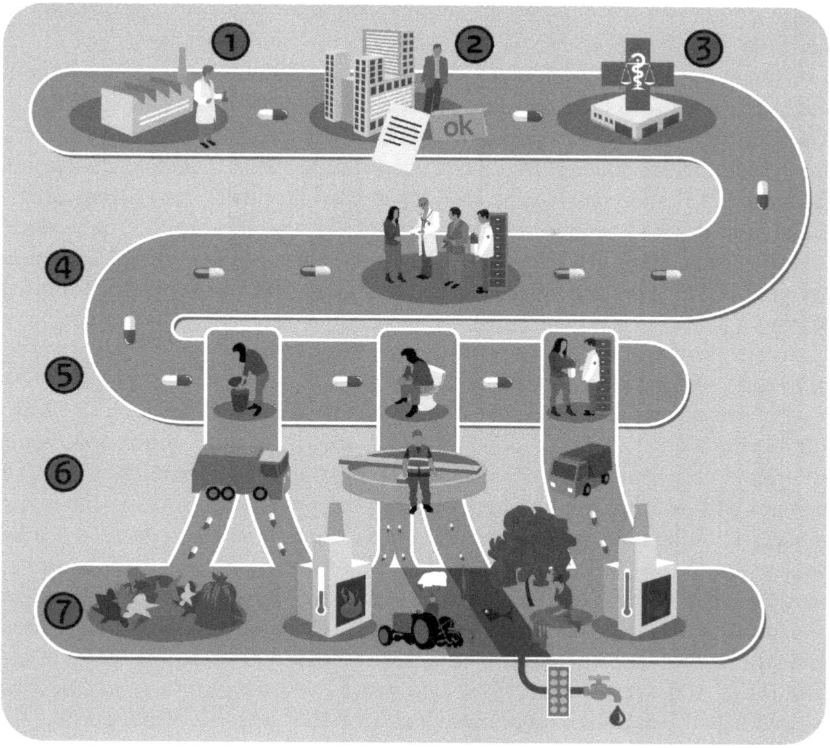

FIG. 1 A simplified representation of the medicinal product chain (Grinten et al., 2016). The numbers in the figure correspond to the following phases: 1. Development of new medicinal product. 2. Registration and market access. 3. Production and distribution of medicinal products. 4. Consumption of medicinal products. 5. Disposal of medicinal products. 6. Treatment of waste containing medicinal residues. 7. Fate of medicinal residues.

Where they are conflicting, decision-making on a higher level may need to consider whether it is appropriate to instigate change. Below, we will follow pharmaceuticals as an example of health-care contaminants along this "medicinal product chain" and consider stewardship at some of the specific phases in the life cycle of a drug.

THE LEGISLATIVE CONTEXT OF HEALTHCARE CONTAMINANTS

Current Legislation

While an environmental stewardship approach can be adopted by legislators themselves, as discussed below, legislation can also be an important driver for other stakeholders, whether directly—because of the wish to operate within the law or the wish to avoid fines—or indirectly; legislation helps shape the debate on what is acceptable and may lead to identification of "industry best practice."

Compounds used in medical devices or as pharmaceuticals are largely exempt from protective legislations such as the Registration, Evaluation, Authorization, and Restriction of Chemicals (REACH; (EC) No 1907/2006) regulation and are covered by separate legislation (see Chapter 13). Measures appropriate to other substances, such as an outright ban or restrictions on use because of environmental concerns, are not seen as appropriate for pharmaceutical compounds because of their importance to human health, and some chemicals controlled in other parts of society may have applications in health-care where benefits of their use may outweigh environmental concerns. Currently, environmental concerns cannot lead to a refusal of a marketing authorization for human pharmaceuticals.

The water environment itself in Europe is addressed by the EU Water Framework Directive (2000/60/EC), which sets maximum concentrations as Environmental Quality Standards or EQS (see also Directive 2008/105/EC) in water bodies for a small number of chemicals defined as priority substances or specific pollutants, as well as assessing the biological elements (e.g., biodiversity and the presence of quality indicator species) directly. In the United Kingdom, the disinfectant triclosan, widely used in personal care products, is a specific pollutant. A proposal (COM(2011) 876) to add three pharmaceuticals to the list—the analgesic diclofenac and the hormones 17α ethinylestradiol (EE2) and 17ß estradiol (E2)—led instead to a decision to place these on a "watch list" to gather monitoring data for the purpose of supporting the determination of appropriate measures to address the apparent risk posed by those substances (European Parliament, 2013). Estrone (E1) and macrolide antimicrobials (azithromycin, clarithromycin, and erythromycin) were later also added (European Commission, 2015). Thus, a small number of specific health care contaminants are monitored and/or controlled in surface waters as specific pollutants or priority substances and limits to environmental concentrations (EQS) apply or may apply in the future. A number of difficulties have however been identified with the ecotoxicity-based approach that currently forms the basis of setting EQS. Firstly, comprehensive ecotoxicity data for many of the thousands of pharmaceutical compounds are not available. Secondly, most pharmaceuticals enter the environment as mixtures, and mixture toxicity is poorly understood. Thirdly, there is emerging evidence of therapeutic or "beneficial" effects on aquatic species, such as reduced mortality (Klaminder et al., 2014), which cannot be measured by standard ecotoxicological methods based on toxic end points and may nevertheless have substantial ecological consequences through disruption of the food web.

While biological assessment matrices have been the subject of much debate, the direct assessment of biological quality of the WFD has as a key advantage that it provides at least some protective regulation in the absence of accurate risk assessments or a detailed understanding of how and why whole ecosystems are affected by complex and variable mixtures of pollutants. Compliance with such regulation then devolves the considerable task of identifying effective mitigating interventions to more local levels of governance, where uncertainties persist. In situations where there is an indication of risk, but a paucity of data, the concept of stewardship may be particularly useful.

Stewardship at Regulatory Level

While the legislative context of health-care contaminants has been outlined above, we may consider how the stewardship approach may be implemented by decision-making in the processes that precede laws being passed or marketing authorizations being granted. This deviates slightly from the order of the "medicinal chain" in Fig. 1 as it considers environmental legislation as an a priori factor that sets the scene for drug designers and manufacturers. Environmental stewardship at the regulatory level, in the absence of comprehensive scientific evidence of environmental risks, may involve application of the precautionary principle, a much-discussed principle that has been the basis for European environmental law since 1992 (see Art. 191, European Commission, 2012). It poses that if there is the possibility that a given policy or action might cause harm to the public or the environment and if there is still no scientific consensus on the issue, the policy or action in question should not be pursued. Though seemingly widely supported, the precautionary principle has attracted controversy in the past when critics have interpreted "precautionary" decisions as thinly veiled forms of trade protectionism: for example, decisions to ban beef imports because of the use of growth hormones (Foster et al., 2000). Proponents of the precautionary principle argue that environmental concern can stimulate innovation and economic development. Its interpretation has varied from a demand for absolute proof of safety before a policy is adopted to a much softer interpretation whereby a lack of scientific certainty cannot be used to delay cost-effective measures, hence giving precedence to cost-benefit analysis over the precautionary principle (Foster et al., 2000). The balance between a precautionary approach and an emphasis on regulatory impact on business has been described as a "regulatory pendulum" (Löfstedt 2004), and the discussion continues in recent debates between scientists on the regulation of endocrine-disrupting chemicals (EDCs) (Abbas et al., 2013; Dietrich et al., 2013; Bergman et al., 2013a) and nanomaterials (Haum et al., 2004; Van Broekhuizen and Reijnders, 2011); both of which are used in the health-care context. In REACH, EDCs are considered of similar regulatory concern as substances of very high concern (European Commission, 2016b), but scientific criteria for what constitutes an EDC have yet to be agreed, and as an interesting example of the "regulatory pendulum," the European Commission recently lost a court case brought by Sweden for failing to do so. The court ruled that criteria for such chemicals should have been identified by the agreed deadline and, in addition, that there was no requirement to carry out an impact assessment, which was nevertheless completed by the commission. In a statement, the commission argued that impact assessment is essential to guide decision-making on the criteria and implies that diverging views among scientists contributed to delays for the identification of criteria (European Commission, 2016a). The case exemplifies that while environmental stewardship is on the one hand enshrined in European legislation, uncertainty and conflicting interests with financial or economic stewardship—in the form of regulatory impact assessment—continue to contribute to delays in regulation. As far as its relevance for health-care contaminants is concerned, it should be noted that once agreement has been reached on what constitutes an EDC, such substances will not be automatically restricted in medicinal devices or indeed pharmaceuticals because of aforementioned exemptions, although undoubtedly the future criteria will influence debate in the health-care setting.

The introduction of the "watch list" mechanism in response to a proposal to introduce EQS for pharmaceuticals can be interpreted in the same balance between economic and environmental considerations. The environmental effects of hormones are relatively well researched compared with other pharmaceutical compounds (e.g., see EE2 entries in the online database WikiPharma, www.wikipharma.org), and the "possibility of harm" referred to in the precautionary principle has thus been established, although uncertainties remain in particular over ecosystem effects. At the same time, it is acknowledged that advanced wastewater treatment technologies such as activated carbon or UV filtration, which could remove pharmaceutical residues, carry their own environmental footprint—chiefly due to energy consumption—and large financial costs (Jones et al., 2007; PILLS, 2012; Gardner et al., 2012). The purpose of monitoring these substances via the watch list was "...supporting the determination of appropriate measures to address the risk posed by those substances" (2011/0429 (COD)). For EE2, the question appears to be not so much whether concern is warranted as what to do about it.

Against this legislative background, pharmaceutical compounds are designed, manufactured, prescribed, used, and eventually released into the environment.

STEWARDSHIP APPROACHES ALONG THE MEDICINAL PRODUCT CHAIN

Product Design and Production

The first phase in the medicinal product chain (Fig. 1) comprises the sphere of product design and manufacturing. This phase is directly affected by the next—registration and market access—and both are considered in this section. In this context, it has been suggested that there is a need for "greener pharmacy": the design of pharmaceutical products and processes that eliminate or reduce the use and generation of hazardous substances (EEA, 2010) and generally have a more environmentally friendly profile, as more extensively discussed in Chapter 12. This challenge to the pharmaceutical industry should not be underestimated but is a promising strategy in the long run (BIO Intelligence Service, 2013). Drug development is notoriously expensive; to bring a new drug to the market has been estimated at $403 million (DiMasi et al., 2003), whereas to bring a new pharmaceutical molecule to market could cost as much as $1.8 billion (Paul et al., 2010). The lengthy process of pre-clinical testing, clinical testing, and approval, which can take over a decade, implies that any new criteria for evaluation must be very carefully considered to avoid wasted resources. Criteria could include biodegradability, the extent to which active metabolites are formed, and (eco) toxicity of both the parent compound and any active metabolites, factors that directly impact phase 7 in the medicinal product chain (Fig. 1). However, generic pharmaceuticals tend to be much cheaper than newly developed, inpatent drugs, and in any case, new pharmaceutical products do not typically replace older ones (Taylor, 2010a), but rather add strings to the physician's bow. Without changes in regulation and policy, the availability of green pharmaceuticals does therefore not necessarily lead to reduced environmental impact.

Nevertheless, as part of a long term strategy, stewardship in this area could be driven by policy, for example, by changes to the patenting system such as extension of the patent duration (EEA, 2010) or lowering of fees and taxes (BIO Intelligence Service, 2013) for drugs that are "benign by design." Development could also be driven by developing a label for "green medicinal products" that would ensure marketing advantages for the less or nonhazardous products (BIO Intelligence Service, 2013) and help raise awareness among prescribers and patients. Finally, as the understanding of environmental effects of pharmaceuticals is expanding, these could be more actively considered in marketing authorizations, if not in the form of a refusal, then perhaps in the form of appropriate restrictions.

Manufacturing

By far the biggest contributor to the procurement carbon footprint for the UK National Health Service is the use of pharmaceuticals, which incur this footprint largely before they reach the pharmacy (Clark et al., 2010). It is generally assumed that emissions from manufacturing overall may be a minor source of pharmaceutical pollution compared with excreted residue (Murray-Smith et al., 2012). Nevertheless, extremely high levels of pharmaceuticals have been detected in effluents from treatment plants receiving manufacturing wastewaters (Larsson et al., 2007; Li et al., 2008) in Asia. Concentrations in the milligram-per-liter range have also been found in discharges from manufacturing sites in the United States and Europe (Larsson, 2014). Risk from pharmaceuticals entering the environment following human consumption tends to be mitigated by dilution: small pharmaceutical loads excreted by individuals are accompanied by a certain volume of wastewater (the "toilet flush"), which limits the concentrations released to sewer. By contrast, this is not the case for emissions from manufacturing; these have the potential to form local "hot spots" (Murray-Smith et al., 2012) and can be highly variable as production campaigns or maintenance activities can lead to peaks in concentrations (Larsson, 2014). Currently, legal controls on pharmaceutical emissions are limited. Emissions of active compounds from manufacturing are not usually subject to regulatory reporting requirements and self-reporting is rare (Larsson, 2014). Pharmaceutical manufacturers in Europe and those exporting to Europe do have to comply with the good manufacturing practice (GMP), but this framework currently does not cover environmental emissions beyond safe storage of wastes. A case study in Sweden demonstrates that environmental stewardship in manufacturing can significantly reduce emissions: a specifically designed wastewater treatment plant operated by AstraZeneca, treating wastewater from two research and manufacturing sites, was found to be able to maintain concentrations of active ingredients at below the low microgram range via a multistage system of treatment processes (although for some drugs this only equated to low environmental risk because of the high environmental dilution factor available in that particular location) (Taylor, 2010b).

Two important drivers for environmental stewardship in manufacturing are frequently cited more generally by advocates of environmental management. One is that financial savings can be achieved through waste avoidance. The pharmaceutical industry have indeed argued that significant discharges of active ingredients from manufacturing facilities is unlikely, in part because the extremely high value of drugs would prevent their release to the environment for economic reasons (Larsson, 2014); an example of a case where environmental and financial stewardship are aligned. The second driver relates to a reduction of reputational risk, or indeed positive reputational effects that may be gained through proactive environmental management, which can bring a marketing advantage. Here, it is important to reflect on the nature of the pharmaceutical industry, which has developed in two separate ways. On the one hand, there is the innovation sector, characterized by high risks and high profits and dominated by a small number of well-known brands such as AstraZeneca, GSK, Pfizer, and others; on the other, the generic manufacturers, who produce the bulk of existing medicinal products and whose names are largely unknown to the public (Taylor, 2010a). Although reputation is important for the innovators and likely to drive stewardship in this sector, this may be less of a consideration for the much more anonymous generics sector, which sells 80%–90% (by volume) of medicines taken by patients. This dichotomous description of the pharmaceutical industry does however not describe all the complexities of a global industry; for example, the innovators increasingly outsource manufacturing and so subcontract to more anonymous manufacturers. Environmental stewardship can—and perhaps should—also be driven by a sense of ethics and corporate responsibility, which may be as likely to arise among generics manufacturers as among innovators, but can be driven by social and political pressures, which vary from region to region. The issue of environmental and social governance and responsibility in complex global industries has been explored through the interpretative framework of "global production networks"; (Henderson et al., 2002; Levy, 2008). Norms and codes of conduct regarding labor or environmental standards exist on multiple levels of governance (Levy, 2008). A closer examination of where and how such norms and standards develop within the pharmaceutical industry and how this is influenced, encouraged, or constrained by hierarchical control (e.g., through outsourcing and other contractual arrangements), power, and economic relationships within the both the generics and innovator sector of the industry may help shed light on how they can be influenced. As strategies, the establishment of ecolabels (which can cover manufacturing and design), introduction of codes of conduct, exposure of malpractice, and increasing visibility have had some influence in other areas, but their strategic potential in the pharmaceutical industry has not yet been evaluated. As a possible way forward, however, an amendment to the EU's GMP regulations to include environmental considerations has been suggested (Pruden et al., 2013), which would drive stewardship in facilities operated or controlled by the innovators and the generics industry alike at least for the market share destined for Europe.

Manufacturers could contribute to a reduction in wasted pharmaceuticals by making available a larger range of pack sizes: to some extent, overprescription (discussed below) results from the fact that a pack size appropriate to the needs of the patient is not always available (BIO Intelligence Service, 2013), and in a survey among members of the public on disposal of unused medicines, an excess number of pills in the package was the most frequently cited reason in several EU countries for medicine not being used (HCWH, 2013). The availability of smaller pack sizes could relatively easily be driven via the marketing authorization. Finally, it has also been suggested that expiry dates tend to underestimate the shelf life of medicines, resulting in disposal when in fact the medicine is still effective (BIO Intelligence Service, 2013). Overcautiousness to protect human health—which can also be seen as careful stewardship but toward different objectives—so has negative consequences for the environment.

The Healthcare Sector

Stewardship at National, Regional, and Institutional Levels of Governance: The Example of Plastics Additives

As we travel further down the "medicinal product chain," we enter the domain of the health-care sector, which includes health-care managers, hospital administrations, prescribers, and patients, and ought to take a look at various levels of governance that control this sector. An illustration of stewardship at these levels can be found in decision-making on phthalates and bisphenol A (BPA). Decision-making on other issues, including prescribing policies or source segregation measures as discussed below, may be expected to occur at similar levels of governance. Phthalates and BPA are both widely used plastics additives; both have been identified as endocrine disruptors (Bergman et al., 2013a,b). In health care, a key concern is the use of phthalates in plastic tubing used for intravenous administration of liquids, particularly on neonatal wards. As exposure toxicity relates to body weight, neonates are at particular risk from EDCs through leaching from intravenous tubing (Mallow and Fox, 2014; Duty et al., 2013). Phthalate-free alternatives are available on the market. In the absence of EU legislation, decisions to reduce or eliminate the use of phthalates and BPA have been made at national, regional, and institutional levels of governance (Table 1).

Clearly, taking decisions at higher levels of governance facilitates the production of guidance and information, which is important to avoid substitutions that deliver no benefits or even disbenefits.

A more extensive study would be required to investigate exactly why decision-makers in these specific cases decided to write policies to reduce EDCs. Concerned individual members of staff or lobbyists may have instigated meetings, raised agenda points, or in some other way set in motion a change. However, it is interesting to note that in most of these cases, interventions are directed exclusively or primarily at neonates. This appears to

TABLE 1 Decision-Making at Different Levels of Governance[a]

Level of Governance	Intervention	Actor
National	Guidelines for reduction; procurement requirements; list of alternatives	Denmark
	Ban on use in certain hospital departments	France
	Recommendation on reduction of hospital use with certain population groups	Germany
Regional	Decision to phase out all use (including in health care)	Stockholm County Council; Sweden
Local	PVC or phthalate-free policies	Vienna Hospital Association, Austria
Institutional	PVC or phthalate-free policies	Westfries Gasthuis, Netherlands

[a] *Examples from Amaral (2015); list not exhaustive.*

indicate that the drivers for these decisions may primarily lie in the protection of human health, a key guiding principle of health-care institutions. While the aquatic environment will benefit from interventions in neonatal wards, softeners will continue to be used widely in society, including in packaging, toys, till receipts, safety equipment, and electronics, and so, potential for environmental gains through reductions in the health-care sector alone is limited. In the sphere of health care, however, there is a unique alignment of human health stewardship and environmental stewardship that does not exist as strongly in other societal contexts, where barriers such as increased costs or the lack of available alternatives may not be overcome as readily.

Pharmaceuticals: Prescribing and Consumption

In this health-care domain—phase 4 in the medicinal product chain (Fig. 1)—a key stage of a drug's life cycle is its consumption by a patient either as a prescribed drug or purchased over the counter. For pharmaceuticals, the stewardship concept in interactions between the prescriber and the patient is particularly relevant in the absence of regulatory restrictions; Daughton and Ruhoy (2013) argue for closer collaboration between the health-care and environmental science professions to develop strategies that would benefit both health and environmental outcomes.

The extent to which drugs are controlled by the medical profession varies by drug and by country; even within the relatively homogenous area of the EU, some drugs can be bought in supermarkets, others only in pharmacies, yet others must be prescribed. In some countries, antimicrobials are only available on prescription, whereas in others, they can be bought over the counter. Sometimes, the distinction is blurred, for example, when a prescription is issued as an online consultation service, which can involve just a few brief questions.

Notions of Health and Illness

Before a decision is made to medicate, a crucial judgment is made on illness or any other term to describe something is amiss, versus health, or a "normal," satisfactory state of wellbeing, or at least an expectation to return to such a state without medical or pharmaceutical intervention. This decision is taken either by an individual (e.g., in the case of over-the-counter medicine) or via an interaction between an individual and their physician. Concepts of health and illness, however, can be subjective, both culturally and personally, and can be influenced and challenged. Pharmaceutical marketing strategies have been implicated in increased demand for medicinal products (Moynihan et al., 2002; House of Commons Health Committee, 2005; BIO Intelligence Service, 2013; Triggle, 2007). Direct-to-consumer advertising of over-the-counter products clearly results in increased pharmaceutical use—if not, pharmaceutical companies would not spend money on it—but prescribers may also be influenced, for example, by educational materials provided by the pharmaceutical industry or conferences sponsored by the industry (Moynihan et al., 2002). Over a decade ago, a UK report (House of Commons Health Committee, 2005) identified as a major issue "the increasing medicalisation of society," which it describes as a trend not created by the pharmaceutical industry but one that has been encouraged by it. It noted the industry was increasingly focused on marketing-based approaches and identified a pervasive and persistent influence of the industry not only on clinical medicine but also on patients, regulators, media, civil servants, and politicians.

It has been proposed that the boundaries of some real health conditions are extended into the normal range, leading to unnecessary drug treatment and ignoring sometimes more appropriate nondrug treatments (Triggle, 2007; House of Commons Health Committee, 2005). For example, it can be argued that while antidepressants can be effectively and successfully used for many patients, "...unhappiness is part of the spectrum of human experience, not a medical condition" (House of Commons Health Committee, 2005). While few would take issue with the fact that depression can be a serious and debilitating condition requiring medical care, it would not be unreasonable to query the fourfold increase in the number of antidepressant prescriptions in Scotland between 2003 and 2012 (Helwig et al., 2016). Baldness, irritable bowel syndrome (IBS), social phobia (shyness), osteoporosis, female sexual dysfunction, and attention-deficit hyperactivity disorder (ADHD) have also been cited as examples of conditions where the boundaries of what requires pharmaceutical treatment appear to have shifted (Triggle, 2007; Moynihan et al., 2002) in articles aiming to provoke debate.

This changing discourse on illness and wellness, which partly takes place in interactions between patients and prescribers, has consequences not only for the affordability of health care (Triggle, 2007) but also for the environment. The role of stewardship in this complex context is hard to define but could be conceived as a more proactive deliberation on how society sees its relationship with medicines and, within that, an awareness of environmental effects. It could also include, or is at least entangled with, notions of stewardship toward one's own health: a desire or responsibility to keep in good health and so avoid the need for medication. On a more practical level, both patients and prescribers could adopt a more discerning attitude when deciding whether to medicate a minor ailment or whether to accept a certain level of inconvenience.

Alternatives to Medicines

A variety of forces within the systems of health care often lead to a consideration of pharmaceuticals as the first line of treatment (Daughton and Ruhoy, 2013). In a 2004 UK survey, more than 80% of practitioners self-reported overprescribing antidepressants to their patients with a quarter calling for increased access to psychological services instead (BBC, 2004). Many health conditions can be prevented or controlled by lifestyle changes, such as diet and exercise. A more methodical, holistic approach may ultimately prove more efficacious, with fewer downstream effects (Daughton and Ruhoy, 2013). Significant proportions of both doctors and patients are open to non-drug alternatives (noPILLS, 2015). In a qualitative study on attitudes to medicines, some members of the population indicated a desire to make healthy lifestyle choices, a preference for natural remedies to "chemicals," and a general belief that individuals can take greater levels of control and responsibility over their own health (noPILLS, 2015).

Overprescription and Dosage

While on the one hand, there is a general sense that there is an overconsumption of pharmaceuticals (noPILLS, 2015), there is equally a common assumption that prescription medicines are "necessary," and a consequent and generally justified reluctance—by environmental scientists and the general public alike—to query prescribers' decisions. This assumption has extended to both the choice of drug and the prescribed dose. However, it has been proposed that in fact dosage has been somewhat neglected from risk mitigation strategies (Daughton and Ruhoy, 2013). Daughton and Ruhoy (2013) draw attention to the fact that while the traditional optimization of the trade-off between therapeutic effect and adverse side effects is complex, this complexity is largely lost in day-to-day prescribing and decisions are guided by on-label dosages, as physicians have neither the time nor the expertise to consider a more individualized prescription. Overprescribing has been reported in several EU countries (noPILLS, 2015), despite the fact that within the medical sphere, there has been attention to the fact that sometimes "less is more." In their review, Daughton and Ruhoy (2013) identify as many as 53 common drugs from medical literature where dosage below the on-label guidelines can be effective, typically at one-half to one-quarter of the minimum on-label dose. Of these, 38 were drugs that have been identified in the water environment. It would appear that in some cases at least, a shift toward reduced prescribing would therefore be aligned with the objectives of careful medical stewardship and environmental stewardship and therefore could constitute "low-hanging fruit." The question is how to drive such a change.

Learning from Antimicrobial Experience

In the context of health care, the term stewardship has most frequently been used to describe efforts to reduce the unnecessary prescription of antimicrobial drugs (e.g., Dixon and Duncan, 2014). While not always defined explicitly, the term is generally used to imply "careful management" by both prescribers and patients, with a view to a minimal prescribing regime and preferential prescription of narrow-spectrum over broad-spectrum antimicrobials and with patients completing the prescribed course and—in case of, for example, an adverse reaction—appropriate disposal, such as by returning leftover drugs to pharmacy. Interest in antimicrobial stewardship has undoubtedly arisen due to a widespread awareness that overprescription of antimicrobial drugs can lead to increased occurrence of antimicrobial resistance (AMR; see Chapter 8 for a discussion of this topic). However, other positive outcomes have since been identified and are now also recognized as drivers of antimicrobial stewardship: cost savings, improved patient (health) outcomes, and toxicological reduction. How stewardship in AMR prescribing can be engendered has been the focus of a number of research efforts (e.g., Dellit et al., 2007). Due to a combination of high-volume use and low threshold values for ecotoxic effects (indicating high toxicity), the therapeutic group of antimicrobial medicines includes some of the drugs posing highest risk to the environment (Helwig et al., 2016). Changes to consumption patterns as a result of stewardship approaches are therefore likely to affect ecotoxicological outcomes in our aquatic environments. The nature of these has to our knowledge not yet been the subject of research; while a reduction of consumption will result in a reduction of risk, this may not be the case of substitution, for example, where a broad-spectrum drug is substituted by a narrow-spectrum one or by a combination of several narrow-spectrum drugs. However, the example of AMR prescribing evidences that a stewardship approach can indeed be used to drive real and significant changes in pharmaceutical consumption patterns.

While the global health threat of AMR is uniquely linked to antimicrobials, benefits of cost savings, improved patient outcomes, and a toxicological reduction may apply equally to other therapeutic groups, and inappropriate prescribing has indeed been described for other therapeutic groups (e.g., Goulding, 2004). Some publications have investigated ways to reduce prescribing of other therapeutic drug groups (e.g., Von Korff et al., 2016), which would have the potential to deliver similar benefits, but not nearly as frequently as in the context of AMR. It may therefore be useful to reflect on some of the findings from the AMR experience to identify how stewardship has been engendered or reductions achieved.

In the hospital setting, increased involvement of pharmacists in prescribing decisions may be useful (Broom et al., 2015). The authors identified that pharmacists' expertise on best-practice decision-making around antimicrobials (e.g., regarding issues of choice, dosage, duration, and mode of delivery) was often not utilized by medical professionals. Interprofessional and organizational barriers posed challenges for pharmacists considering intervention in the prescription process, and organizational (policy) change would be required to encourage the use of pharmacists' expertise in order to drive prescribing improvement. To facilitate antimicrobial stewardship, Dellit et al. (2007) recommended involvement of hospital administration and medical staff leadership, obtaining appropriate authority from the hospital administration, negotiation of compensation and expected outcomes, and tracking of prescriptions. Suggested elements of strategic interventions included education, guidelines, antimicrobial order forms, streamlining or de-escalation of therapy, and dose optimization. However, there was also evidence that education alone, without active intervention, was only marginally effective (Dellit et al., 2007).

Drugs in many therapeutic groups are prescribed primarily by general practitioners, with only a small fraction prescribed by specialists. In the noPILLS project, it emerged that the role of community pharmacists in different countries varies; in Germany, the pharmacist indeed acts as an advisor to the physician, and in the Netherlands, a relatively close cooperation between pharmacist and prescriber exists. In Luxembourg, it is the pharmacist that is held responsible should the medication have adverse side effects, and he or she can modify a prescription by calling the prescribing physician if deemed necessary (noPILLS, 2015). By contrast, in a qualitative Spanish study on antimicrobial prescribing, participating physicians cited indiscriminate dispensing of antimicrobials by pharmacies as a reason why they sometimes prescribed antimicrobials (antimicrobials) unnecessarily (Vazquez-Lago et al., 2012). Proactive involvement of the pharmacist may allow for more appropriate dosage prescriptions but may require organizational change, and the relationship between physicians and pharmacists and their relative responsibilities toward prescribing may be a key factor in prescribing outcomes. The issue of control over prescribing not only relates to pharmacists; in an Indian study (Kotwani et al., 2010), physicians pointed out that "quacks" (nonqualified doctors) contribute to the problem (of indiscriminate antimicrobial prescribing), indicating that perhaps wider availability contributes to a sense of powerlessness among physicians to address overprescription.

The studies by Kotwani et al. (2010) and Vazquez-Lago et al. (2012) share a remarkable number of findings, given they were carried out in completely different health-care systems and cultures. Both cite patient expectations as a reason for prescribing antimicrobials unnecessarily. However, already in 1997, Cockburn and Pit (1997) concluded that although patients brought expectations to the consultation regarding medication, the doctors' opinions about their expectations were the strongest determinants of prescribing. Several studies (Butler et al., 1998; Himmel et al., 1997) also indicated some time ago that patient satisfaction is not necessarily related to receiving prescriptions. Interviews with people in the community in the context of the noPILLS project found that generally, people would prefer to take no more medication than they "need" and that a certain wariness over taking "too many" drugs exist. This cautiousness extended to (and could be even stronger when relating to) dependents in their care, such as children or elderly relatives. If patient expectations are an important factor in the prescription and consumption of pharmaceuticals, education may engender stewardship. In a study (Macfarlane et al., 2002), an information leaflet (on the natural course of their condition and the advantages and disadvantages of antimicrobial use) issued to patients who were given a prescription for antimicrobials to use *at their own discretion* reduced the uptake of the prescription. Little et al. (2004)

logically suggest that if *perceived* patient pressure is a stronger predictor of all doctor behaviors, unnecessary resource use and adverse effects can be limited if doctors directly ask patients about their expectations, when management decisions are not thought to be medically needed. Both Kotwani et al. (2010) and Vazquez-Lago et al. (2012) also report physicians' fear of getting "blamed" (vs. "peace of mind"), indicating that while there may be an acceptance that reducing pharmaceutical use in general constitutes health-care stewardship, "erring on the side of caution" in individual cases may result in overprescription. This transferal from the general to the individual may be the most challenging aspect of applying stewardship in prescribing.

Unlike the issue of AMR, pharmaceuticals as emerging environmental contaminants are still mostly considered an environmental issue rather than a health concern and as such perhaps less obvious or more controversial to be accepted by prescribers as something that should affect their prescribing decisions. Strategically, a starting point could therefore be to investigate closely how and where reduced lower dosage or nonpharmaceutical therapy improves patients' outcomes, so that a reduction in prescriptions can be driven by stewardship within the profession's own frame of reference. Pharmacists may be able to play an important role in this process. An awareness and consideration of environmental effects of drug therapy can and should begin to feed into this frame of reference over time; training, professional education, and labeling can all contribute to this process.

Wasted Pharmaceuticals

A large amount of both prescribed and OTC pharmaceuticals are disposed of unused. In the United Kingdom, it is estimated that community care medicine wastage is in the order of £300M per year (noPILLS, 2015). This includes over-the-counter medicines held "in stock" that have gone beyond their expiry date or medicine obtained because of a specific condition that was cured before it was all used. Sometimes, treatment is interrupted sooner than expected, such as due to the death of a patient or because of a change of medicine following an adverse reaction or ineffectiveness (Mackridge and Marriott, 2007). As mentioned above, changes in pack sizes could go some way to address these issues. Not only stocking up on drugs such as painkillers and cough medicines but also overprescribing and misprescribing (including higher than necessary dose strengths or longer than needed durations (Daughton and Ruhoy, 2013)) can contribute to accumulation of drugs in households, unauthorized distribution of drugs to others, self-medication, and unintended poisoning of children in particular, as well as higher than necessary costs to patients and health-care systems.

Patients dispose of unused pharmaceuticals mainly via one of three routes: through take-back facilities at pharmacies, via solid waste (to landfill or incineration) or via the sink or toilet (some aspects of pharmaceutical waste management are covered in the previous chapter of this book). The choice of route determines environmental outcomes, and patients' disposal behavior is therefore a key area of interest. For pharmaceuticals that are extensively metabolized by the patient, disposal of unused pharmaceuticals may be a major contributor to environmental risk posed by the parent compound. When larger amounts are disposed incorrectly, measurable peak concentrations may occur; indeed, the relative concentrations of parent compounds and metabolites have been used to identify incorrect disposal events (Petrie et al., 2016). Even if for other drugs incorrect disposal is likely to be a relatively minor pathway compared with the excreted fraction, addressing this issue is a relatively uncontroversial way to reduce environmental concentrations, and efforts have been made to persuade people to take their medicines back to the pharmacy. The qualitative noPILLS study exposed quite complex and sometimes unexpected factors that influenced participants' disposal behavior. In an example of careful stewardship directed toward the safety of other people, participants disposed of their medicines down the toilet as they felt this was a safer alternative to putting them in the bin. Dissolved drugs were seen to "melt away" where nobody could access them. While many people thus expressed a desire to "do the right thing," this did not always lead to the most desirable behavior. Importantly, many participants that did return drugs to the pharmacy thought they were being asked to take drugs back, so they would be reused and were keen to contribute to cost savings for the health service or perceived distribution to "poor people." Consequently, one participant felt there was no point in taking back drugs that had expired (noPILLS, 2015).

A rudimentary understanding of environmental effects of pharmaceuticals appeared to exist among some members of the general public, in particular on hormones affecting fish, with some expressing a notion that effects on humans via the food chain may exist too. However, there was also evidence of confusion and misconceptions, perhaps not surprising given the complexity of the issue (noPILLS, 2015). It is therefore encouraging that while people who take pharmaceuticals back to the pharmacy cannot be assumed to do so for environmental reasons, it appears that people who see themselves as "living an environmentally friendly life" are more likely to take their unused pharmaceuticals back to pharmacy than those who do not (Bound et al., 2006). A combination of facilitation (by providing convenient disposal facilities) and an appeal to the public's sense of environmental stewardship and responsibility may therefore have significant potential.

The noPILLS study concluded that "...with a clear message in an effective format, it may well be possible to 'redirect' people's well-intended behaviour towards more desirable disposal habits" and that "raising awareness may indeed [yield] some 'low hanging fruit' in terms of achieving a reduction of pharmaceuticals in the environment" (noPILLS, 2015).

Source Segregation of Urine

There is a possibility for intervention between consumption by the patient and entry into the sewer: source segregation of urine. EU research project noPILLS, focused on the prevention of pharmaceutical inputs, investigated the acceptability of "road bags" for patients that had received iodinated contrast agents. In the study, outpatients that were given this medication were asked to collect their urine in special collection bags filled with absorbent granules that subsequently go for incineration, for a period of 24h during which most of the compound is excreted. This pilot resulted in a measurable reduction in contrast agents in hospital effluent. The project concluded that with a little extra available time per patient, urine separation could be included in routine procedures. A notable finding was that active staff involvement was a critical factor in engaging and motivating patients, pointing to a crucial need to educate and inform medical staff (noPILLS, 2015).

"Road bags" would be suitable for the reduction of other compounds of specific concern but is probably limited to a small number of individually targeted compounds. Urine segregation toilets require more investment but would remove all contaminants in urine at minimal inconvenience to users and have been well accepted by most users (Larsen et al., 2009), offering the additional advantage of enabling nutrient recovery. At the moment, segregation systems are mainly implemented in pilot projects and have yet to be incorporated in mainstream developments.

Wastewater Treatment

The final opportunity to intercept pharmaceuticals before they reach the natural environment is in wastewater treatment facilities (phase 6 in Fig. 1). While commonly applied technologies such as activated sludge treatment do not removal all pharmaceutical residues, advanced technologies capable of removing all or nearly all pharmaceutical residue are available and have the advantage of removing other micropollutants as well. It should be noted that this option of "last resort" is usually not available for the reduction of residues of veterinary medicines and that a proportion of raw wastewater often bypasses treatment altogether in times of high rainfall when drainage systems cannot cope with the volumes of storm water in combined sewer systems. In some regions of the world, even basic treatment technologies are not commonly used, and advanced treatment technologies are not currently a realistic prospect. Also important to understand, as outlined above, is that such removal technologies tend to come at a high cost (ultimately to water bill payers) and have a significant environmental footprint themselves, as they tend to be energy intensive. Nevertheless, as for most pharmaceuticals excretion into the sewer is the primary pathway into the environment, this option must be considered. Again, there is evidence of environmental stewardship at various levels of governance. In 2014, Switzerland became the first country in Europe to introduce a law with a target of reducing the release of micropollutants (which include pharmaceuticals and EDCs) into the environment. The law primarily impacts on around 100 of the larger wastewater treatment plants; the selected plants treat about half of the country's wastewater and will work to a target of 80% removal of micropollutants. Three-quarters of the cost of the upgrades will be paid for a nationwide wastewater tax of CHF 9 (around £7) per person per year (EAWAG, 2014). In Germany, such a legal driver does not yet exist, but some treatment plants have already constructed additional steps for micropollutants, and further feasibility studies are being conducted. In the Netherlands, the Foundation for Applied Water Research (STOWA), a demand-driven organization serving regional water managers, together with one regional waterboard, gathered information on cost implications for the Dutch situation, noting that in the absence of a legal driver, the main question is cost.

Separate treatment of highly concentrated effluents, such as those from hospitals, may be worth considering (e.g., Verlicchi et al., 2010). While the overall fraction of pharmaceutical residue released from hospitals is likely to be small compared with the fraction emitted by the community, treating hospital wastewater can remove a relatively large pharmaceutical load by treating a relatively small amount of wastewater. Furthermore, in specific geographic situations—such as the presence of a large hospital in a rural area—the hospital's contribution to the load arriving at a wastewater treatment plant may be considerable (Herrmann et al., 2015). In a small number of hospitals, so far mostly as pilot facilities in the context of research projects (see, e.g., www.pills-project.eu), decentralized treatment has indeed been implemented at the hospital level. Nevertheless, the environmental stewardship demonstrated by these participating institutions has already yielded benefits of new insights into the utilization of advanced treatment technologies under the very specific operating conditions of hospitals.

Generally, within the water industry, wastewater treatment managers and operators have to work within a regulatory framework that imposes restrictions on certain discharge parameters. Should they exceed the limits set in their permit, fines may be issued. Building in sufficient margin to ensure the limits can be met under any conditions is expensive. It is not unique to the wastewater industry that it is therefore understood and accepted by management that occasionally, permit conditions are broken and fines incurred; this may be as the result of a formal business risk analysis but more likely because there is an acceptance that sometimes, "these things happen." Equally, there can be a sense of "getting away with it" if such unauthorized discharges escaped the regulator's controls. A qualitative study with risk managers in the United Kingdom, Canada, and America found that a risk management culture remains elusive in many wastewater utilities (Allan et al., 2013). The authors identified that out of seven core risks from interviews with the utilities' risk managers—after financial risk, environmental, public health, regulatory, and safety risk—financial risk was most frequently identified as a "core risk," and that this mostly revolved around financial consequences of risk rather than about financial risk per se. The authors also identified that the level of investment into the utilities fell far short of that required to improve water service standards in line with customer expectations and pose that operational and systems risk will be more likely in underinvested systems (Allan et al., 2013). This underlines that not all countries are at the stage where investment in tertiary treatment is likely to be implemented soon.

DISCUSSION

It is undisputed that use of medicines brings benefits to society, not only in terms of health outcomes but also by making a significant contribution to the economy, both through commercial activity in the pharmaceutical sector and, for example, by reducing loss of productivity through sickness. However, unintended adverse effects include risk or harm to humans and the environment. This is echoed in the economic sphere; economic benefits may need to offset against the cost of adverse human health effects, the cost of mitigation of pharmaceutical residues in waters, or, on a larger scale, loss of ecosystem services through a reduction in water quality or even the cost of possible wider health effects such as a hypothetical link between EDCs (which include some pharmaceuticals) and a decline in male fertility (Jobling et al., 2009). Thus, a new balance requires to be made. As a society, it is necessary to reexamine our relationship with pharmaceuticals. We see evidence of stewardship where despite a lack of comprehensive understanding of environmental effects, awareness of the issue of micropollutants appears to be driving interest in advanced treatment technologies in the wastewater sector. Nevertheless, with current financial constraints and perhaps more importantly concern over the environmental footprint of such technologies themselves, this solution is seen by many as at best part of a way forward. In pharmaceutical design and manufacturing, financial drivers, such as through reputational benefits, can drive environmental stewardship in parts of the industry, but this is not the case across the entire sector, and in a market-driven industry, incentives or procurement controls could help to better align financial and environmental drivers. The health-care sector appears to offer the best opportunities for a stewardship approach, as health outcomes, financial outcomes, and environmental outcomes could all be enhanced through stewardship in prescribing. The antimicrobial experience teaches us that information, education, facilitation, and an active program of interventions promote the best chance to move from good intentions to real change. From a patient's perspective, with respect to disposal, a general preparedness to "do the right thing" may be channeled toward environmental stewardship through a combination of awareness raising and facilitation. With respect to consumption, there is a need to reexamine our relationship with pharmaceuticals. While we have not considered new drug development in detail, in the context of the AMR debate, it has been proposed that decoupling the link between profitability and sales volume could enable reward for the development of new antimicrobial drugs (O'Neill, 2016). This coupling is the basis of the pharmaceutical industry's market-based approach that has been identified as an important influence on patients and prescribers leading to increased medicalization and prescription. This leads to the question whether alternative funding mechanisms, not only for antimicrobial development but also for other parts of the pharmaceutical industry, would allow a profitable innovation sector that is less reliant on sales volumes and could thus reduce some of the marketing pressures on prescription and consumption behavior. This, however, is beyond the scope of this chapter.

CONCLUSION

In the absence of a full understanding of environmental effects and risks, stewardship approaches have much to offer, in particular when environmental outcomes can go hand in hand with improved patient outcomes and reduced costs. As outlined above, the concept can be applied along the entire medicinal product chain and at all levels of governance; while legislators themselves can certainly adopt a precautionary approach, it is in particular in the absence of legislation that stewardship approaches can make a real difference. There is good evidence that this is often the case in the health-care sector. To effect real and lasting change, environmental stewardship needs to be more than just good intentions and be driven by policy and regulation, embedded in health-care practice through facilitation, education, and optimized organizational structures and cultures, and integrated in our thinking about health and health care.

REFERENCES

Abbas, A., Ahrens, L., Andersson, A.M., Andersson, P., et al., 2013. The 2013 Berlaymont Declaration on Endocrine Disrupters.

Allan, R., Mauelshagan, C., Luis, A.M., Jeffrey, P., Pollard, S., 2013. Making risk management stick: Reflections on risk governance in water utilities. In: Water Contamination Emergencies: Managing the Threats. The Royal Society of Chemistry, Cambridge, UK, pp. 33–46. https://doi.org/10.1039/9781849737890-00033.

Amaral, M.J., 2015. Non-Toxic Healthcare: Alternatives to Phthalates and Bisphenol A in Medical Devices. Healthcare Without Harm. Brussels, Belgium.

BBC, 2004. Depression Pills Too Accessible. BBC News Channel. Available from: http:/news.bbc.co.uk/1/hi/health/3579635.stm. Accessed 6 June 2016.

Bergman, Å., Andersson, A.-M., Becher, G., van den Berg, M., et al., 2013a. Science and policy on endocrine disrupters must not be mixed: a reply to a 'common sense' intervention by toxicology journal editors. Environ. Health 12 (1), 69. https://doi.org/10.1186/1476-069X-12-69.

Bergman, Å., Heindel, J.J., Jobling, S., Kidd, K.A., Zoeller, R.T., 2013b. State of Science of Endocrine Disrupting Chemicals 2012: Summary for Decision-Makers. Report Prepared for UNEP/World Health Organisation.

BIO Intelligence Service, 2013. Study on the Environmental Risks of Medicinal Products. Final Report Prepared for Executive Agency for Health and Consumers.

Bound, J.P., Kitsou, K., Voulvoulis, N., 2006. Household disposal of pharmaceuticals and perception of risk to the environment. Environ. Toxicol. Pharmacol. 21 (3), 301–307. https://doi.org/10.1016/j.etap.2005.09.006.

Van Broekhuizen, P., Reijnders, L., 2011. Building blocks for a precautionary approach to the use of nanomaterials: positions taken by trade unions and environmental NGOs in the European nanotechnologies debate. Risk Anal. 31 (10), 1646–1657. https://doi.org/10.1111/j.1539-6924.2011.01615.x.

Broom, A., Broom, J., Kirby, E., Plage, S., Adams, J., 2015. What role do pharmacists play in mediating antibiotic use in hospitals? A qualitative study: table 1. BMJ Open 5 (11). e008326. https://doi.org/10.1136/bmjopen-2015-008326.

Butler, C.C., Rollnick, S., Pill, R., Maggs-Rapport, F., Stott, N., 1998. Understanding the culture of prescribing: qualitative study of general practitioners' and patients' perceptions of antibiotics for sore throats. BMJ [Br. Med. J.] 317, 637.

Clark, J.H., Breeden, S.W., Summerton, L., 2010. Green(er) pharmacy. In: Kümmerer, K., Hempel, M. (Eds.), Green and Sustainable Pharmacy. Springer-verlag, Berlin, p. 41.

Cockburn, J., Pit, S., 1997. General practice prescribing behaviour in clinical practice: patients' expectations and doctors' perceptions of patients' expectations—a questionnaire study. BMJ [Br. Med. J.] 315 (7107), 520–523.

Daughton, C.G., 2003. Cradle-to-cradle stewardship of drugs for minimizing their environmental disposition while promoting human health. Environ. Health Perspect. 111 (5), 757–774.

Daughton, C.G., Ruhoy, I.S., 2013. Lower-dose prescribing: minimizing "side effects" of pharmaceuticals on society and the environment. Sci. Total Environ. 443, 324–337. https://doi.org/10.1016/j.scitotenv.2012.10.092.

Dellit, T.H., Owens, R.C., McGowan, J.E.J., Gerding, D.N., Weinstein, R.A., Burke, J.P., Huskins, W.C., Paterson, D.L., Fishman, N.O., Carpenter, C.F., Brennan, P.J., Billeter, M., Hooton, T.M., 2007. Guidelines for developing an institutional program to enhance antimicrobial stewardship. Clin. Infect. Dis. 44 (1), 159–177. https://doi.org/10.1097/IPC.0b013e318068b1c0.

Dietrich, D.R., Von Aulock, S., Marquardt, H., Blaauboer, B., et al., 2013. Scientifically unfounded precaution drives European Commission's recommendations on EDC regulation, while defying common-sense, well-established science and risk assessment principles. ALTEX 30 (3), 381–382. https://doi.org/10.1016/j.tem.2012.12.005.

DiMasi, J.A., Hansen, R.W., Grabowski, H.G., 2003. The price of innovation: new estimates of drug development costs. J. Health Econ. 22, 151–185.

Dixon, J., Duncan, C.J., 2014. Importance of antimicrobial stewardship to the English National Health Service. Infect. Drug Resist. 2014 (7), 145–152. Available from: https://www.dovepress.com/importance-of-antimicrobial-stewardship-to-the-english-national-health-peer-reviewed-fulltext-article-IDR#.

Duty, S.M., Mendonca, K., Hauser, R., Calafat, A.M., Ye, X., Meeker, J.D., Ackerman, R., Cullinane, J., Faller, J., Ringer, S., 2013. Potential sources of bisphenol a in the neonatal intensive care unit. Pediatrics 131 (3), 483–489. https://doi.org/10.1542/peds.2012-1380.

EAWAG, 2014. Wastewater Treatment Plant Extension Underway. News Portal. Available from: http://www.eawag.ch/en/news-agenda/news-portal/news-detail/news/der-klaeranlagen-ausbau-laeuft/?tx_news_pi1%5Bcontroller%5D=News&tx_news_pi1%5Baction%5D=detail&cHash=6a856ddab073ff2e13e9f25bc860fa1a. Accessed 26 June 2016.

EEA, 2010. Pharmaceuticals in the Environment. Results of an EEA Workshop. EEA Technical Report Series 1725-2237. https://doi.org/10.2800/31181.

EPA Innovation Action Council, 2005. Everyday Choices: Opportunities for Environmental Stewardship. Report for United States Environmental Protection Agency, Washington, p. 19.

European Commission, 2016a. Commission Action to Comply With Judgement in Case T-521/14: Sweden vs the Commission. EP Plenary – Commission Statement 2 February 2016. Available from: https://ec.europa.eu/commission/2014-2019/andriukaitis/announcements/ep-plenary-commission-statement-commission-action-comply-judgement-case-t-52114-sweden-vs-commission_en. Accessed 20 June 2016.

European Commission, 2016b. Endocrine Disruptors. Available from: http://ec.europa.eu/environment/chemicals/endocrine/index_en.htm. Accessed 20 June 2016.

European Parliament, 2013. Surface Waters: 12 New Controlled Chemicals, Three Pharmaceuticals on Watch List. Press Release.

European Commission, 2015. Commission Implementing Decision (EU) 2015/495 of 20 March 2015 establishing a watch list of substances for Union-wide monitoring in the field of water policy pursuant to Directive 2008/105/EC of the European Parliament and of the Council. Off. J. Eur. Union 78, 40.

European Commission, 2012. The treaty on the functioning of the European Union. Off. J. Eur. Union C326, 132.

Foster, K.R., Vecchia, P., Repacholi, M.H., 2000. Science and the precautionary principle. Science 288, 979–981.

Gardner, M., Comber, S., Scrimshaw, M.D., Cartmell, E., Lester, J., Ellor, B., 2012. The significance of hazardous chemicals in wastewater treatment works effluents. Sci. Total Environ. 437, 363–372. https://doi.org/10.1016/j.scitotenv.2012.07.086.

Goulding, M.R., 2004. Inappropriate medication prescribing for elderly ambulatory care patients. Archives of Internal Medicine 164 (3), 305–312. https://doi.org/10.1001/archinte.164.3.305.

Grinten, E.V.D., Breure, A., Lambooij, M., Lette, M., 2016. Towards Balancing the Benefits of Pharmaceutical Care and Minimizing Its Environmental Harm: Identification of Potential Levers in the Medicinal Product Chain. RIVM, Utrecht.

Haum, R., Petschow, U., Steinfeldt, M., Von Gleich, A., 2004. Nanotechnology and Regulation within Framework of the Precautionary Principle. Institut für Ökologische Wirtschaftsforschung, Berlin.

HCWH, 2013. Unused Pharmaceuticals Where Do They End Up? A Snapshot of European Collection Schemes. Available from: https://noharm-europe.org/documents/unused-pharmaceuticals-where-do-they-end-snapshot-european-collection-schemes.

Helwig, K., Hunter, C., McNaughtan, M., Roberts, J., Pahl, O., 2016. Ranking prescribed pharmaceuticals in terms of environmental risk: inclusion of hospital data and the importance of regular review. Environ. Toxicol. Chem. 35 (4), 1043–1050. https://doi.org/10.1002/etc.3302.

Henderson, J., Dicken, P., Hess, M., 2002. Global production networks and the analysis of economic development. Rev. Int. Polit. Econ. 9 (3), 436–464. https://doi.org/10.1080/09692290210150842.

Herrmann, M., Olsson, O., Fiehn, R., Herrel, M., Kümmerer, K., 2015. The significance of different health institutions and their respective contributions of active pharmaceutical ingredients to wastewater. Environ. Int. 85, 61–76. https://doi.org/10.1016/j.envint.2015.07.020.

Himmel, W., Lippert-Urbanke, E., Kochen, M., 1997. Are patients more satisfied when they receive a prescription? The effect of patient expectations in general practice. Scand. J. Prim. Health Care 15 (3), 118–122.

House of Commons Health Committee, 2005. The Influence of the Pharmaceutical Industry: Fourth Report of Session 2004-05. vol. 1 The Stationary Office Limited, London.

Jobling, S., Burn, R.W., Thorpe, K., Williams, R., Tyler, C., 2009. Statistical modeling suggests that antiandrogens in effluents from wastewater treatment works contribute to widespread sexual disruption in fish living in english rivers. Environ. Health Perspect. 117 (5), 797–802. https://doi.org/10.1289/ehp.0800197.

Jones, O.A.H., Green, P.G., Voulvoulis, N., Lester, J.N., 2007. Questioning the excessive use of advanced treatment to remove organic micropollutants from wastewater. Environ. Sci. Technol. 41 (14), 5085–5089. https://doi.org/10.1021/es0628248.

Klaminder, J., Jonsson, M., Fick, J., Sundelin, A., Brodin, T., 2014. The conceptual imperfection of aquatic risk assessment tests: highlighting the need for tests designed to detect therapeutic effects of pharmaceutical contaminants. Environ. Res. Lett. 9, 1–7. https://doi.org/10.1088/1748-9326/9/8/084003.

Von Korff, M., Dublin, S., Walker, R.L., Parchman, M., Shortreed, S.M., Hansen, R.N., Saunders, K., 2016. The impact of opioid risk reduction initiatives on high-dose opioid prescribing for patients on chronic opioid therapy. J. Pain 17 (1), 101–110.

Kotwani, A., Walttal, C., Katewa, S., Joshi, P.C., Holloway, K., 2010. Antibiotic use in the community: what factors influence primary care physicians to prescribe antibiotics in Delhi, India? Fam. Pract. 27 (6), 684–690.

Larsen, T.A., Alder, A.C., Eggen, R.I.L., Maurer, M., Lienert, J., 2009. Source separation: will we see a paradigm shift? Environ. Sci. Technol. 43 (16), 6121–6125. https://doi.org/10.1021/es9010515.

Larsson, D.G.J., 2014. Pollution from drug manufacturing: review and perspectives. Philos. Trans. R. Soc. B 369, 20130571. https://doi.org/10.1098/rstb.2013.0571.

Larsson, D.G.J., de Pedro, C., Paxeus, N., 2007. Effluent from drug manufactures contains extremely high levels of pharmaceuticals. J. Hazard. Mater. 148 (3), 751–755. https://doi.org/10.1016/j.jhazmat.2007.07.008.

Levy, D.L., 2008. Political contestation in global production networks. Acad. Manag. Rev. 33 (4), 943–963. https://doi.org/10.5465/AMR.2008.34422006.

Li, D., Yang, M., Hu, J., Ren, L., Zhang, Y., Li, K., 2008. Determination and fate of oxytetracycline and related compounds in coxytetracyclin production wastewater and the receiving river. Environ. Toxicol. Chem. 27, 80–86.

Little, P., Dorward, M., Warner, G., Stephens, K., Senior, J., Moore, M., 2004. Importance of patient pressure and perceived pressure and perceived medical need for investigations, referral, and prescribing in primary care: nested observational study. BMJ [Br. Med. J.] 328 (7437), 444.

Löfstedt, R.E., 2004. The swing of the regulatory pendulum in Europe: from precautionary principle to (regulatory) impact analysis. J. Risk Uncertain. 28 (3), 237–260. https://doi.org/10.1023/B:RISK.0000026097.72268.8d.

Macfarlane, J., Holmes, W., Gard, P., Thornhill, D., Macfarlane, R., Hubbard, R., 2002. Reducing antibiotic use for acute bronchitis in primary care: blinded, randomised controlled trial of patient information leaflet. BMJ [Br. Med. J.] 324, 91–94.

Mackridge, A.J., Marriott, J.F., 2007. Returned medicines: waste or a wasted opportunity? J. Public Health 29 (3), 258–262. https://doi.org/10.1093/pubmed/fdm037.

Mallow, E.B., Fox, M.A., 2014. Phthalates and critically ill neonates: device-related exposures and non-endocrine toxic risks. J. Perinatol. 34 (12), 892–897.

Moynihan, R., Heath, I., Henry, D., 2002. Selling sickness: the pharmaceutical industry and disease mongering. BMJ (Clinical Research ed.) 324 (7342), 886–891. https://doi.org/10.1136/bmj.324.7342.886.

Murray-Smith, R.J., Coombe, V.T., Grönlund, M.H., Waern, F., Baird, J.A., 2012. Managing emissions of active pharmaceutical ingredients from manufacturing facilities: an environmental quality standard approach. Integr. Environ. Assess. Manag. 8, 320–330. https://doi.org/10.1002/ieam.1268.

noPILLS, 2015. Interreg IV B NEW Project Partnership 2012–2015: noPILLS Report. Emschengenossenschaft, Essen.

O'Neill, J., 2016. Tackling Drug-Resistant Infections Globally: Final Report and Recommendations. The Review on Antimicrobial Resistance. Wellcome Trust/HM Government, London.

Paul, S.M., Mytelka, D.S., Dunwiddie, C.T., Persinger, C.C., Munos, B., Lindborg, S.R., Schacht, A.L., 2010. How to improve R&D productivity: the pharmaceutical industry's grand challenge. Nat. Rev. Drug Discov. 9, 203–214. https://doi.org/10.1038/nrd3078.

Petrie, B., Youdan, J., Barden, R., Kasprzyk-Hordern, B., 2016. New framework to diagnose the direct disposal of prescribed drugs in wastewater – a case study of the antidepressant fluoxetine. Environ. Sci. Technol. 50 (7), 3781–3789. https://doi.org/10.1021/acs.est.6b00291.

PILLS, 2012. Pharmaceutical Input and Elimination From Local Sources: Final Report of the European Co-Operation Project PILLS. Emschergenossenschaft, Essen.

Pruden, A., Larsson, D.J., Amézquita, A., Collignon, P., Brandt, K.K., Graham, D.W., Lazorchak, J.M., Suzuki, S., Silley, P., Snape, J.R., Topp, E., Zhang, T., Zhu, Y.-G., 2013. Management options for reducing the release of antibiotics and antibiotic resistance genes to the environment. Environ. Health Perspect. 121 (8), 878.

Taylor, D., 2010a. Ecopharmacostewardship – a pharmaceutical industry perspective. In: Kümmerer, K., Hempel, M. (Eds.), Green and Sustainable Pharmacy. Springer-verlag, Berlin, p. 106.

Taylor, D., 2010b. Ecopharmacostewardship – a pharmaceutical industry perspective. In: Kümmerer, K., Hempel, M. (Eds.), Green and Sustainable Pharmacy. Springer-verlag, Berlin, p. 114.

Triggle, D.J., 2007. Treating desires not diseases: a pill for every ill and an ill for every pill? Drug Discov. Today 12 (3–4), 161–166. https://doi.org/10.1016/j.drudis.2006.12.001.

Vazquez-Lago, J.M., Lopez-Vazquez, P., Lopez-Duran, A., Taracido-Trunk, M., Figueiras, A., 2012. Attitudes of primary care physicians to the prescribing of antibiotics and antimicrobial resistance: a qualitative study from Spain. Fam. Pract. 29 (3), 352–360. https://doi.org/10.1093/fampra/cmr084.

Verlicchi, P., Galletti, A., Petrovic, M., Barceló, D., 2010. Hospital effluents as a source of emerging pollutants: an overview of micropollutants and sustainable treatment options. J. Hydrol. 389 (3–4), 416–428. https://doi.org/10.1016/j.jhydrol.2010.06.005.

Chapter 12

Management of Environmental Contaminants From Health Care: Sustainable Pharmacy

Klaus Kümmerer
Leuphana University of Lüneburg, Lüneburg, Germany

INTRODUCTION

As discussed in the previous chapters in this book, pharmaceuticals often are not fully metabolized or mineralized in the body of the target organism after their administration. Consequently, parent compounds and metabolites are excreted into wastewater. Organic diagnostic agents are designed for not being metabolized. They are excreted unchanged. Disinfectants applied for surface disinfection and laundry detergents, cleaning agents, personal care products, and others enter also effluents. Advanced effluent treatment does often remove only a minor share of the pharmaceutically active compounds (APIs), adjuvants and metabolites, and other chemical compounds—if sewage treatment is in place at all. Furthermore, often, advanced treatment does neither result in a full removal (Margot et al., 2013), let alone complete mineralization of the compounds, that is, break down to carbon dioxide, water, and inorganic salts. In contrary, the advanced oxidative treatments often result in the formation of many unknown products, so-called transformation products (TPs) of unknown fate, toxicity, and risk. Some of them have been shown to be even more toxic as their parent compounds (Mitch and Sedlak, 2002; Schmidt and Brauch, 2008; Tootchi et al., 2013; Li and Lin, 2015; Marti et al., 2015; Zeng and Mitch, 2015). Accordingly, all these compounds and their respective TPs are detectable in the environment (Kümmerer, 2010a; Lambropoulou and Nollet, 2014; Rozas et al., 2016). Risk assessment for these transformation products is tedious—if possible at all—as data on their fate and effects in the environment are hardly available and as they are numerous. Ecopharmacology classifies the undesirable presence of pharmaceuticals in the environment as an unwanted side effect of medical treatment (Kümmerer and Velo, 2006). Furthermore, in the medical care sector, many

Environmental Contaminants, Vol. 11. https://doi.org/10.1016/B978-0-444-63857-1.00012-7

different materials including medicinal products, laboratory consumables, and others are used that end up as waste. These consist of many different materials that are mixed and are therefore difficult to recycle.

In this chapter, some of the opportunities that are provided by green and sustainable pharmacy to manage environmental contaminants from health care will be presented. According to the principles of sustainable pharmacy and chemistry (Kümmerer and Hempel, 2010; Kümmerer, 2010a,b; Clark et al., 2010; Kümmerer and Clark, 2016), the whole life cycle of a compound or a material has to be taken into account to identify opportunities for risk management and risk reduction. Some opportunities to reduce the introduction of pharmaceuticals into the environment by the different stakeholders are summarized in Table 1 (Kümmerer, 2010a,b). Similar ideas apply for the case of chemicals and medicinal products and waste, other than pharmaceuticals.

SUSTAINABLE CHEMISTRY AND PHARMACY

Nowadays, it is widely accepted that focusing on end of the pipe treatment of environmental problems is not sustainable and will not be successful in the long run. It was learned that the main unwanted input into the environment in developed countries is the products themselves that cause the problems through their mere presence in the environment. Input sometimes happens on account of proper use. It was also learned that if this issue is to be tackled properly, solutions should not only be helpful for some few developed countries; rather, a solution that works everywhere is needed. It was in this context that a broader approach was already addressed in the mid-1990s (EC, 1996). As for pharmaceuticals, for example, it was also learned in the 1990s that for synthesis of 1 kg of an active pharmaceutical compound, the amount of waste generated is up 50–100-fold the one generated within the synthesis of a bulk chemical (Sheldon, 1992, 2007). The concept of green chemistry was developed subsequently and began to gain momentum within chemistry (Anastas and Warner, 1998; Clark and Smith, 2005; Clark, 2006; Lapkin and Constable, 2008, https://www.epa.gov/greenchemistry/basics-green-chemistry#definition). Within this framework, the focus is not only on the use or the synthesis of a chemical. Instead, the full life cycle of a chemical and its impact on the environment is looked at. This includes raw materials used, synthesis, manufacturing, use, and end of life. Nowadays, this concept has further been developed including also legal, ethical, and economical aspects and new business models (Kümmerer and Clark, 2016; Kümmerer 2017). In analogy, this should also apply to pharmacy—green and sustainable pharmacy (Kümmerer, 2010a,b; Clark et al., 2010). Only recently, the United Nations Environmental Assembly (http://www.unep.org/about/sgb/cpr_portal/Portals/50152/UNEA2%20RES/7.pdf) named sustainable chemistry as an important building block in the context of the sustainable development goals. The same holds for sustainable pharmacy.

TABLE 1 Opportunities to Reduce the Input of Pharmaceuticals into the Environment

Who	Possible Measures and Activities
Pharmaceutical companies	Publication of data relevant for environmental assessment Publication of analytic methods and results
	Offering suitable package sizes
	Integration of environmental aspects into the development of new APIs and new therapies
	Dedication to green and sustainable pharmacy
	Fewer over-the-counter products
	Establishment of take-back systems wherever not already present
	Establishment of corporate sustainability
	Research into sustainable pharmacy
Patients	Improvement of compliance
	Intake of APIs only if necessary and only after prescription by a medical doctor
	Outdated drugs should not; instead, they should be returned to the pharmacy if a take-back system is established or into household waste if appropriate (check with local authorities and pharmacies)
	Proper hygiene
Pharmacists	Information to patients
	Participate in take-back systems if appropriate (check with local authorities)
Hospitals	Integration of the delivering pharmacy/wholesaler into the handling of outdated drugs
	Information to doctors and patients
	Proper hygiene
Medical doctors	Asking for appropriate package size
	Prescription according to environmental criteria if alternatives are available
	Information of patients
Health insurance	Requesting/paying for appropriate package size
	Keeping the necessary medical standards and demonstration of reduction potential and economical benefits
	Information to doctors and patients

Continued

TABLE 1 Opportunities to Reduce the Input of Pharmaceuticals into the Environment—Cont'd

Who	Possible Measures and Activities
Veterinary medicine	Restrictive prescription Improvement of compliance Improvement of hygiene Less exchange of animals between flocks and farms Information to farmers
Wastewater handling and treatment	Reduction of input by broken sewerage/piping Reduction of total water flow to be treated (separate piping of wastewater and rain water) and thereby increasing concentration of APIs
Drinking water treatment	Extended monitoring Advanced treatment if necessary Information to the general public
Authorities	Initiation and backup of communication between all stakeholders Development of limits/thresholds for APIs in different environmental compartments and drinking water
Banks	Inclusion of sustainability aspects into the rating of companies (manufacturers, retailers, etc.)
Politics	Inclusion of APIs in environmental legislation More restrictive connection between environmental properties and authorization of pharmaceuticals Improvement of legislation for the management of outdated drugs Incentives for greener and sustainable pharmaceuticals Promoting research and education in sustainable pharmacy

From Kümmerer, K. 2008. Strategies for reducing the Input of pharmaceuticals into the environment. In: Kümmerer, K. (Ed.) Pharmaceuticals in the Environment. Sources, Fate, Effects and Risks. Springer, Berlin Heidelberg, pp. 411–418.

A pharmaceutical can be "green" in terms of the quality and quantity of waste generated during its synthesis according to the principles of green chemistry. However, the pharmaceutical may persist or accumulate in the environment after excretion. As for medicinal products, "green" materials might not be recyclable because they contain too many different compounds or are mixed with others after their respective use. Or a renewable feedstock might have been used. However, reproduction of the renewable feedstock needs much water and fertilizer, may be in competition with food production, or depends on an endangered species. Then, it cannot be called sustainable. The same holds if the compound itself is "green" or even sustainable; however, the material flows connected to its production, distribution, and usage are very large (in terms of quantity and/or in terms of e.g. such as toxicity) or depend on nonrenewable resources. The application of greener products or chemicals may become less green or even nonsustainable if they are produced and applied in high volumes. Fig. 1 presents some opportunities for reduction of the environmental impact caused by the use of pharmaceuticals, contrast media, disinfectants, and medicinal products along their life cycle.

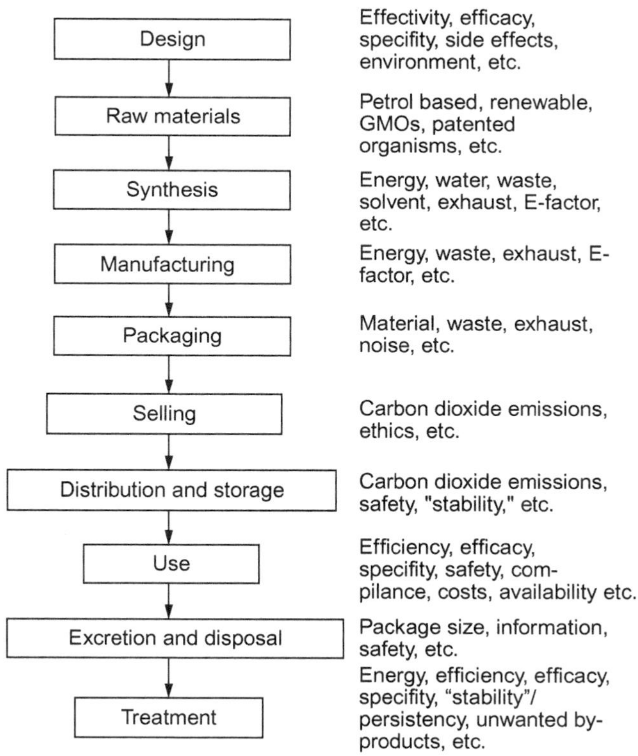

FIG. 1 The life cycle of a pharmaceutical and some points that are relevant for sustainability. *Reproduced with permission from Kümmerer, K., 2010a. Pharmaceuticals in the environment. Annu. Rev. Environ. Resour. 35, 57–75; Kümmerer, K., 2010b. Why green and sustainable pharmacy? In: Kümmerer, K., Hempel, M. (Eds.), Green and Sustainable Pharmacy. Springer, Berlin Heidelberg, pp. 3–10 as for* E-*factor, see Sheldon, R.A., 1992. Organic synthesis—past, present and future. Chem. Ind. 23, 903–906; Sheldon, R.A., 2007. The E factor: fifteen years on. Green Chem. 9, 1273–1283.*

THE END OF THE PIPE'S END

Given the poor prognosis for end of the pipe approaches such as effluent treatment and waste incineration or dumping, it is profitable to shift the focus to "the beginning of the pipe"; that is the reasons for the use of products of pharmaceutical and medical industries—the active ingredients, adjuvants in case of pharmaceuticals, disinfectants, and contrast agents and medicinal products and their constituents. Avoiding unnecessary incentives for the use of pharmaceuticals, disinfectants, and other products not only just reduces the material effluent from a hospital but also reduces the environmental burden at several other stations of the life cycle of pharmaceuticals, chemicals, and other products.

NEW BUSINESS MODELS

We use pharmaceuticals, disinfectants, and other chemical and medical products not just "for fun." We use them because they offer a certain function and a certain service. Therefore, the proper understanding of the broader context of the application of such products is the starting point instead of just thinking of applying a certain product. This different view also opens up opportunities for new business models and new business opportunities.

A practical example from health care in this context is about disinfection (https://www.dbu.de/OPAC/ab/DBU-Abschlussbericht-AZ-26035.pdf). A disinfectant offers the function to kill pathogens. However, that is only part of the full picture. What we really want when we apply a disinfectant is a certain standard of hygiene. From this point of view, the application of a proper disinfectant is just one building block among others such as establishing procedures that avoid (cross) contamination or the wrong handling of devices, using proper furniture, thinking about where to disinfect, and supervising and educating staff properly—to mention just a few measures that contribute to a certain standard of hygiene.

The traditional business model is that a company produces and sells a disinfectant—the more, the better. The user, for example, the hospital, has the opposite interest—to buy or more correctly to pay as little as possible. The result of such a situation is that both do not really talk about the needs and how a certain standard of hygiene can be met—especially including measures that do not need a disinfectant. However, if the hospital is not paying for a certain amount of disinfectant but instead is maintaining a predefined standard of hygiene, the situation is completely different. The vendor will fully bring in his complete knowledge, including education, regulation, and investigating carefully where disinfection is really needed and where other measures are more appropriate. The vendor also will notice that if less disinfectant is needed, that will save him costs, for example, for buying the raw materials and manufacturing the disinfectant and transporting it to and handling it in the hospital. Instead, the vendor will make some of his profit by consultation, education, and supervision. In fact, both parties pursue then the same goal; both have the same interest, that is, meeting the needed standard of hygiene. This improves the exchange of knowledge, fosters cooperation, and reduces the usage and subsequent introduction of disinfectants into the environment. Furthermore, the hospital gets rid of the responsibility, the handling of dangerous substances, the organizing of supervisions and education, etc. In fact, the staff can focus on their main business—medical treatment and medical care. In short, a win-lose situation has turned into a win-win-situation. Many other examples are known (http://www.chemicalleasing.org/docs/20160310_10%20Years%20Chemical%20Leasing%20Report%20and%20Strategy%20Outlook_FV.pdf), however not many yet from the medical area. Proper nutrition, physical exercise, healthy lifestyle, or acupuncture for the treatment of pain have been shown to be opportunities to reduce the usage of pharmaceuticals and will help in the reduction of the environmental burden of pharmaceuticals, disinfectants, and medicinal products along their life cycle and thereby also the one of health care in general.

STAKEHOLDERS

The framework of sustainable chemistry and pharmacy does not just deal with the chemicals, pharmaceuticals, and materials. It also asks to have a view on the role of stakeholders. In general, a better understanding of the individual role of involved people (patients, doctors, and pharmacists (Götz and Deffner, 2010; Castensson and Ekedahl, 2010; Wennmalm and Gunnarsson, 2010)) can contribute to the usage of less pharmaceuticals, disinfectants, and other products in the health-care sector. Proper information, education, and communication will contribute to the reduction of environmental burden related to health-care activities (Kümmerer and Schramm, 2008).

BENIGN BY DESIGN

Another beginning of the pipe solution is to consider the products and molecules themselves. There is a tendency in the development toward new APIs that possess increasingly lower activity thresholds (Triggle, 2010). The total tonnage of pharmaceuticals emitted into the environment could therefore be expected to decrease in the future. However, lower activity thresholds may lead to a higher risk as the substances might be more toxic to environmental organisms. Furthermore, the diversity of compounds and products is expected to increase. This would also mean more monitoring, more TPs, and more effort, time, and money being spent for risk assessment. Therefore, it is questionable whether such developments do really contribute to a reduction of related emissions into the environment. A high rate of metabolism will reduce the presence of an API in the environment. However, if the metabolites are of higher activity as the parent and/or if they are of different toxicity, the situation may be worse. Often, metabolites are more polar than the parent compounds. That renders them more mobile in the aquatic environment. Improvement of the share of an administered drug that is taken up within the body of the target organisms or the human body will reduce the share of a drug that is excreted unchanged. Furthermore, if the compounds are transported more effectively to the target organ or the compound is more specific, the amount of an API that is needed to exert the effect wanted and the input into the environment is lower. Another approach would be to design molecules of APIs in such a manner that the remainder that is not taken up and stays, for example, within the human gut is inactivated or, even better, fully degraded before excretion.

Some medicines contain molecules based on proteins and peptides, the so-called biopharmaceuticals or bioceuticals. Biopharmaceuticals are medical drugs produced using biotechnology by means other than direct extraction from a native (i.e., nonengineered) biological source. They may be retrieved from organisms or produced by organisms. They are often big molecules (molecular weight about 15,000–150,000 Da). Some smaller ones are chemically synthesized. Some are natural proteins; others can be different—similar to natural proteins and peptides (Leader et al., 2008). Hormones and monoclonal antibodies are prominent groups. The first and well-known example was recombinant human insulin. A newer generation of biopharmaceuticals is the proteins that are chemically modified, that is, by linking them to long-chain polyethylene glycols resulting in so-called PEGylated proteins (Webster et al., 2007). One goal for protein drugs is to render them more stable (Fang, 2015).

Bioceuticals have neither been in focus of environmental research nor risk management. Often, it is just assumed that bioceuticals will be readily biodegradable in the environment as they are "natural compounds." However, there is no real data basis that would back up this assumption. Not all natural compounds are readily biodegradable. It is still open to discussion whether bioceuticals are safer for the environment than conventional small molecules that are used as APIs. The environmental significance of bioceuticals is not yet clear. First results indicate that nonmodified proteins should be degradable in sewage treatment paints (Straub, 2010). However, the basis of knowledge is still very little. PEGylated proteins have a long biological half-life, and excretion is delayed. The modification of the natural structure of proteins has a clear effect on biodegradability—it is lower (Straub, 2010). Bioceuticals have only a limited scope of application, up to now mainly in anticancer treatment, and they are often very expensive. For the time being, bioceuticals do not provide a substitute of conventional active ingredients in general.

In current discussions on green and sustainable pharmacy, improvements in synthesis are very prominent (Summerton et al., 2016), whereas the environmental properties of the molecules do not get enough attention.

Environmentally fully mineralized pharmaceuticals would avoid all the follow-up problems associated with persistence (Cairns and Mount, 1990), effluent treatment (Fatta-Kassions et al., 2016a), water pollution, effects against nontarget organisms in the environment, and challenges of risk assessment including the related investments that could be allocated to other parts of the health system. Furthermore, it would contribute to several sustainable development goals (SDGs, https://sustainabledevelopment.un.org/?menu=1300) especially in countries where there is a water shortage and therefore an urgent need for (waste)water reuse (Fatta-Kassions et al., 2016b).

This approach, called benign by design, means that ready mineralization after use is taken into account even before a pharmaceutical is synthesized (Fig. 2). The targeted design of pharmaceuticals is common practice in pharmaceutical industries. Environmental properties have to be included in the same way pharmaceutical lead structures are already improved with respect to their pharmacokinetic and pharmacodynamic profile. If fast and full mineralization is included, the resulting molecules are "benign by design." However, the inclusion of environmental aspects is neither as yet a topic at top management level nor in the education of medical doctors, pharmacists, and medicinal chemists.

Conventional wisdom assumes that an API has to be stable to reach the market. That is what chemists, medical chemists, and pharmacists are taught since decades. The link between structure and properties of a chemical is at the core of chemistry and pharmacy. Therefore, it can be used for benign design. The physical chemical conditions vary along the life cycle of an API and a pharmaceutical excipient. Typical conditions of importance that differ are moisture, access and type (e.g., spectrum and intensity) of light, temperature, oxygen concentration, and pH (Kümmerer, 2007, 2010a,b, 2012). They differ, for example, during storage, in the human body and municipal sewage, sewage treatment, or surface water. Bacterial diversity

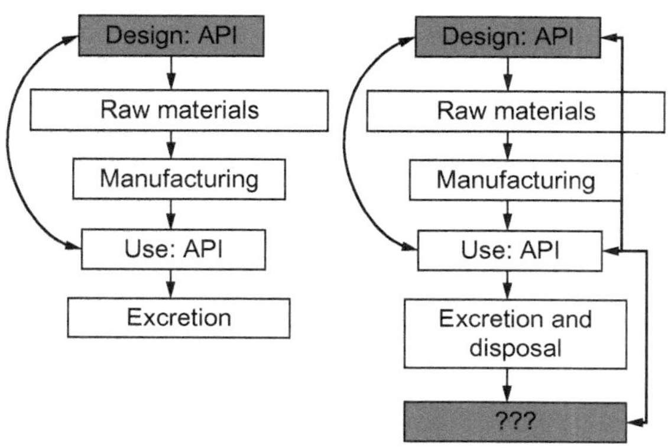

FIG. 2 Conventional (*left*) and sustainable approach for the design of new pharmaceuticals (*right*) (Kümmerer, 2010a,b).

and bacterial density differ in these different environments. The metabolic diversity and the sheer number of bacteria and their potential and pathways for the breakdown of molecules (i.e., number and species of enzymes) differ.

If we are aware of those facets that are of significance for the "stability" of a chemical in the environment, this knowledge can then be used to design chemicals not only for optimized performance during their application but also within the latter stages of their life cycle. In fact, stability seems to be a marketing buzzword, whereas reactivity is a sound scientifically based concept that allows for benign design.

The full, truely sustainable functionality of a molecule includes the properties necessary for good performance within all its life stages—not only during application. An analysis of data from the author's own biodegradability database (OECD test series 302 and 301; in total about 2200 different chemicals, about 500 have been tested in two or more tests) shows that currently 979 chemicals in the market are readily or inherently biodegradable (36%), whereas the number is 29% for pharmaceuticals from different chemical and therapeutic classes (in total 28 out of 96; Kümmerer, 2010a,b). Furthermore, it is well known that some APIs are sensitive against light or they are not stable enough at ambient temperature. Therefore, some pharmaceuticals such as ciprofloxacin are stored under brown glass or in the fridge (e.g., some ß-lactams after reconstitution in water).

Strategies on how to design pharmaceuticals and other chemicals for complete mineralization in the environment, tools needed, and examples demonstrating the proof of concept are already available (Leder et al., 2015; Rastogi et al., 2015a,b, 2014; Kümmerer, 2007). Such new approaches result in new active molecules and products and will therefore also present new business opportunities.

However, there are arguments against the benign by design concept in case of pharmaceuticals. An ethical one is that it is brought forward by the pharmaceutical industry and that we should not deny anyone access to a new pharmaceutical. However, many drugs have not fully or not at all been developed or manufactured by industries for economical reasons, for example, orphan drugs, new antibiotics, drugs for the treatment of children, malaria, and AIDS. Some new drugs, for example, for the treatment of cancer, most often bioceuticals, are very expensive and therefore not affordable for most people. So the question is do we discuss real ethical issues or just economical ones. Furthermore, it is argued that drug development is very expensive, and including biodegradability would make it even more expensive. However, there are no data available. As for convectional drugs, some estimates say that about two-thirds of the costs are just marketing costs to introduce new drugs. There is also the fear that integrating environmental biodegradability will result in a shortage of new compounds as further regulation (environment) will result in less compounds. However, there is already more than one compound on the market for most groups and medical indications. Furthermore, it has been shown in several studies and can be seen from the history of pharmacy and toxicology that new requirements and regulations result in the longer run in new and better compounds, that is, they are a driver of progress and are opening up new innovation space (see, e.g., toxicological requirements after the Contergan (thalidomide) case). Another argument brought forward against biodegradable drugs is that the requirement of drugs being environmentally biodegradable refers to competition of drug-developing companies with the ones manufacturing and selling generics. However, competition is already the case and always was. In fact, competition is another driver of progress. If a new compound is better than the old one in terms of efficacy, efficiency, and unwanted side effects, that will not be an issue; instead, it will be an advantage for innovative companies if customers ask for it and if regulations are in place calling for it.

REFERENCES

Anastas, P.T., Warner, J.C., 1998. Green Chemistry: Theory and Practice. Oxford University Press, New York/Oxford.

Cairns, J., Mount, D.I., 1990. Aquatic toxicology 2. Environ. Sci. Technol. 24, 154–158.

Castensson, S., Ekedahl, A., 2010. Pharmaceutical waste: the patient ole. In: Kümmerer, K., Hempel, M. (Eds.), Green and Sustainable Pharmacy. Springer, Berlin Heidelberg, pp. 179–200.

Clark, J.H., 2006. Green chemistry: today (and tomorrow). Green Chem. 8, 17–21.

Clark, J.H., Breeden, S.W., Summerton, L., 2010. Green(er) pharmacy. In: Kümmerer, K., Hempel, M. (Eds.), Green and Sustainable Pharmacy. Springer, Berlin Heidelberg, pp. 37–59.

Clark, J.H., Smith, P., 2005. Basic principles of green chemistry. Innov. Pharm. Technol. 16, 94–97.

EC, 1996. EC Council Directive 96/61/EC of 24 September 1996 concerning integrated pollution prevention and control. Off. J. L 257, 0026–0040. 10/10/1996.

Fang, J., 2015. Reliability of machine learning based algorithms for designing protein drugs with enhanced stability. Drug Des. 4, e130.

Fatta-Kassions, D., Dionysiou, D.D., Kümmerer, K., 2016a. Long-term strategies for tackling micropollutants. In: Fatta-Kassions, D., Dionysiou, D.D., Kümmerer, K. (Eds.), Advanced Treatment Technologies for Urban Wastewater Reuse. The Handbook of Environmental Chemistry, vol. 45. Springer, Cham, Heidelberg, New York, Dordrecht, London, pp. 291–299.

Fatta-Kassions, D., Dionysiou, D.D., Kümmerer, K. (Eds.), 2016b. Wastewater reuse and current challenges. In: The Handbook of Environmental Chemistry, vol. 44. Springer, Cham, Heidelberg, New York, Dordrecht, London.

Götz, K., Deffner, J., 2010. Options for a more environmentally friendly handling of pharmaceuticals. In: Kümmerer, K., Hempel, M. (Eds.), Green and Sustainable Pharmacy. Springer, Berlin Heidelberg, pp. 149–164.

Kümmerer, K., 2007. Sustainable from the very beginning: rational design of molecules by life cycle engineering as an important approach for green pharmacy and green chemistry. Green Chem. 9, 899–907.

Kümmerer, K., 2010a. Pharmaceuticals in the Environment. Annu. Rev. Environ. Resour. 35, 57–75.

Kümmerer, K., 2010b. Why green and sustainable pharmacy? In: Kümmerer, K., Hempel, M. (Eds.), Green and Sustainable Pharmacy. Springer, Berlin Heidelberg, pp. 3–10.

Kümmerer K. (2012) Drugs. In: Anastas P. (Ed.), Handbook of Green Chemistry. vol. 9 (Boethling R., Voutchkova A. (Volume Eds.), Designing Safer Chemicals). Wiley-VCH, Weinheim, pp. 251–279.

Kümmerer, K., 2017. Angew. Chem. Int. Ed. 56, 16420–16421.

Kümmerer, K., Clark, J.H., 2016. Green and sustainable chemistry. In: Heinrichs, H., Martens, P., Michelsen, G., Wiek, A. (Eds.), Sustainability Science. Springer Science+Business Media, Dordrecht, pp. 43–59.

Kümmerer, K., Hempel, M. (Eds.), 2010. Green and Sustainable Pharmacy. Springer, Berlin Heidelberg.

Kümmerer, K., Schramm, E., 2008. Arzneimittelentwicklung: Die Reduzierung von Umweltbelastungen durch gezieltes Moleküldesign. Umweltwiss. Schadst. Forsch. 20, 249–263.

Kümmerer, K., Velo, G., 2006. Ecopharmacology: a new topic of importance in pharmacovigilance. Drug Saf. 29, 371–373.

Lambropoulou, D.A., Nollet, L.M.L. (Eds.), 2014. Transformation Products of Emerging Contaminants in the Environment: Analysis, Processes, Occurrence, Effects and Risks. Wiley.

Lapkin, A., Constable, D. (Eds.), 2008. Green Chemistry Metrics: Measuring and Monitoring Sustainable Processes. Wiley-Blackwell, Oxford.

Leader, B., Baca, Q.J., Golan, D.E., 2008. Protein therapeutics: a summary and pharmacological classification. Nat. Rev. Drug Discov. 7, 21–39.

Leder, C., Rastogi, T., Kümmerer, K., 2015. Putting benign by design into practice-novel concepts for green and sustainable pharmacy: designing green drug derivatives by non-targeted synthesis and screening for biodegradability. Sustain. Chem. Pharm. 2, 31–36.

Li, S.W., Lin, A.Y.C., 2015. Increased acute toxicity to fish caused by pharmaceuticals in hospital effluents in a pharmaceutical mixture and after solar irradiation. Chemosphere 139, 190–196.

Margot, J., Kienle, C., Magnet, A., Weil, M., Rossi, L., de Alencastro, L.F., Abegglen, C., Thonney, D., Chèvre, N., Schärer, M., Barry, D.A., 2013. Treatment of micropollutants in municipal wastewater: ozone or powdered activated carbon? Sci. Total Environ. 461–462, 480–498.

Marti, E.J., Pisarenko, A.N., Peller, J.R., Dickenson, E.R.V., 2015. N-nitrosodimethylamine (NDMA) formation from the ozonation of model compounds. Water Res. 72, 262–270.

Mitch, W.A., Sedlak, D.L., 2002. Formation of N-nitrosodimethylamine (NDMA) from dimethylamine during chlorination. Environ. Sci. Technol. 36, 588–595.

Rastogi, T., Leder, C., Kümmerer, K., 2014. Designing green derivatives of β-blocker metoprolol: a tiered approach for green and sustainable pharmacy and chemistry. Chemosphere 111, 493–499.

Rastogi, T., Leder, C., Kümmerer, K., 2015a. Re-designing of existing pharmaceuticals for environmental biodegradability: a tiered approach with β-blocker propranolol as an example. Environ. Sci. Technol. 49, 11756–11763.

Rastogi, T., Leder, C., Kümmerer, K., 2015b. A sustainable chemistry solution to the presence of pharmaceuticals and chemicals in the aquatic environment-the example of re-designing β-blocker atenolol. RSC Adv. 5, 27–32.

Rozas, O., Vidal, C., Baeza, C., Jardim, W.F., Rossner, A., Mansilla, H.D., 2016. Organic micropollutants (OMPs) in natural waters: oxidation by UV/H_2O_2 treatment and toxicity assessment. Water Res. 98, 109–118.

Schmidt, C.K., Brauch, H.J., 2008. N,N-dimethylsulfamide as precursor for N-nitrosodimethylamine (NDMA) formation upon ozonation and its fate during drinking water treatment. Environ. Sci. Technol. 42, 6340–6346.

Sheldon, R.A., 1992. Organic synthesis—past, present and future. Chem. Ind. 23, 903–906.
Sheldon, R.A., 2007. The E factor: fifteen years on. Green Chem. 9, 1273–1283.
Straub, J.O., 2010. Protein and peptide therapeuticals: an example of 'benign by nature' active pharmaceutical ingredients. In: Kümmerer, K., Hempel, M. (Eds.), Green and Sustainable Pharmacy. Springer, Berlin Heidelberg, pp. 127–133.
Summerton, I., Sneddon, H.F., Jones, L.C., Clark, J.H. (Eds.), 2016. Green and Sustainable Medicinal Chemistry. Methods, Tools and Strategies for the 21st Century Pharmaceutical Industry. RSC Publishing, London.
Tootchi, L., Seth, R., Tabe, S., Yang, P., 2013. Transformation products of pharmaceuti-cally active compounds during drinking water ozonation. Water Sci. Technol. Water Supply 13, 1576–1582.
Triggle, D.J., 2010. Pharmaceuticals in society. In: Kümmerer, K., Hempel, M. (Eds.), Green and Sustainable Pharmacy. Springer, Berlin Heidelberg, pp. 23–35.
Webster, R., Didier, E., Harris, P., Siegel, N., Stadler, J., Tilbury, L., Smith, D., 2007. PEGylated proteins: evaluation of their safety in the absence of definitive metabolism studies. Drug Metab. Dispos. 35, 9–16.
Wennmalm, A., Gunnarsson, B., 2010. Experiences with the Swedish classification system. In: Kümmerer, K., Hempel, M. (Eds.), Green and Sustainable Pharmacy. Springer, Berlin Heidelberg, pp. 243–249.
Zeng, T., Mitch, W.A., 2015. Contribution of N-nitrosamines and their precursors to domestic sewage by greywaters and blackwaters. Environ. Sci. Technol. 49, 13158–13167.

FURTHER READING

Kümmerer, K., 2008. Strategies for reducing the input of pharmaceuticals into the environment. In: Kümmerer, K. (Ed.), Pharmaceuticals in the Environment. Sources, Fate, Effects and Risks. Springer, Berlin Heidelberg, pp. 411–418.

Chapter 13

Policy, Regulations, and Risk Assessment

Richard Murray-Smith* and David Taylor†
*Regulatory Science Associates, Kip Marina, Inverkip, Renfrewshire, Scotland
†wca, Brunel House, Volunteer Way, Faringdon, Oxfordshire

INTRODUCTION

The provision of health care is very diverse, taking place in many different environments. These include large regional hospitals; health clinics; primary care facilities such as doctors, dentists, opticians, and podiatrists; nursing homes; residential homes; and domestic houses. As would be expected, the environmental impact of these activities is diverse as are the policies and regulations that attempt to address them.

In most cases, the most significant environmental impacts will arise from energy and water use, transport, and liquid and solid waste disposal. The operators will manage these impacts by compliance both with legal requirements and corporate policies. In most cases, the impact of liquid and solid wastes will be addressed by both internal policies and external regulations, whereas reducing the environmental impact of transport and the consumption of energy and water will be primarily driven by corporate polices. For example, the 2008 Climate Change Act in the United Kingdom is addressed to the government and includes no specific requirements on business; nevertheless, many businesses have developed policies to reduce their carbon emissions toward meeting the overall national targets.

Most large organizations will have policies on environmental protection including sustainability that are used to develop internal standards and improvement targets. For example, in the United Kingdom, the National Health Service has a 5-year sustainable development strategy (NHS, 2016) including targets for carbon emissions, water use, and waste going to landfill. These targets are very challenging with, for example, NHS England attempting to achieve by 2050 an 80% reduction in its carbon emissions compared with 1990. Kaiser Permanente, the large private US Health-care provider, has an

Environmental Contaminants, Vol. 11. https://doi.org/10.1016/B978-0-444-63857-1.00013-9

environmental stewardship program with similar goals including a target to reduce water consumption by 30% (Kaiser Permanente, 2016).

In the waste field, regulations are usually the principal driving force, and these can be complex, especially in larger institutions like hospitals where the diversity of wastes arising often requires segregation and separate management. In the United Kingdom, the Department of Health has produced a comprehensive, 187-page, guidance document (HMSO, 2013) that details the myriad regulations that cover all aspects of waste management and regulation in health-care facilities.

In smaller establishments delivering health care, such as residential and nursing homes, both liquid and solid wastes are largely controlled by conventional regulations also applicable to domestic and commercial waste management. For example, collection of domestic solid waste will be undertaken by local authorities with, where appropriate, provisions for segregation and recycling. Similarly, liquid waste will be discharged to the domestic sewerage system and treated in the same way as conventional domestic waste streams. However, even in very small institutions such as surgeries, there will also be regulatory requirements for the segregation of some specific clinical waste categories that will need to be dealt with separately. For example, in the case of a dental practice, mercury amalgam wastes and sharps will need special provision.

In larger institutions such as hospitals, liquid wastes are still likely to be discharged to the local sewer, but more controls are likely to be required to meet the demands of the sewage work operator, who has in turn to meet a wide range of environmental regulations. Hospital wastewater can contain a wide range of hazardous substances, such as pharmaceutical residues, hazardous substances, pathogens, and radioisotopes. These substances can be found in a wide range of concentrations due to the size of a hospital, the bed density, the number of inpatients and outpatients, the number and type of wards, the number and types of services, the country, and the season.

Consequently, the health-care facility management will need to reach agreement(s) with the sewage work operator relating to the discharge of body fluids (other than urine or feces), X-ray contrast media, radioactive waste from diagnosis and intensive radiotherapy, cytotoxic or cytostatic medicines, and laboratory wastes. Some countries have produced general guidance for health-care managers to assist in these discussions (Water UK, 2014).

Solid waste produced in health-care facilities can be divided into two types: nonclinical waste and clinical waste. Nonclinical waste will be subject to the local regulations normally applicable for the collection of domestic solid waste with provisions for appropriate segregation and recycling. In the EU member states, these provisions will also require the separate collection of waste electric and electronic equipment (EU, 2016a).

Clinical waste can be defined as any waste that consists wholly or partly of the following:

- Human or animal tissue
- Blood or bodily fluids
- Excretions
- Swabs or dressings
- Drugs or other pharmaceutical products
- Syringes, needles, or other sharp instruments that, unless rendered safe, may prove hazardous to any person coming into contact with it
- Any other waste arising from medical, nursing, dental, veterinary, pharmaceutical or similar practice, investigation, treatment, care, teaching or research, or the collection of blood for transfusion, being *waste that may cause infection* to any person coming into contact with it

In some countries, all clinical waste is considered to be hazardous waste and subject to appropriate regulations; in others, clinical waste is further subdivided into hazardous and nonhazardous.

It should be recognized that regulations governing the disposal of hazardous wastes tend to be very stringent. In the EU, this requires registration and licensing of both the waste producer, the waste disposal contractor, and any persons involved in transport or storage of the waste. In addition, the "duty of care" requirements mean that the original producer of the waste retains a legal obligation to ensure that the ultimate disposal is satisfactory.

Any of the wastes arising from health-care facilities (hospitals, hospices, surgeries, etc.) may affect the environment, and regulations and policies already exist in most cases to manage any adverse impact. However, the biggest area of public concern currently relates to the potential impact of waste pharmaceuticals. Waste cytotoxic and cytostatic drugs are always treated as hazardous clinical waste although other medicines may not be. This is primarily because of the problems this causes for collection and storage in take-back schemes. In the United States, the regulations governing waste medicines from health-care facilities are currently being revised (USEPA, 2015).

In addition to the regulations governing health-care facilities, regulations are also in force to manage other parts of the pharmaceutical supply chain from manufacture to use by patients and eventual disposal.

PHARMACEUTICAL MANUFACTURING

Effluent discharges from pharmaceutical manufacturing plants are regulated in most parts of the world. Sometimes, this is done using specific regulation, but often, the pharmaceutical manufacturer is simply regulated by the same legal instruments that affect all industry. The majority of local or national regulators responsible for the environment have powers to limit the amount of material that may be released from a manufacturing process both into the atmosphere and local watercourses. In some cases, where the factory effluent is discharged into a sewerage system, this control may be exerted indirectly via limits on the eventual release by the sewerage system operator.

In the United States, a comprehensive document was published by the Environmental Protection Agency in 1998 (USEPA, 1998) outlining best available treatment options and the regulatory treatment of effluents arising from pharmaceutical manufacture. However, the most comprehensive control system is currently that provided in the European Union by the Directive on Industrial Emissions (EU, 2010), and this is becoming a model for the development of similar legislation across the world.

The purpose of the directive is to achieve integrated prevention and control of pollution arising from those industries considered to pose the largest risk to the environment (these activities are listed in Annex I of the directive). It lays down measures designed to prevent or, where that is not practicable, to reduce emissions to the air, water, and land from these activities, including measures concerning waste, in order to achieve a high level of protection of the environment taken as a whole. Research and development is excluded from the provisions of the Directive, but pharmaceutical manufacturing is specifically included in Annex 1 of the Directive under Section 4.5 "Production of pharmaceutical products including intermediates."

Article 11 of the Directive sets out eight basic principles that must be applied to all installations. The key principles from the perspective of the pharmaceutical sector are contained in paragraphs (a) all the appropriate preventive measures are taken against pollution, (d) the generation of waste is prevented in accordance with Directive 2008/98/EC, and (e) where waste is generated, it is prepared for re-use, recycled, recovered or, where that is technically and economically impossible, it is disposed of while avoiding or reducing any impact on the environment.

The operator needs to obtain a permit to operate the facility that has to demonstrate that "best available techniques" (BAT) are in use for all the products being manufactured. Specific guidance on what is considered to be BAT for the manufacture of pharmaceuticals has been provided by the European Commission (EU, 2006), and simply stating that the process being used conforms to BAT is not sufficient; the applicant must document both the rationale together with a description of the alternatives that have been rejected. In addition, unlike its predecessor, the integrated pollution prevention and control (IPPC) directive (EU, 1996) and the industrial emissions directive (EU, 2010) require that all new manufacturing facilities must as a minimum meet the BAT standards contained in the BAT reference document (EU, 2006), while existing facilities must be improved to meet this standard.

The permit granted to the operator will include a range of conditions limiting what may be discharged into air and water. These limits can relate either to integrated parameters such as pH or biological oxygen demand or to specific substances such as copper. In all cases, the operator will be expected to demonstrate compliance with regular monitoring.

In the past, competent authorities have not felt it necessary to set specific limits on individual pharmaceutical active ingredients, but with the growing interest and public concern about the presence of pharmaceutical residues in drinking water, some specific limits on APIs have started to appear (Irish EPA, 2004), and the number of specific limits imposed on the release of active ingredients from manufacturing facilities can be expected to increase.

In addition to the general provisions of the Industrial Emissions Directive, Article 18 may turn out to be of much greater significance. This article requires measures to be taken beyond BAT if this is necessary to prevent an environmental quality standard from being exceeded.

> *Where an environmental quality standard requires stricter conditions than those achievable by the use of the best available techniques, additional measures shall be included in the permit, without prejudice to other measures which may be taken to comply with environmental quality standards.*

Since an increasing number of precautionary environmental water quality standards are being introduced by member states as part of the implementation of the EU Water Framework Directive (EU, 2016b), this is likely to provide additional pressure to reduce emissions from pharmaceutical manufacturing facilities.

Until 2007, there was a general assumption that discharges of active pharmaceutical ingredients (APIs) from manufacturing facilities would be trivial because production volumes are generally low (compared with other types of chemicals) and process efficiencies were generally assumed to be high. However, Larsson and his coworkers (Larsson et al., 2007) looked at the discharge of APIs from manufacturing plants in India and found that in some cases, effluents contained milligram-per-liter concentrations of individual APIs. Since then, several other similar emissions have been identified in China (Li et al., 2008) and Norway (Thomas and Langford, 2010). Thus, although the effluent discharges from many pharmaceutical production plants are usually stringently controlled, this is by no means universal and may not be applicable to APIs.

The discovery of these emissions of APIs, particularly antibiotics, from manufacturing sites, is causing disquiet among stakeholders and demands for action to be taken (O'Neill, 2015). However, enforcing state-of-the-art pollution control on a global scale is practically and politically very difficult. National environmental regulations are at different stages of development, and harmonization takes a long time even when possible. However, suggestions are being made that a modification to an existing set of globally applicable regulations that apply specifically to the pharmaceutical industry might offer a way forward. A key concern of medicine regulators such as the European Medicines Agency (EMA) and the US Federal Drug Administration (FDA) is preserving rigorous quality control to ensure the safety of pharmaceuticals in use. As a consequence, medicine regulators now insist that pharmaceuticals cannot be marketed in their areas of jurisdiction unless they have been manufactured in compliance with the global manufacturing practice (GMP) regulations. One solution to the problem of unregulated effluent discharges is seen by some as amending the GMP regulations that apply to all pharmaceutical production facilities, to incorporate appropriate effluent discharge controls. This was first suggested in 2002 by Velagaleti et al. (2002) and is now being championed within the European Union by the Swedish Government (Swedish MPA, 2011). If approved, this would incorporate mandatory emission standards to the aquatic environment for individual pharmaceuticals within the GMP requirements. However, this apparently simple change may prove difficult to implement in practice.

PHARMACEUTICAL USE

Patient use of medicines and subsequent excretion to wastewater are generally regarded as being the largest contributing sources of pharmaceutical residues in the environment, by overall volume (Taylor, 2016). Resulting concentrations in the environment tend to be low but widespread. Once excreted by the patient, the pathway to the environment is the same for pharmaceuticals as for any other household chemical or personal care product that is ultimately released to domestic wastewater (i.e., down the toilet, sink, or drain in individual households). Depending on where people live, domestic wastewater will either be discharged to septic tanks or direct to surface water (rivers/streams, etc.) or, most likely, collected via sewer systems and treated in a municipal wastewater treatment plant before treated water is released to the environment. In the EU, ~65%–95% of homes are connected to a municipal wastewater treatment plant (EEA, 2016). Wastewater treatment will often remove a proportion of chemicals from the waste stream, but invariably, a residual amount of most chemicals will remain, which will enter either the aquatic environment via the discharge of treated wastewater or the terrestrial environment via application of treated sludge or wastewater (through irrigation) to land, often as a fertilizer. While they are no different in principle to other chemicals, pharmaceuticals have received significant attention over the past two decades as they are known to be biologically active and they are also found in the environment. Perhaps, most significantly from an environmental perspective was the discovery in the 1990s (Sumpter and Jobling, 1995) that estrogenic compounds, including 17β-ethinylestradiol (EE2), could induce vitellogenin synthesis in male fish at extremely low (ng/L) concentrations, fueling concerns about these compounds being potential "endocrine disruptors." This fact, together with the knowledge that many other pharmaceuticals could be found in the environment, led to a rapid increase in research efforts and an increase in regulatory requirements, to better understand the environmental risks associated with pharmaceuticals in the environment.

Pharmaceuticals can be found in most aquatic environments at concentrations typically ranging from low nanogram-per-liter to low microgram-per-liter levels (frequently <0.01 μg/L and occasionally >0.1 μg/L, Taylor, 2016), depending on the pharmaceutical, how widely it is used, whether or not it is excreted and/or removed during wastewater treatment, and the nature of the local receiving environment. Most pharmaceuticals are in continuous use (on average in a given area), with some exceptions such as seasonal antihistamines and other medicines such as analgesics and decongestants that might "peak" during a flu epidemic, for example. All will tend to be found at higher concentrations where there is a high-inhabitant population and/or low dilution in the environment.

Regulatory Frameworks

Assessment of the environmental risks associated with pharmaceutical use by patients is the legal responsibility of the European Medicines Agency (EMA) and the Federal Drug Agency (FDA), in the EU and the United States, respectively. In the EU, an environmental risk assessment (ERA) must be submitted as part of any new marketing authorization application (MAA) and any subsequent type II variations or extension applications (typically changes to strength, form, or route of administration of the active substance) that significantly increases patient use. Similarly, in the United States, a new drug application (NDA) or supplemental application (sNDA) must include an appropriate environmental assessment (EA) or a request for a categorical exclusion from the need to prepare an EA, in which case justification must be provided.

In both the EU and the United States, comprehensive guidance exists on how to undertake an ERA and EA and under what circumstances they are required, including reference to the relevant regulations and directives (EMA, 2006, 2015; FDA, 1998, 2016). Currently, no equivalent guidance exists in other regions in the world, and few other regions require any formal consideration of environmental risks—those that do often accept cross-reference to EU regulatory ERAs or United States EAs. Future ERA requirements may be anticipated in other regions, possibly within 1–2 years in the case of Canada and possibly Japan though timings are unclear.

For established medicines that predate current regulatory requirements, there is currently no mechanism under the EU Medicines Directive 2001/83/EC to address any emerging environmental risks, unless a company makes a new submission for a new product containing an old active pharmaceutical ingredient (API). Thus, there are an estimated 2000 legacy pharmaceuticals used in ca. 3000 older products for which data are sparse or nonexistent and that are currently unlikely to be assessed under existing medicine legislation in the EU and the United States (Mudgal et al., 2013). This is recognized as a knowledge gap that is currently being explored within a collaborative Innovative Medicines Initiative (IMI) between the European Commission and the pharmaceutical industry (IMI, 2015).

Regulatory ERA and EA Approaches

The basic concepts of regulatory ERA and EA are the same in the EU and the United States. Both approaches follow a tiered testing strategy with an initial concentration threshold below which no testing is needed, but after which testing gradually becomes more intensive until the risks can be adequately characterized. Both concepts are based on a comparison of exposure concentration with effect concentration; if the predicted concentration in the environment exceeds the concentration that is predicted to cause an adverse effect, then a potential risk is identified, which may trigger further testing to refine the risk assessment. However, while EU ERA and US EA are similar concepts, there are several differences between their approaches. Table 1 summarizes some of the key similarities and differences, and these are discussed in more detail in the following paragraphs.

A key difference in the ERA and EA approaches is the "action limit" used to trigger different levels of assessment and laboratory testing. For example, in the United States, an EA is required if the concentration in the receiving environment is expected to exceed 0.1 μg/L. In the EU, the equivalent trigger is 0.01 μg/L, a factor of 10 lower. This reflects a different level of precaution and results in more compounds having to be tested and assessed in the EU than in the United States. Another difference is that when an EA is triggered in the United States, different tests are required in comparison with the requirements for an ERA in the EU. The EU has the most stringent requirements in this respect, with chronic (long-term) ecotoxicity studies typically being required for most pharmaceuticals that enter "Phase II Tier A," the entry level for testing in the EU (EMA, 2006). In contrast, in the United States, entry level testing begins with acute (short-term) testing but has higher assessment factors (safety factors) built into the process as a result. In both the EU and the United States, if concerns (potential risks) are identified, further testing is normally required, and at the highest level, the ecotoxicological data required to refine the risk assessment would be similar, if not the same, in both the EU and the United States.

With respect to exposure assessment, the EU and United States approaches are also very similar in the way that concentrations are calculated for risk assessment purposes; however, the EU also requires a "persistence, bioaccumulation, and toxicity" (PBT) assessment, which means that more emphasis is placed on understanding the hazardous properties, irrespective of the actual exposure in the environment. If a compound is "P," "B," and "T" (i.e., fulfills the specific criteria for all three properties), then it is automatically deemed hazardous to the environment. Furthermore, if a compound fulfills the criteria for "very persistent" (vP) and "very bioaccumulative" (vB), then it will also be deemed hazardous to the environment, irrespective of predicted levels of exposure and toxicity. The basis of PBT and vPvB assessment is the "precautionary principle" that is much more embedded in EU legislation (as a result of the OSPAR convention; OSPAR, 1992) than in the United States.

TABLE 1 Comparison of US EA and EU ERA Approaches

Feature	US Approach (FDA, 1998, 2016)	EU Approach (EMA, 2006, 2015)	Comment/Comparison
EA and ERA triggers and exclusions			
Regulatory triggers	NDA[a], sNDA[b]	MAA[c], type II variation[d], extension application[e]	Essentially equivalent
Normally out of scope or excluded from the need to prepare a full EA/ERA[f]	Applications that do not increase the use of the active moiety Applications for research, development, or clinical trials Applications for substances that occur naturally in the environment[g] Applications for biologic products License renewals Medicines containing GMOs	Applications that do not increase the use of the active moiety Applications for research, development, or clinical trials Applications for biologic products License renewals Medicines containing GMOs	Essentially equivalent. The US EA has more specific exclusion criteria for naturally occurring substances
PEC/MEEC triggers requiring a full ERA/EA	If maximum expected environmental concentration (MEEC) > 0.1 μg/L (equivalent to market volume of 44,000 kg/year in the United States, based on the FDA calculation used)	If predicted environmental concentration (PEC) > 0.01 μg/L (equivalent to therapeutic dose of 2 mg/day, based on the EMA calculation used)	EU trigger is a factor of 10 lower. Few products exceed the US MEEC, but many products exceed the EU PEC action limit

Continued

TABLE 1 Comparison of US EA and EU ERA Approaches—Cont'd

Feature	US Approach (FDA, 1998, 2016)	EU Approach (EMA, 2006, 2015)	Comment/Comparison
Other triggers requiring a full ERA/EA	If extraordinary circumstances exist, the MEEC trigger may not apply. Following recent guidance (FDA, 2016), special consideration must be given to compounds with estrogenic, androgenic, and thyroid activity to judge whether the MEEC trigger is appropriate	Some drug substances may affect the reproduction of vertebrate or lower animals at concentrations lower than 0.01 μg/L. These substances should enter phase II, and a tailored risk assessment strategy should be followed that addresses its specific mechanism of action (EMA, 2006)	Both approaches are now essentially similar following recent FDA guidance (FDA, 2016). In the EU, medicines that directly interact with endocrine systems are often assessed regardless of the PEC
	No PBT/vPvB trigger	Includes PBT/vPvB trigger—if log K_{ow} is >4.5, a full PBT assessment is needed following ECHA principles and guidance, irrespective of PEC	US EA risk-based EU ERA risk+hazard (PBT/vPvB)-based
Testing requirements			
Entry level ecotoxicity testing	Acute tests on fish, daphnia, and algae to recognized guidelines (e.g., OECD 201, 202, and 203) Activated sludge, respiration inhibition test OECD 209	Chronic tests on fish, daphnia, and algae to recognized guidelines (e.g., OECD 210, 211, and 201) Activated sludge, respiration inhibition test OECD 209	For US EAs, chronic studies are required if the lowest EC50 from the acute tests is <10 times the MEEC

Entry level phys-chem testing	Water solubility Dissociation constant(s) Octanol/water partition coefficient Vapor pressure or Henry's law constant	None specified	US EA phys-chem studies required to recognized guideline (e.g., OECD) and GLP
Entry level fate testing	Hydrolysis Aerobic biodegradation Soil biodegradation Photolysis Sorption/desorption (if $K_{ow} > 3$)	Sorption/desorption (OECD 106, OECD 121, or OPPTS 835.1110) Ready biodegradation (OECD 301) Water-sediment transformation (OECD 308, aerobic only)	US EA—exact studies not specified but should be to recognized guidelines (e.g., OECD) and GLP. Purpose is to quantify removal mechanisms. EU studies more prescriptive
Environmental compartments considered	Surface water Wastewater treatment plants Terrestrial environment Atmospheric environment	Surface water Wastewater treatment plants Terrestrial environment	Atmospheric environment rarely considered in detail as pharmaceuticals is usually (but not always) nonvolatile
Higher-tier studies required	Yes—if potential risks identified May include chronic ecotox, bioaccumulation, and terrestrial studies as needed	Yes—if potential risks identified May include chronic ecotox, bioaccumulation, and terrestrial studies as needed	Essentially similar in concept but different triggers and considerations (see EMA 2006; FDA, 1998 for details)
Factors considered in the PEC assessment			
Depletion mechanisms considered for MEEC and PEC calculations	Wastewater treatment, removal via biodegradation and adsorption to biosolids	Wastewater treatment, removal via biodegradation and adsorption to biosolids	Essentially similar but slightly different calculations/criteria

Continued

TABLE 1 Comparison of US EA and EU ERA Approaches—Cont'd

Feature	US Approach (FDA, 1998, 2016)	EU Approach (EMA, 2006, 2015)	Comment/Comparison
Human metabolism taken into account	Yes. If the pharmacological potency of the metabolites is known, then the PEC can be adjusted accordingly (assumes ecotoxicity proportional to pharmacological activity)	No. A "total residue approach" is used—assumes 100% excretion as parent compound as a worst case. If a risk is identified, then metabolism may be taken into account if the ecotoxicity of the metabolites is known	EU approach more precautionary, requires metabolite testing if taking metabolism into account to reduce the PEC
Key factors used in PEC/MEEC calculation	Wastewater 375 L/day per capita[h] Kilogram per day emission active substance Dilution factor 10	Wastewater 200 L/day per capita Kilogram per day emission active substance Dilution factor 10	Essentially similar but slightly different calculations and wastewater volumes
Factors considered in the PNEC assessment			
Assessment factor (AF)	AF = 1000 applied to lowest acute EC50 from three trophic levels[i] If three chronic NOECs are available, AF = 10 (same as EU approach)	AF = 10 applied to lowest chronic NOECs from three trophic levels (fish, daphnia, and algae)	Note that the AF of 1000 is the same as the EU approach for general chemicals when only acute data are available

[a] *New drug application.*
[b] *Supplementary NDA.*
[c] *Marketing authorization application.*
[d] *EU Regulation (EC) No 1234/2008.*
[e] *Extension applications typically required for changes to strength, form, or route of administration of the active substance applications (not considered a variation).*
[f] *Strictly, some of these are not "out of scope," but due to the nature of the substances, they are assumed to be low risk and hence do not require any testing.*
[g] *As long as the product does not alter significantly the concentration or distribution of the substance, its metabolites or degradation products in the environment.*
[h] *Current estimate from current US population, and wastewater total wastewater volume of 1.214×10^{11} L/day.*
[i] *In the FDA guidance, the EC50/100 is compared with the MEEC, which is the same as comparing the EC50/1000 with the EU PEC equivalent after dilution in the environment.*

In general, medicines that could pose a potential risk to the environment will either be high volume or very toxic to aquatic organisms or a combination of both. Those that are very toxic often include antimicrobials, which are typically toxic to microorganisms (as they are designed to be) and often toxic to blue-green algae. They also include some medicines that interact with endocrine pathways and that can affect reproductive capability in aquatic organisms, particularly fish, at low concentrations. Such compounds include potent estrogens and antiestrogens that are often selected for their ability to interact endocrine pathways in humans. EMA guidance (EMA, 2006) states that such compounds should be tested irrespective of the amount entering the environment. In other words, such compounds may automatically require a tailored testing strategy that addresses the potential reproductive toxicity of the compound. In practice, this normally means undertaking longer-term tests that are more likely to capture potentially sensitive stages in the reproductive cycle. Reproductive toxicity is not always linked to endocrine disruption; however, the endocrine system is particularly sensitive, and so, compounds that interact with it, either directly or indirectly, may do so at extremely low concentrations that are typically much lower than concentrations needed for other toxicological modes of action.

As mentioned above, in the United States, EAs are not normally required for pharmaceuticals that do not exceed a "maximum expected environmental concentration" (MEEC) of 1 μg/L at the point of entry into the environment (equivalent to 0.1 μg/L in the receiving environment). There is, however, a requirement to consider whether "extraordinary circumstances" apply, and while guidance was previously lacking as to what this means in practice, recent FDA guidance (FDA, 2016) now requires special consideration of whether "extraordinary circumstances" may apply for compounds that have estrogenic, androgenic, or thyroid activity. Any pharmaceutical with these properties may need to be assessed regardless of the MEEC and appropriate testing undertaken to address the specific activity of concern. This is essentially similar in principle to the EU approach (EMA, 2006), and any laboratory testing to address a specific activity of concern would need to satisfy both US and EU requirements.

Apart from the EU emphasis on hazard assessment, the measures of risk in the EU and the United States are similar in concept. In the EU, the risk assessment is based on a comparison of the predicted environmental concentration (PEC) and the predicted no-effect concentration (PNEC), the so-called PEC/PNEC ratio. In the United States, the EA is based on a comparison of the MEEC with the lowest EC50 divided by an appropriate assessment factor. The difference is that in the EU, the PNEC is normally based on chronic data with an assessment factor of 10, while in the United States, the EC50 is normally based on acute data but a larger assessment factor is applied. In the United States, the lowest EC50/1000 is essentially taken to be equivalent to the EU PNEC. This is also identical to the way the PNEC is derived from acute data for general chemicals according to ECHA guidance in the EU (Ref). In practice, if an applicant is planning to market a medicinal product in both the EU and the United States, then the studies undertaken to satisfy EU requirements can also potentially be used to satisfy US requirements. For example, it should be possible to avoid having to undertake acute ecotoxicity studies for a US EA if chronic studies are already available for EU ERA purposes.

In the EU, if the PEC/PNEC is >1, then some form of risk refinement is generally required. This will typically involve refining the PEC or the PNEC. Since introduction of the EMA 2006 ERA guideline, all new medicines entering a phase II assessment (the point at which entry level testing commences) will normally have three chronic end points for aquatic species, so opportunities to refine the PNEC may be limited. In contrast, the PEC is often based on conservative assumptions (see Table 1), particularly with respect to degradation during sewage treatment where the default position is generally to assume no removal, that is, 100% emission of the active ingredient to wastewater and/or sewage sludge. This is because most pharmaceuticals fail the ready biodegradation test (RBT), which is usually the initial criteria used to determine removal during sewage treatment. The RBT is a very stringent test; however, there may be more opportunity for PEC refinement by undertaking more realistic studies to better simulate biodegradation during sewage treatment.

Similar principles apply in the United States, though there may also be the opportunity to generate chronic ecotoxicity data, since these are not initially required. In addition, FDA guidance allows human metabolism to be taken into account to lower the PEC, if the pharmacological activity of the metabolites is known. If the pharmacological activities of the metabolites are not known, then they are assumed to have the same pharmacological activity as the parent compound. In the EU, human metabolism can potentially also be taken into account (EMA, 2015), but actual ecotoxicity data are required for the human metabolites. Such ecotoxicity data are rarely available for human metabolites, so in practice, the PEC is normally always based on the so-called "total residue" approach, that is, assuming 100% excretion from humans as the main active ingredient for the purpose of calculating the PEC.

In both the EU and the United States, certain classes of chemicals are exempt from the need to undertake an EA or ERA, provided relevant criteria are met. These tend to include naturally occurring substances and biological medicines such as proteins and vitamins that are assumed to break down rapidly in the environment.

Medicines containing GMOs must also consider risks to the environment; however, traditional ERA approaches do not generally apply. There is dedicated guidance for this in the EU (EMA, 2005) and the United States (FDA, 2015), most of which focusses on the need to demonstrate either zero release to the environment or nonviability of GMOs once they are in the environment.

Risk Management

With virtually all chemicals across all industry sectors, EU regulations exist, which require some form of ERA to be undertaken as part of the marketing authorization process. This will normally involve a PEC/PNEC and a PBT/vPvB assessment. If risks are identified during these assessments (PEC/PNEC >1 or the chemical is identified as a potential PBT/vPvB), further studies are typically required to refine the ERA. If after appropriate refinement a risk still remains, a policy decision needs to be taken as to whether the benefits of continued use of the chemical outweigh the risk to the environment. Within REACH, the EU legislation covering general chemicals, this step involves undertaking a socioeconomic analysis (SEA). While the process can be complex, the output is essentially a comprehensive risk-benefit assessment upon which the European Chemicals Agency (ECHA) can authorize (or not) specific uses of a chemical. For pesticides, the equivalent process is managed by the European Food Safety Authority (EFSA), and other industry sectors also have equivalent processes (e.g., biocides and cosmetics) under different parts of EC legislation.

For pharmaceuticals, the situation is somewhat different since the outcome of the ERA cannot be a barrier to approval of a medicine for human use. The EU Medicines Directive 2001/83/EC states that "The environmental impact should be assessed and, on a case-by-case basis, specific arrangements to limit it should be envisaged. In any event this impact should not constitute a criterion for refusal of a marketing authorisation." In effect, for the purposes of granting marketing approval in the EU, the benefit of medicines to humans is presumed to outweigh any risks to the environment.

Nevertheless, all stakeholders generally agree that the environmental risks should be assessed and managed appropriately. The directive specifically states that "The environmental impact should be assessed and, on a case-by-case basis, specific arrangements to limit it should be envisaged."

Very few individual pharmaceuticals have actually been shown to pose a significant risk to the environment (Mattson et al., 2007), so the adequacy or otherwise of the current EU policy with respect to implementing appropriate risk management measures has arguably not really been put to the test. Current ERA guidance (EMA, 2006) recommends that all medicines include a statement on the patient information leaflet regarding safe disposal. Apart from that, the only known example of an additional environmental risk management measure for a human medicine was for a contraceptive patch containing EE2, where a disposal sachet was also included with the product with instructions for patients to dispose in the household waste bin and not flush down the toilet (Holm et al., 2013).

National regulators do theoretically have the power to withdraw marketing approval in their country if the environmental risks are seen as unacceptable (EC, 2004). However, chemicals released into the environment do not tend to stay within national borders—the River Rhine flows through several countries, for example, so management of risks at a national level can be problematic if different countries adopt different ideas about what is or is not "acceptable" risk. Hence, there is good reason for EMA managing this at the EU level.

To date, there has never been a case of a medicine being withdrawn from the market because of an identified risk to the environment from human use. Diclofenac, which caused serious damage to the Asian vulture population in India and Pakistan, was being abused as a veterinary pharmaceutical rather than a human medicine; it is now very heavily regulated in that region (Cuthbert et al., 2011).

Environmental Legislation

That few environmental risk management measures that have been applied under the current medicine legislation do not however reflect the whole picture with respect to current concerns about potential risks associated with pharmaceuticals in the environment.

There are a few pharmaceuticals where questions remain concerning environmental risk, and these are now being addressed to some extent within the EU Water Framework Directive. The most well-known and debated case is that of the steroidal estrogens, particularly ethinyloestradiol (EE2), as mentioned above, where environmental concentrations are known to be close to and in some cases above the PNEC. In theory, it would be possible to upgrade domestic sewage treatment plants to reduce levels in the environment to below the PNEC; however, there are significant cost implications of doing so. In the United Kingdom, it has been estimated that the cost could be in the region of €32–37 bn for England and Wales (Owen and Jobling, 2013). As a result of this high cost, attention is now focused on trying to further understand the scale of the problem and to reduce uncertainties in the risk assessments.

In many cases, the precautionary approach helps to address these kinds of uncertainties, particularly when the cost of action is low. However, when the cost of action is high, there is, understandably, a need to be more sure of the benefit. For example, a pharmaceutical manufacturing facility may apply sufficient on-site controls so that effluent discharges to the environment do not exceed a PNEC. However, it does not necessarily follow that it is appropriate to upgrade thousands of sewage treatment plants across Europe to meet the same PNEC standard. In the first case, it may be relatively easy to accept the uncertainty in the PNEC and the costs of implementing appropriate controls, while in the second case, the costs are much higher, so it is in society's interest to be much more sure that the benefits will outweigh the costs of any action.

In order to address some of these uncertainties, EE2, estradiol, and diclofenac have been placed on an EU "watch list" under the Water Framework Directive (WFD). EU member states are required to monitor concentrations of these chemicals in the aquatic environment for a period of 2 years, after which a decision will be made as to whether or not any of these chemicals should become priority substances and have a formal environmental quality standard (EQS) assigned. Under the WFD, concentrations in the environment cannot exceed the EQS. As discussed above, for patient use of medicines, source control is extremely difficult if not impossible without restricting patient use, so the focus of attention is likely to be on improved wastewater treatment if concentration in the environment needs to be reduced to meet an EQS. Due to the potential cost implications of this, as discussed above, the watch list is intended as a mechanism to gather more evidence to reduce uncertainty in the risk assessment. This evidence will provide information on environmental concentrations in surface waters across Europe and may, as a result, identify "hot-spot" areas that exceed the PNEC.

In reality, this issue is more complex. The wider goals of the Water Framework Directive are to achieve "good ecological status" and to ensure that chemicals are not present at harmful levels. Upgrading a sewage treatment plant will almost certainly remove lots of other chemicals and pharmaceuticals, for example, so the cost-benefit assessment is not necessarily substance-specific and should consider the wider implications for the ecological health of such action. In practice, this requires a coordinated effort to relate the measured concentrations in the environment (as being collected under the watch list monitoring campaigns) with information on the ecological health in those areas where samples are being taken. Without such coordination, interpreting the ecological significance of data from the watch list is likely to be challenging.

Relationship Between Medicines and Environmental Legislation

In the EU and the United States, medicinal products are developed, assessed, and approved by their respective medicine agencies. In principle, the environmental risk assessment is based on an individual product basis. It does not account for any other products by other companies, possibly for other therapeutic indications, which might contain the same active ingredient. Thus, while the risk associated with the use of an individual product may be well characterized under medicine legislation, the risk resulting from the total active ingredient residue from all products is not assessed as part of the regulatory approval process. Some initiatives have addressed this issue within voluntary schemes. For example, in the Swedish Fass.se scheme (FASS, 2012), medicines are classified against ERA criteria, and the PEC component is based on the total residue in the environment resulting from all medicinal products containing the same active ingredient that are marketed in Sweden.

From an environmental perspective, the EU Water Framework Directive EQS concept is also substance, not product-specific. For any priority substance, an EQS applies to the total concentration of the substance from all sources and products containing the substance. This means that if an EQS is exceeded, regulators can theoretically choose from a number of options to mitigate the risks posed, either by restricting use or by imposing emission limits. However, since for human medicines the ERA cannot be a barrier to approval of a medicinal product, it is theoretically possible to have a situation where a product is approved for human use under the Medicines Directive but is not allowed to be discharged to wastewater under the Water Framework Directive. This explains some of the difficulty of managing the potential risks associated with patient use of medicines and why it has been such a contentious issue, particularly in the EU.

Partly in response to the challenges identified above, in 2013, the European Commission, under the framework of Directive 2013/39/EU, was asked to develop a strategic approach to pollution of water by pharmaceutical substances by the end of 2015. To inform the strategy, a study was carried out for the commission to highlight the issues and possible solutions (Mudgal et al., 2013). Progress on this has been slow; however, several issues remain unresolved. A recently published Commission Roadmap (EU, 2017) indicates that further work is needed to "identify remaining knowledge gaps and uncertainties," and to "explore how to address the challenge to protect the environment (and human health via the environment) and at the same time safeguard access to effective and appropriate pharmaceutical treatments for human patients and animals, considering inter alia the opportunities for innovation." Further studies are now in progress.

UNUSED AND LIFE EXPIRED MEDICINES

Ensuring a continuous supply of pharmaceuticals to all patients when they are needed is a very difficult task especially where any lack of supply is considered to be unacceptable. As a consequence, although companies try to minimize wastage, unused and life-expired medicines will inevitably occur in the supply chain from manufacturer to distributer and to retailer. However, waste disposal is not an issue here since manufacturers operate reverse distribution systems to recover this material for destruction.

Potential problems do exist after the pharmaceutical has been prescribed for or purchased by the patient. In this case, unused and life-expired material can result from a number of causes: The patient's medication may be inappropriate and subsequently changed by their medical practitioner, too much material may have been prescribed or purchased, and the patient may get better or die. Consequently, some surplus medication will need to be discarded. In a hospital or other institution in the EU, pharmaceuticals should be treated as "clinical waste" and disposed of in line with the relevant national waste regulations (see Introduction). However, no regulations currently apply to disposal by individual patients although guidance is provided in most countries.

Until the discovery of pharmaceutical residues in surface water in the 21st century, the major risk presented by the disposal of unused and life-expired pharmaceuticals by patients was considered to be the potential exposure of children and animals if they were discarded in the domestic waste. Consequently, common guidance was to flush such material into the domestic sewerage system. Today, it is recognized that this is inappropriate since it leads to potential contamination of watercourses and current guidance now takes two different forms.

In some countries such as the United States and Germany where domestic waste is either incinerated or confined in secure landfill sites, advice on disposal of surplus medicines is to mix them with undesirable materials to ensure that they cannot be reused and then deposit them in the household waste (USFDA, 2016).

However, many countries have developed "take-back schemes" and request patients to return unused and life-expired pharmaceuticals to their local pharmacy that will then ensure appropriate destruction. There are well-established schemes in some European countries such as France, Sweden, and the United Kingdom (Taylor and Poulmaire, 2008), and in 2004, the European Union issued a directive (EU, 2004) that specifically required that "all member states shall ensure that appropriate collection systems are in place for medicinal products that are unused or have expired." This requirement is now slowly being implemented within the 28 current EU member states.

In the United States, "take-back" schemes have been inhibited by the requirements of the US Drug Enforcement Agency (DEA). They insist on very strict and onerous oversight requirements because of their concern that "take-back" schemes might be used to divert drugs of abuse back to the population at large. However, a number of individual state and county jurisdictions are currently attempting to institute take-back schemes of their own with a requirement that the pharmaceutical industry should pay the operating costs. This is currently being challenged through the US court system (Silverman, 2015).

REFERENCES

Cuthbert, R., Taggart, M.A., Prakash, V., Saini, M., Swarup, D., 2011. Effectiveness of action in India to reduce exposure of gyps vultures to the toxic veterinary drug diclofenac. PLoS ONE 6 (5), e19069.

EC, 2004. Regulation (EC) No 726/2004 of the European Parliament and of the Council of 31 March 2004 laying down Community procedures for the authorisation and supervision of medicinal products for human and veterinary use and establishing a European Medicines Agency. Article 18;4. Off. J. L 136, 0001–0033. 30/04/2004.

EEA, 2016. Urban Waste Water Treatment. Available from: http://www.eea.europa.eu/data-and-maps/indicators/urban-waste-water-treatment/urban-waste-water-treatment-assessment-3. Accessed 3 January 2016.

EMA, 2005. Environmental Risk Assessments For Medicinal Products Containing, or Consisting of, Genetically Modified Organisms (GMOs) (Module 1.6.2). EMEA/CHMP/BWP/135148/2004. Available from: http://www.ema.europa.eu/docs/en_GB/document_library/Scientific_guideline/2009/09/WC500003806.pdf. Accessed 4 October 2016.

EMA, 2015. Questions and Answers on 'Guideline on the Environmental risk Assessment of Medicinal Products for Human Use'. EMA/CHMP/SWP/44609/2010 Rev. 1. Available from: http://www.ema.europa.eu/docs/en_GB/document_library/Scientific_guideline/2015/03/WC500185159.pdf. Accessed 4 September 2016.

EMA, 2006. Guideline on the Environmental Risk Assessment of Medicinal Products for Human Use. European Medicines Evaluation Agency Guidance Note EMEA/CHMP/SWP/4447/00. Available from: http://www.ema.europa.eu/docs/en_GB/document_library/Scientific_guideline/2009/10/WC500003978.pdf. Accessed 4 September 2016.

EU, 1996. Directive 1996/61/EC on integrated pollution prevention control. Off. J. L 257, 26–40.

EU, 2004. Directive 2004/27/EC of the European Parliament and of the Council of 31 March 2004 amending Directive 2001/83/EC on the Community code relating to medicinal products for human use. Off. J. L 136, 34–57.

EU, 2006. Reference Document on Best Available Techniques for the Manufacture of Organic Fine Chemicals. Brussels, European Commission.

EU, 2010. Directive 2010/75/EU of the European Parliament and of the Council of 24 November 2010 on industrial emissions (integrated pollution prevention and control). Off. J. L 334, 17–119.

EU, 2016a. Waste Electrical & Electronic Equipment (WEEE). Available from: http://ec.europa.eu/environment/waste/weee/index_en.htm. Accessed 4 October 2016.

EU, 2016b. The EU Water Framework Directive. Available from: http://ec.europa.eu/environment/water/water-framework/index_en.html. Accessed 4 October 2016.

EU, 2017. Commission Roadmap: Strategic Approach to Pharmaceuticals in the Environment. EC Ref. Ares (2017)2210630. Available from: https://ec.europa.eu/info/law/better-regulation/initiatives/ares-2017-2210630_en. Accessed 7 October 2017.

FASS, 2012. Environmental Classification of Pharmaceuticals at www.fass.se. Guidance for Pharmaceutical Companies. Farmaceutiska Specialiteter i Sverige (FASS), Stockholm. Available from: http://service.lif.se/Deltagarservice/Sidinnehall/Deltagare_Dokument/Regulatory/Milj%C3%B6/Environmental%20classification%20of%20pharmaceuticals%20att%20wwwfassse%20120816.pdf. Accessed 4 September 2016.

FDA, 1998. Guidance for Industry: Environmental Assessment of Human Drug and Biologics Applications. US Center for Drug Evaluation and Research, Washington, DC. CMC 6 (Rev 1). Available from: http://www.fda.gov/downloads/Drugs/GuidanceComplianceRegulatoryInformation/Guidances/ucm070561.pdf. Accessed 4 September 2016.

FDA, 2015. Determining the Need for and Content of Environmental Assessments for Gene Therapies, Vectored Vaccines, and Related Recombinant Viral or Microbial Products. U.S. Department of Health and Human Services, Food and Drug Administration Center for Biologics Evaluation and Research, Washington, DC. Available from: http://www.fda.gov/downloads/BiologicsBloodVaccines/GuidanceComplianceRegulatoryInormation/Guidances/CellularandGeneTherapy/UCM439273.pdf. Accessed 4 October 2016.

FDA, 2016. Environmental Assessment: Questions and Answers Regarding Drugs With Estrogenic, Androgenic, or Thyroid Activity. Guidance for Industry. US Center for Drug Evaluation and Research, Washington, DC. Available from: http://www.fda.gov/downloads/Drugs/GuidanceComplianceRegulatoryInformation/Guidances/UCM444658.pdf. Accessed 4 September 2016.

HMSO, 2013. Health Technical Memorandum 07-01: Safe management of healthcare waste. UK Department of Health. Available from: https://www.gov.uk/government/uploads/system/uploads/attachment_data/file/167976/HTM_07-01_Final.pdf. Accessed 16 October 2017.

Holm, G., Talbot, J., Taylor, D., Snape, J.R., Murray-Smith, R., Sörme, P., 2013. Implementing ecopharmacovigilance in practice: challenges and potential opportunities. Drug Saf. 36, 533–546.

IMI, 2015. Intelligence Led Assessment of Pharmaceuticals in the Environment. Available from: http://i-pie.org/. Accessed 3 January 2016.

Irish EPA, 2004. Integrated Pollution Control License. 673 Wexford, Irish Environmental Protection Agency.

Kaiser Permanente, 2016. Environmental Stewardship Programme. Available from: http://share.kaiserpermanente.org/article/environmental-stewardship-overview/. Accessed 4 October 2016.

Larsson, D.G.L., de Pedro, C., Paxeus, N., 2007. Effluent from drug manufactures contains extremely high levels of pharmaceuticals. J. Hazard. Mater. 148, 751–755.

Li, D., Yang, M., Hu, J., Ren, L., Zhang, Y., Li, K., 2008. Environ. Toxicol. Chem. 27, 80–86.

Mattson, B., Nasmen, I., Strom, J., 2007. Sweden's voluntary environmental drug classification system. RAJ Pharma, 138–153.

Mudgal, S., De Toni, A., Lockwood, S., Salès, K., Backhaus, T., Halling Sorensen, B., 2013. Study on the Environmental Risks of Medicinal Products. Final Report Prepared for Executive Agency for Health and Consumers. BIO Intelligence Services, Paris.

National Health Service, 2016. Sustainable Development Strategy for the Health and Social Care System 2014–2020. Available from: http://www.sduhealth.org.uk/policy-strategy/engagement-resources.aspx. Accessed 4 October 2016.

O'Neill, J., 2015. The Review on Antimicrobial Resistance—Antimicrobials in Agriculture and the Environment: Reducing Unnecessary Use and Waste. HMSO, London.

OSPAR, 1992. Convention for the Protection of the Marine Environment of the North-East Atlantic (The "OSPAR Convention").

Owen, R., Jobling, S., 2013. Ethinyl oestradiol in the aquatic environment. In: Late Lessons From Early Warnings II. European Environment Agency, Copenhagen (Report No. 1/2013). (Chapter 13).

Silverman, E., 2015. Pharmaceutical Industry Must Pay for Drug Take-Back Programs. Available from: http://blogs.wsj.com/pharmalot/2015/05/26/pharmaceutical-industry-must-pay-for-drug-take-back-programs/. Accessed 4 October 2016.

Sumpter, J.P., Jobling, S., 1995. Vitellogenesis as a biomarker of oestrogenic contamination of the aquatic environment. Environ. Health Perspect. 103, 173–178.

Swedish MPA, 2011. Platform to Enable the Initiation of a Revision of EU Legislation on Good Manufacturing Practice, GMP, in Order for Legislation Also to Comprehend Environmental Considerations. Swedish Medical Products Agency, Stockholm.

Taylor, D., 2016. The pharmaceutical industry & the future of drug development. Issues Environ. Sci. Technol. 41, 1–33.

Taylor, D., Poulmaire, M., 2008. In: An initial survey of unused & expired medicine take-back schemes in the European Union, pharmaceutical products in the environment: towards lowering occurrence and impact. KNAPPE International Conference, Nimes France.

Thomas, K., Langford, K.H., 2010. In: Point sources of human pharmaceuticals into the aquatic environment. In: Kummerer, K., Hempel, M. (Eds.), Green & Sustainable Pharmacy. UK Department of Heath (2013), Springer, Berlin.

USEPA, 1998. Development Document for Final Effluent Limitations Guidelines and Standards for the Pharmaceutical Manufacturing Point Source Category. US Environmental Protection Agency, Washington, DC. EPA-821-R-98-005.

USEPA, 2015. Proposed Rule: Management Standards for Hazardous Waste Pharmaceuticals. Available from: http://www.epa.gov/hwgenerators/proposed-rule-management-standards-hazardous-waste-pharmaceuticals. Accessed 4 October 2016.

USFDA, 2016. How to Dispose of Unused Medicines. Available from: http://www.fda.gov/ForConsumers/ConsumerUpdates/ucm101653.htm. Accessed 4 October 2016.

Velagaleti, R., Burns, P.K., Gill, M., Prothro, J., 2002. Impact of current good manufacturing practices and emission regulations and guidances on the discharge of pharmaceutical chemicals into the environment from manufacturing, use, and disposal. Environ. Health Perspect. 110, 213–220.

Water UK, 2014. National Guidance for Healthcare Waste Water Discharges. Available from: http://www.water.org.uk/publications/water-industry-guidance/national-guidance-healthcare-waste-water-discharges. Accessed 2 May 2016.

FURTHER READING

Department of Health, 2013. Safe Management of Healthcare Waste. Health Technical Memorandum 07-01. Department of Health, London. Available from: https://www.gov.uk/government/uploads/system/uploads/attachment_data/file/167976/HTM_07-01_Final.pdf. Accessed 4 October 2016.

HMSO, 2008. Climate Change Act 2008 (c.27). HMSO, London.

USEPA, 2016. Management of Pharmaceutical Hazardous Waste. Available from: http://www.epa.gov/hwgenerators/management-pharmaceutical-hazardous-waste. Accessed 4 October 2016.

Index

Note: Page numbers followed by "*f*" indicate figures, and "*t*" indicate tables.

Q

R

S

Z

Kolibri 360 GmbH
Ferdinand-Jühlke-Straße 7
99095 Erfurt, Deutschland
produktsicherheit@kolibri360.de